**The
Broken
Land**

The Broken Land

Adventures in
Great Basin Geology

Frank L. DeCourten

The University of Utah Press

Salt Lake City

The Defiance House Man colophon is a registered
trademark of the University of Utah Press.
It is based upon a four-foot-tall ancient Puebloan
pictograph near Glen Canyon.

Library of Congress Cataloging-in-Publication Data

DeCourten, Frank.
   The broken land : adventures in Great Basin geology /
Frank L. DeCourten.
      p. cm.
Includes bibliographical references and index.
   ISBN 978-0-87480-751-6 (pbk. : alk. paper)
1. Geology—Great Basin. I. Title.
   QE79 .D43 2003
   557.9—dc21
                                        2003002468

Contents

Leaning forward over the guitar in my lap, I raise my hands toward the campfire to warm fingers that have lost some flexibility to the cold night air. A shaft of smoke rises from the fire in a billowing spiral that leans a bit to the south before it bleeds away into the sparkling blackness of a western Utah night sky. The clatter of laughter moves like waves through the twenty-five faces surrounding the fire. Some of those faces are young, with hopeful, excited eyes darting everywhere, but mostly directed toward an unknowable future. Other faces, furrowed from life, hold quieter eyes reflective of an unchangeable past. Superimposed on each face is the elation of the moment: a warm blush of life, as difficult to describe as it is to miss, shining from the ring of bright smiles. Several days of roaming free through the exquisite emptiness of Utah's West Desert, delighting in the natural majesty around every bend in the dusty gravel roads, has somehow inspired a collective euphoria among the small band of explorers. This sagebrush wonderland has simultaneously stretched our eyes, minds, and souls. My campers all seem like children again, delighting in their personal discoveries of a world so different from the one they normally occupy. I glance over at Tom, my old pal and the best damn partner any cowboy ever had. Beneath his tilted hat brim and partly shaded from the flickering firelight, he's got that faraway look that I know from our many years together means he's not really here at all. He's out there riding his old Paint through the desert night 100 years ago, searching, as cowboys like us always do, for that something that we can never define. Tom's prone to such ethereal drifting. It's one of the things I like best about him.

As I prepare to join him on that ride, I let my eyes drift beyond the smiling faces illuminated by the fire, out to the edge of the amber radiance. I can barely see the ragged junipers standing like shadowy sentries in the distance. Farther away, the saw-tooth silhouette of the distant mountains is defined only by the absence of stars near the horizon. The ragged canyon that twists into the Confusion Mountains above our camp is completely invisible in the sublime blackness around us. Watching the shimmering firelight wither into the darkness beyond, I ponder the paradox of being in the midst of a virtually limitless terrain, but at the same time enclosed in a tiny world defined by the circular glow of burning wood. Yet the dissonance between the expansive darkness and confined brightness is oddly comfortable. I feel protected and secure here, far removed from the din and demands of a life that bounds uncontrollably, like some wild mustang, across the alien landscape of the modern world. Here in the desert, where the antiquity is almost palpable, I welcome the mystical tendrils that seem to rise from the sand like shadscale branches, wrap around my boots, and hold me still. Feeling riveted to this special place, and firmly braced in my own time, some of the internal commotion produced by my drifter's soul finally seems quieted. For someone who has had lifelong difficulty calling anyplace "home," this small ringlet of campfire light is probably as close as I'll ever get. Welcome to the Yellow Circle, my refuge.

Out in the desert beyond the Yellow Circle, life goes on as it always has on nights like this. The coyotes are there, silent for the time being, going about their canine business. The jackrabbits are holding tight. I can almost feel the glistening eyes of a cougar trained on the Yellow

The Yellow Circle

Circle from some stone ledge in the dark mountains high above us. The nighthawks and owls sail over us; sometimes, if it's quiet enough, you can hear their wings rip the air. The bats flutter about, occasionally dipping into the light to perform one their remarkable mid-air maneuvers. They vanish in an instant. Once in a while, a kangaroo rat flirts with the edge of the Yellow Circle just long enough to cause someone to jump. "What the hell was that?" Laughter breaks out and fabricated stories of snakes, scorpions, and other, more mythical, desert beasts follow. The soil beneath our boots is alive with insects, but we only notice them when they crawl up inside the legs of our pants, which is not often. During lulls in the campfire banter, the breeze carries a soft melody of assorted squeaks, chirps, buzzes, hoots, and calls that emanate from somewhere out in the dark abyss surrounding us. We are in the middle of a magical parade of nocturnal life, and we are part of it. Some would say that the Yellow Circle is an intrusion on the natural world, that it doesn't belong here, and that it disrupts the rhythm of the desert ecosystem. But it never seems that way to me. Our luminous niche is full of life, as is the darkness beyond it, and the sense of belonging to it all enhances the joy of living in each of us. For all of us huddled around the fire tonight, life is a miraculous adventure. We marvel at the ways of the creatures around us, and we delight in the thrill of being immersed in something much bigger than ourselves. It wouldn't surprise me if, over the many years that I have gathered with people around the Yellow Circle, some life has even been conceived under its influence. But, tonight, somewhere out in the desert shadows, a few lives will end, some lives will begin, and other lives will simply continue. The pageant of life continues to flow all around us, and I don't think that the Yellow Circle will change any of that very much.

Something else is alive out here, too. Beneath the Yellow Circle, about fifteen miles below the embers in the fire, are rocks made soft under the influence of heat and pressure. These pliant rocks are alive with motion. To be sure, the motion is slow, perhaps a few millimeters per year, but it is motion nonetheless. Forces within the

earth are gradually pulling this layer of ductile material into a thin sheet, undermining the support of the thick blocks of solid stone above. Stresses caused by these subsurface movements produce fractures in the brittle rock near the surface. As the earth's crust is broken into enormous blocks along these fractures, they heave about on the underlying rock mush. Some blocks rise to become mountains, while others sink to form the foundations of valleys, like the one we are camped in. The blocks are constantly shifting and the landscape around us is continually changing as the mountains ascend at their own imperceptible pace. From within the Yellow Circle, we can't watch them rise any more than we can see the observant cougars perched on their ledges. But they *are* rising.

As the mountains soar, layer upon layer of rock is exposed to view along their steep flanks. Some of these layers contain fossils of primitive sea creatures that lived on the floor of some vanished ocean hundreds of millions of years ago. Other layers consist of welded and fused ash particles produced during ancient volcanic cataclysms. Still other rock layers record the development of inland lakes and raging floods. The legacy of the land is written in these layers of rock; and, we know from examining this record that the history of the Great Basin has been deep, rich, and sometimes convulsive. The desert may seem timeless from within the Yellow Circle tonight, but the permanence of the surrounding landscape is only an illusion. Even the land is alive here, constantly changing and evolving, as do all living things. Nothing here is "finished." Everything around us—the plants, the animals, the rocks, and the terrain itself— is in the process of becoming something else. And, of course, so are we. None of us around the campfire tonight are the same people we were yesterday, or will be tomorrow.

Too soon, morning steals the darkness. As the sun peeks over the eastern horizon, its rays of light flood across the desert, washing away the Yellow Circle to reveal the grand panorama of western Utah. I peek out of my sleeping bag with squinting eyes that struggle to focus in the bright glare. Though it is still low in the east, the sun has painted the Confusion Mountains a bright yellow-gray streaked with shadowy canyons and dotted with vivid green clumps of junipers. I notice a brown lump near the burned-out campfire. It's just Tom tucked inside his bedroll. I wonder if he's back from his ride yet. In any case, I'll have to make the coffee this morning, as usual. So, I pull on my boots, find the old black coffee pot, fill it with grounds and water, and walk toward the firepit. The refuse of the night before lies scattered around—a can here, a bottle there, my guitar propped up against the trunk of a shaggy juniper tree. I notice a small spiral-bound notebook, half buried in the sand. On the open page, names and addresses are scrawled in writing that is barely legible. Three days ago, these people were all total strangers, now they are exchanging personal information like long-lost lovers. The Yellow Circle never fails to connect people— to each other, to the coyotes, to the sagebrush, to the clouds, to all those things wild and raw that live around us and within us. I feel a little warm glow inside after seeing the notebook; it reminds me of nature's incomparable power to weave beautiful human tapestries from even the most dissimilar threads. Over by the fire, Tom is beginning to stir; his horse must be played out.

A few minutes later, people are starting to gather around the smoldering fire to sip from steaming cups and nibble on what's left of our provisions. This is our last morning in the West Desert, and soon it will be time to break camp and depart. But no one seems very enthusiastic about that prospect. A few campers are slowly gathering up their gear, rolling sleeping bags, and packing tents with measured effort and with, I suspect for some, deliberate delay. Several folks hover around trading stories about the events that transpired in the Yellow Circle just hours earlier. Desert legends can be made overnight, it seems, and it happens out here all the time. Few people ever return from the Confusion Mountains without a story to tell.

Two hours pass, after which our caravan of trucks crawls slowly down the rough dirt road leading back to Highway 50, raising a cloud of yellowish dust that drifts across the sagebrush plain. As the sounds of the rumbling trucks fade into the distance, the last few dust grains settle on the rocks surrounding the firepit back at our vacant campsite. In our absence, a pinyon jay descends from the limb of a juniper tree nearby and lands a few feet from the pile of ashes. The bird jumps when the last ember in the firepit pops. The bit of burning wood glows bright orange for a few seconds, then fades to black. The Yellow Circle has vanished, and silence returns to our abandoned campsite, except for the soft whisper of the breeze through the junipers and rabbitbrush.

We make one final stop on the way home near Sevier Lake. With the trucks pulled over along the shoulder of the road, I give my final lecture to the group. It's my usual "This Land is Your Land" sermon, ending with an invitation to return to this part of Utah, considered by many a wasteland of little value. I try to remind everyone of the special feeling of the desert. I speak of the absence of fences, the howl of the wind, the smell of dust, the rumble of tires over rough roads, the delicious desolation, and the incomparable sensation of wildness in the heart. We review the lessons of the rocks and all the things they have taught us about the history of this land. We chat about the absurdity of human "ownership" and how vain it is to assert that we can truly possess anything in nature, including parcels of this desert. We look back to the west, toward the Confusion Mountains, for a silent moment until someone says, "Well, back to the real world."

The *real world?* No, I think to myself, the desert we are leaving is more real than anything in our fabricated urban habitat. We are returning to a world of illusion where humans have obscured things truly meaningful and genuine with such delusions of relevance as banks, courts, stock markets, shopping malls, churches, legislative houses, and universities. Walking to my truck, I think of all the conventions, habits, and institutions by which humans isolate themselves from nature, and I realize that, for the last few days anyway, we have escaped them all. My fondest hope is that the people who sit for a night or two in the Yellow Circle will take some of it with them back to the "real" world. I hope they will remember that all the things we saw in the desert will still exist, whether or not we are around to experience them. Likewise, the things we *felt* in the wilderness are still there deep inside each of us, and can be retrieved on demand, if we only remember to do so. In the clamorous, confining, and counterfeit world that most of us inhabit, we tend to lose our spiritual perspectives, suppress the wildness in our souls, and tune out the songs that nature sings. Deaf to the music and insensitive to the wonder that surrounds us, daily minutia becomes our master. Too many of us limp along in our self-designed prison with unconnected and fragmented souls, distracted from the pain we inflict on ourselves by the smoke and mirrors of a consumer culture promulgated by corporate tyrants and government co-conspirators. We desperately seek some external elixir to restore meaning and purpose to life, a search that I believe is doomed to failure without an emotional connection to the living world around us. We so seldom realize that the serenity we crave can be achieved simply through a deeper intimacy with the land and life that enfolds us. The emotions that come so effortlessly around the Yellow Circle—that extraordinary feeling of being ferociously in love with the world—can provide the kind of fulfillment we all seek, if we can only sustain it against the suffocating influence of a dispassionate life solely geared toward the acquisition of wealth and status. As I watch the campers climb back into the trucks, I wonder how many of them understood the words. Did anyone grasp the message? Will any of them ever think again about the smell of the sage, the sound of the nighthawks, or the taste of the dust? Do they still *feel* the desert in their hearts?

A few hours later, back in the city, the trucks are being unloaded as night falls in the neon wilderness. Under the pallid orange lights in the parking lot, people are carrying gear from the trucks, packing it into their cars, shaking out blankets, and pouring sand from their boots. Tom and I lean against the back of the Suburban, regretting that another trip has ended. We toast our adventure with the last two bottles of cold beer. A young woman approaches to offer us her thanks for the outing. She had a good time, she said, and she learned so much. She made new friends and had seen so many things that she never expected to encounter in the "god-forsaken" desert. She said she was amazed that there was so much beauty in the barren wastes of western Utah. And, oh, by the way, she liked the music, too. As she turned to leave, I glanced at Tom, who was giving me a subtle half-grin, as he always does when things like that happen. He's grinning because he saw it too, and he knows the meaning. While she spoke to us, there was a faint but unmistakable sparkle behind the woman's smiling eyes.

It was yellow.

I once shared a flight from Los Angeles to Salt Lake City with a middle-aged woman from Cincinnati named Nancy. Nancy sat in the window seat, while I, seeking relief from my chronic flight anxiety, took an aisle seat on the same row. About an hour after take-off, I had learned that Nancy had been to California to visit relatives and was returning to her native Midwest. In the course of our casual chatter, Nancy asked me what I did and why I was traveling to Utah. In answer to her questions, I informed her that I lived in Salt Lake City, was a geologist, and worked at a natural history museum. Nancy said that I must love to fly because of the great views of mountains, canyons, and plains that an aerial perspective affords. I told her, however, that I'd rather walk on a mountain than look down on one through 30,000 feet of air. She seemed perplexed by my admission, but smiled just the same. A short time later, the plane banked to the left and, as the wing dipped below the horizon, I noticed that we were turning above the Escalante Desert in southwest Utah.

Nancy gazed silently down on the tawny chestnut terrain below. The barren desert of western Utah and eastern Nevada receded into the hazy distance like a vast brown ocean. The isolated mountains stood like waves in this sea, the nearer ones partially obscuring the more distant. As I glanced through Nancy's window, she turned to me and said, "What a wasteland! Is this Death Valley?" I explained that, no, we were northeast of Death Valley by a considerable distance, and that the desert floor below us was actually several thousand feet above sea level. I pointed out to Nancy the Wah Wah and Needles Ranges, separated from each other by Pine Valley, with the snowcapped peaks of the Snake Range in Nevada looming on the distant hori-

zon. I told her of the extensional forces that were at work in the region, ripping blocks of rock apart to create the corrugated terrain. I explained what "basin and range" meant. Nancy was unimpressed. As our brief chat concluded, she said it was too bad that so much land was uninhabitable, but she certainly understood why. "Who would want to live *THERE*?" she exclaimed. I didn't answer, resisting the temptation to say the same thing about Ohio: Isn't that where the rivers catch fire?

At the airport in Salt Lake City, Nancy and I parted company as she hurried away to catch her connecting flight to Cincinnati and I headed for the parking lot. I was still thinking about her derision of the desert lands of western Utah as I drove home along I-80. I excused her contempt as a consequence of an easterner's unfamiliarity with the beauty and spaciousness of western Utah and eastern Nevada. She would undoubtedly change her mind if she had the opportunity to explore the countless geological marvels and other natural wonders that exist in that "wasteland." But disdain and nescience toward the western deserts is not the exclusive domain of easterners ("east" being anywhere to the right of the Rockies). Even many native Utahns commonly perceive the western portion of their state as an empty and worthless expanse. On the opposite side of the Great Basin, I have heard many Californians express similar sentiments about the desert lands east of the Sierra Nevada. A bartender at the Owl Club in Eureka, Nevada, once responded to my praise of the central Nevada desert by exclaiming, "OK, that's it for you." Even among westerners, who should all know better, the common perceptions are that Great Basin landscapes, communities, and culture are far inferior to those of neighboring regions. Few people would express the same admiration for Gabbs, Ely, Elko, Delta, Pioche, or Cedar City that are lavished upon Taos, Moab, Estes Park, Jackson Hole, Park City, or Sun Valley. Without the glitzy entertainment of casinos, the lure of brothels, or the promise of easy wealth from minerals (or misdeeds), the expansive sagebrush plains of the Great Basin seem to offer little appeal in comparison to the scenic coast, the shining mountains, or the verdant valleys. To far too many people, the Great Basin is nothing more than that desolate, god-forsaken desert out west where we bury toxic wastes, place our most hideous works of art, test our malevolent weapons, legalize some forms of "immorality," imprison aliens from other worlds, and conduct all sorts of other unsavory activities. It's a place we might have to drive through (the faster, the better!), hoping that the car doesn't break down until we arrive at a more accommodating destination.

This disparagement of the Great Basin desert is nothing new. It has resonated through the descriptions of the region since the time Anglo-Europeans first set eyes on it. During the late 1840s, Captain John Frémont, originator of the physiographic term for this area, had a particular fondness for such words as "sterility," "barrenness," and "waste" in his published descriptions of these desert lands. Scan the language, documented by C. Elizabeth Raymond in the essay "Sense of Place in the Great Basin," used by early travelers through the region. You'll find "dreary and dismal," "forsaken and cursed," "desolate waste," and "paltry, unimportant ugliness." There were, even in the late nineteenth century, a few voices of admiration for the western deserts, but they were muted beneath the avalanche of negative imagery that set the stage for public perceptions of the Great Basin during the following century. It is not surprising that our attitudes toward the desert lands of the West have traditionally shown little regard for, or even an awareness of, the boundless natural beauty that exists in the arid lands of North America. This reflects, in part, the initial experience of our forebears, who were usually more concerned with mere survival in the brutal desert wilderness. Of course, we can't really blame the early explorers and pioneers for their apparent insensitivity. It is difficult to admire the scenery of the desert while dying of thirst and hunger. But, few of us in modern times confront the same hardships and hazards that the pioneers faced. We normally travel through the Great Basin untroubled by bloody feet, swollen tongues, blistered skin, or dehydration delirium. And yet, many (most?) people still fail to fully, or easily, appreciate the grandeur of the bleak landscape.

Thoreau asserted that people will discover "… as much beauty … in the landscape as we are prepared to behold—not a grain more." So, what is it, precisely, that prepares us to behold beauty in nature? Why are vistas of sagebrush and mahogany-colored mountains thrilling to me but dreary to Nancy? Why will so many people wander wide-eyed for hours among Las Vegas's glittering monuments to greed and corruption, while viewing the desert around them only as an irritating obstacle to travel? I can't answer that question for everyone; but, in preparing to behold beauty in landscapes, there is something very special about seeing the land through geologically enlightened eyes. For me, I think the wonder of nature in general, and my love for the western deserts in particular, stems from understanding, and appreciating, the deeper history of land.

I am a geologist by training and experience; certainly not the best the profession has to offer, but good enough to recognize the significance of what nature reveals to us in the scrappy landscapes of the Great Basin. The rocks, fossils, and geological structures exposed in this magnificent desert allow us to recognize and decipher an immense history. Of course, we don't know it all: secrets locked in stone are sometimes difficult to extract. Even with all the imperfections in our knowledge, however, we know that it has taken well over two billion years to create the terrain that now stands in western Utah, Nevada, and eastern California. It is an amazing legacy, encompassing an incomprehensibly long interval, involving many cataclysmic upheavals punctuated by more serene interludes. The tumultuous geological heritage of the Great Basin is fascinating for its own sake, but I think the historical perspective that geology provides serves a purpose far greater than mere factual enlightenment. Knowing the history of the land makes the modern landscape more meaningful, more worthy of our respect and reverence simply for being what it is. To be sure, the geological evolution of the Great Basin has produced plenty of material wealth in the form of silver, gold, copper, molybdenum, beryllium, and other metals. But the real treasure of the Great Basin is the intellectual benchmark that its history provides for our own existence as a part of the modern landscape. Awareness of the stories embedded in rocks and mountains places a special significance on our own presence within the natural continuum of space and time. We begin to feel like we are a part of the natural rhythms that have resonated through Earth for eons. Learning to read the deep history engraved in cold, gray stone along the walls of some desert canyon in the Great Basin brings us face to face with the realization that the story does not stop in the stone. Our existence here and now is a consequence of the same natural cycles that have created the rocks, mountains, and desert. We stand in the middle of nature's historical panorama, with our feet planted upon the immeasurable past while we survey the astonishing results. This may all be God's plan, or it may be something completely different. In any event, it is the rocks that tell us how the plan was executed: the pace, the pattern, the processes. And we are not merely passive observers of the products of those events—we are part of the whole process.

John Muir once referred to the canyons of the High Sierra as "nature's poems carved on tablets of stones." When I stroll through the desolate lands of the Great Basin, where there is little to conceal the rock, that kind of poetry blooms into a sonnet of powerful revelation, and I can hear the verses more clearly there than anywhere else. In this stark land, emphatic rock exposures proclaim the deep history with such boldness that we can not only decipher it, we can actually *feel* it with every step. The varied terrain of rawboned peaks, swelling mountains, ragged and twisted ravines, and level sagebrush

expanses bespeaks a history that dwarfs our own. Just as our fondness often grows for human acquaintances that we come to know well, our reverence toward the land escalates as we absorb, and ponder, its deep history. My love of the desert lands of the West is, I think, not an expression of any lofty personal ethic, but simply the consequence of struggling to comprehend the depth of their immense history. Geologists are not morally superior to other people because many of us feel such a powerful attachment to the land. It's just that loving the Great Basin desert, and harboring outrage over its destruction and abuse, is an occupational hazard for anyone who is sensitive to, and respectful of, its extraordinary past.

The intent of this book is to examine some of the amazing geological events that have made the Great Basin what it is. At the outset, however, we should understand that no book can completely cover all the details of the geologic history of such a large region, nor will we ever arrive at the "end" of the story. I once asked a elderly miner in a Tonopah bar if he had lived in Nevada his whole life, and, repeating the old joke, he replied, "Not yet!" In a similar way, the geological story of the Great Basin is still unfinished and, for all practical purposes, always will be. Only when the internal, heat-driven churning of the earth ceases will the dynamic processes that have shaped the modern Great Basin expire. In the meantime, for at least the next several billion years, the region will continue to evolve in response to the same fundamental forces that have defined its past. And it is that past, the rich geological legacy of the region, that will constitute the principal focus of this exploration. Among the lengthy procession of events recorded in the rock sequences of the Great Basin, we can only hope to present and comprehend the major steps, those that have had the greatest impact on the geological framework of the region. Of course, painting this historical portrait with such a broad brush means that we will have to leave some of the lesser events, those confined to localized areas or of relatively minor impact, for later contemplation. Nonetheless, we can still attain a solid overall understanding of the geological history of the region by grasping the significance and effects of the episodes explored in the following chapters. To accomplish this task, we will have to explore many specific localities, consider many different geological processes, learn the names and features of many types of rocks, and struggle to appreciate the immense depth of geologic time. This task may sound daunting at first but, like any other exploration of the unknown, each new vista that we absorb brings more coherence to the entire panorama. The intrigue in our survey builds with the realization that the more we learn, the more exciting our venture becomes. Each answered question completes a part of the story, but it also begets additional mystery. Event by event, the history begins to emerge from the rocky archives until the overall picture comes into focus. As we decipher the geological heritage of the Great Basin, we will also, of course, be deepening our love for the landscape. But, every journey of discovery starts somewhere. So, let's begin the voyage by considering what, exactly, is this remarkable place called the *Great Basin*?

THE GREAT BASIN DEFINED

When Captain John C. Frémont approached Utah Lake with his ragged entourage in 1844, he was close to completing a epic journey of exploration throughout the West, one that forever changed public perceptions of the region between Colorado and California. This expedition, his second for the U.S. Bureau of Topographical Engineers, was the greatest escapade of a man whose pioneering exploits and political ambitions were legendary even in his own time. Starting in St. Louis in May of 1843, Frémont and his men had arduously completed a circuit that, in part, led them from Wyoming, across northern Utah, into Idaho and eastern Oregon, down California, and back across southern Nevada and southwest Utah. In the mid-1800s, these regions were mostly unknown tracts of wild desert land with few well-mapped routes of travel. Myth and rumor flourished in the absence of any factual knowledge of the physiography of such a large portion of North America.

As he neared Utah Lake, Frémont was already aware that the fabled waterway linking the Rocky Mountains and the Pacific Coast, one of the principal aims of his journey, did not exist. Other explorers before him had raised doubts about this "Rio Buenaventura" fantasy, but Frémont's expedition was the first to circle the entire region in which it was rumored to exist. As he traveled north through western Utah toward Wyoming on May 23, 1844, Frémont was closing his grand circle of exploration. He had found no trace of the mythical stream that had induced generations of earlier explorers to enter a dangerous and forbidding land. His report, published a year later, would ultimately dispel the legend of the Buenaventura forever and introduce a persistent and alluring regional term:

> The *Great Basin*—a term which I apply to the intermediate region between the Rocky Mountains and the next range, containing many lakes, with their own system of rivers and creeks, (of which the Great Salt is the principal), and which have no connexion with the ocean, or the great rivers which flow into it.

Though Frémont had found no evidence of the Rio Buenaventura, he had observed numerous rivers and streams during the expedition. Some of these streams ran to the coast as expected, but several seemed to turn inland, flowing into an ill-defined interior basin centered on the modern state of Nevada (see Plate 1). Frémont had seen the Bear and Sevier Rivers in Utah turn west and eventually drain into landlocked alkaline lakes, the Great Salt Lake and Sevier Lake, respectively. A similar pattern of drainage was evident to Frémont, from both his own observations and records of earlier explorers, in the mountains south of the Snake and Owyhee Rivers, near the present Idaho-Nevada boundary. In that region, small streams such as the Mary's River and Maggie Creek glide south into the Humboldt River, which in turn vanishes into the thirsty silt of the Humboldt-Carson sink in west-central Nevada. In the area where California, Nevada, and Oregon meet, Frémont observed several streams, such as the Quinn River, that carry water a short distance from isolated highlands into small basins in the

Black Rock Desert that are completely enclosed by higher ground. Crossing over the Sierra Nevada, Frémont could not have missed noting how this mountain system separates the sea-reaching Sacramento and San Joaquin River systems from the headwaters of the Truckee, Carson, Walker, and Owens Rivers, all of which carried water into the great mysterious desert to the east, from which it never reemerged. South of the Sierra Nevada, Frémont encountered the Mojave River and its tributaries, which rise from the San Bernardino Mountains and twist to the northeast across the barren desert before dying in the blistering white alkaline silt of Bristol Lake. From the Mojave region to his encampment at Utah Lake, Frémont again noted small streams flowing northward into the interior abyss, along with a few others, such as the Virgin River, that ran south to meet the mighty Colorado. Frémont's explorations did not lead him through the entire region of internal drainage; in fact, he personally saw only a very small portion of it. Nonetheless, his observations, coupled with those of earlier travelers in the western wilderness, convinced Frémont that there must be a large region between the Rocky Mountains and the Sierra Nevada into which rivers flow, never to meet the sea. Thus, Frémont could not have selected a better term for this vast interior tract that collected and retained water from adjacent highlands—it was truly a *GREAT BASIN*.

Following Frémont's expeditions, as the settlement of the West accelerated, numerous scientific and topographic surveys entered the Great Basin, each diminishing a bit of the mystery of the region. Between 1849 and 1859, six major geographic expeditions entered the Great Basin region. Scattered across the landscape today are natural features that bear names honoring the leaders and participants of those early expeditions: the Stansbury Mountains, Beckwith Pass, Steptoe Valley, Simpson Springs, the Whipple Mountains, to name a few. In the 1860s and 1870s, a second wave of government expeditions entered parts of the Great Basin and added important scientific information to the geographic and topographic work of the earlier explorations. A new set of names found homes

Figure 1.1 Wheeler Peak in Great Basin National Park.

on the landforms of the Great Basin and adjacent regions during this era: King Peak, Wheeler Peak, and Powell Plateau, among others.

As cartographers, geologists, botanists, and zoologists gathered more information on this poorly known expanse, its boundaries could be drawn with greater precision. Since Frémont's original designation of the Great Basin was based on topography and surface drainage, this became the initial criteria for drawing the boundaries of this natural province. But confusion began to emerge as other scientists documented the unique geology, botany, and zoology of the region: the floral, faunal, and geological boundaries did not always coincide. They were close, but not precisely the same as the boundaries confining the region of internal drainage. For example, botanists have identified unique communities of plants that typify the flora of the Great Basin. Big sagebrush, *Artemesia tridentata,* is the floral icon of this assemblage. The intoxicating scent of sage drifts everywhere across the Great Basin landscape; this shrub is so widespread that author Stephen Trimble, in the title of his superb book, refers to the region as the "Sagebrush Ocean." Accompanying the sagebrush are, depending on specific climate and soil conditions, other such characteristic plants as rabbitbrush (*Chrysothamus nauseosus*), shadscale (*Atriplex confertifolia),* ephedra or "Mormon tea" (*Ephedra nevadensis*), greasewood (*Sarcobatus vermiculatus*), juniper (*Juniperus osteosperma*), and pinyon pine (*Pinus monophylla*). Though it is characteristic of the Great Basin region, this plant assemblage also can be found in the plateaus to the east and in the well-

Figure 1.2 Sagebrush, the floral icon of the Great Basin.

Figure 1.3 Great Basin collared lizard.

common kingsnake (*Lampropeltis getula*), among many others. Likewise, a characteristic array of lizards inhabits the Great Basin, including the Great Basin collared lizard (*Crotaphytus bicinctores,* (Figure 1.3), the zebra-tailed lizard (*Callisaurus draconoides*), and the sagebrush lizard (*Sceloporus graciosus*). The mammal fauna includes such emblematic "western" critters as coyotes (*Canis latrans*), badgers (*Taxidea taxus*), black-tailed jackrabbits (actually, hares of the species *Lepus californicus*), pronghorn (*Antilocapra americana;* not truly an antelope, contrary to the common name for this ungulate), mule deer (*Odocoileus hemionus*), mountain lions (*Puma concolor*), and many others. The smaller mammal fauna is dominated by rodents such as the ubiquitous packrat (*Neotoma desertorum*) and the kangaroo rats (at least two species of the genus *Dipodomys*), along with several different kinds of mice, voles, squirrels, and gophers.

The bird fauna of the Great Basin is an extremely varied assortment comprised of native desert dwellers, scores of migratory species, and numerous semipermanent visitors from adjacent regions. Moreover, each local habitat in the Great Basin—sagebrush plains, saline marshes, pinyon-juniper woodlands, parched sandy deserts, conifer forests, and subalpine tundra—supports its own specific coterie of birds. Nonetheless, the most common birds in the Great Basin are the typical "western" species, such as ravens (*Corvus corax*), several types of nighthawks and hawks (most belong to either the genera *Buteo* or *Accipiter*), golden eagles (*Aquila chrysaetos*), turkey vultures (*Cathartes aura*), mourning doves (*Zenaidura macroura*), sagehens (*Centrocercus urophasianus*), and dozens of songbirds.

In addition to these more conspicuous elements of the Great Basin biota, there are equally fascinating, but often overlooked, communities of insects, mollusks, and fish. However, the overall animal assemblages prevalent throughout the Great Basin collectively include many species that range well beyond the region of internal drainage. Black-tailed jackrabbits live throughout the West, as do mule deer. Gopher snakes and collared lizards can be found in the

drained deserts to the south. Thus, the botanical Great Basin, defined by the distribution of unique plant assemblages, is not exactly the same as the topographic (or hydrographic) Great Basin.

In a similar manner, the animals that live in the Great Basin desert constitute a distinctive, but widespread, assemblage that characterizes the region. Zoologists have compiled extensive catalogs of the various species of birds, reptiles, insects, fish, and mammals that inhabit the Great Basin, along with detailed information about the ranges of each. Among the reptiles, for example, the Great Basin rattlesnake (*Crotalus viridis lutosus*), one of eight subspecies of the western rattlesnake, is particularly characteristic of the region. Other common members of the Great Basin serpent brigade include the gopher snake (*Pituophis catenifer*), the long-nosed Snake (*Rhinocheilus lecontei*), and the

mountains and plateaus east of the Great Basin, as well as in many portions of central California to the west. Ravens and coyotes live almost anywhere, even despite brutal attempts by humans to eliminate them where they are not wanted, which is almost everywhere beyond the desert. Zoologists, in attempting to delineate the Great Basin on the basis of characteristic natural assemblages of mammals, birds, insects, fish, and mollusks, are confounded by the fact that so few of the native species are restricted entirely to the region. Thus, defining the Great Basin on the basis of endemic organisms, or the distribution of characteristic assemblages, produces boundaries that are often at odds with, and commonly less definite than, the hydrographic limits used by other scientists.

Even the general geophysical framework of the region—defined by specific types and ages of bedrock, the manner in which the rocks are faulted and contorted, and the structure of the deeper crust—extends in places beyond the lines that enclose Frémont's original province of internal drainage. Likewise, the prehistoric and native human populations of the Great Basin, scrutinized for more than a century by anthropologists, exhibit enough spatial, temporal, and cultural overlap with those of adjacent regions to obscure the definition of the region based on archaeological patterns and the regional history of human occupation. Hence, the boundaries of the Great Basin Province based on botanical, zoological, geological, or archaeological data do not always coincide with the original topographic or hydrographic confines. As naturalists continued to explore the phenomena of the Great Basin throughout the twentieth century, the meaning of the term became obscured by inconsistencies in its definition. Any particular ornithologist, botanist, geologist, physiographer, meteorologist, archaeologist, or zoologist was often describing slightly (or radically!) different things with the term "Great Basin." Because there were so many ways to define it, this historically mysterious region became, once again, ambiguous—even to those who were fascinated by the natural phenomena and tantalizing mysteries of the vast but somewhat vacuous landscape.

Decades of further exploration, intensive mapping, and debate about the Great Basin have still not resolved all of the obfuscation surrounding the region. Today, we still live with these multiple ways of defining what the Great Basin is. The term means somewhat different things to different people, depending on what aspect of this natural wonderland is being considered. There is a floristic Great Basin, a zoological Great Basin, a geophysical Great Basin, and even a climatological Great Basin. Most geologists, perhaps because of our predisposition toward things that can be mapped with some degree of precision and our reverence for historical priorities, use Frémont's original definition: that which is based on the criteria of internal drainage. Any pattern of regional drainage is, of course, really defined by topography, the highs and lows of the landscape, inasmuch as water only runs downhill, and probably always has. In such topographic usage, a *basin* is defined as a low area surrounded by higher terrain. Such a basin will collect and retain water from the adjacent highlands, and would therefore be easily delineated on a map by encircling the upper ends of all the streams and tributaries that drain into it. For this reason, the boundaries of the Great Basin based on drainage can be located with much less ambiguity than limits based on climate, flora, fauna, soil, or bedrock. The periphery of a drainage basin will separate tributaries flowing in opposite directions at some obvious, if not perfectly precise, place. Plant and animal communities, on the other hand, may blend and overlap over a broad area where one assemblage replaces another. Precisely where the faunal-floral boundary is located within the mixed zone may be unclear. In a similar way, climatic zones may also have broad areas of transition from one type to another, as temperate zones blend into steppes, which in turn merge into true deserts. Drainage patterns are not permanent features of the landscape, nor are they always easy to map, but they do provide the most consistent way to define the vast desert region we are discussing.

When defined and mapped on the basis of internal drainage, the Great Basin encompasses more than 160,000 square miles. This

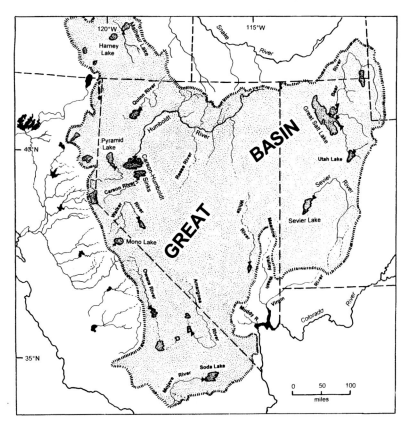

Figure 1.4 The Great Basin Province, defined on the basis of surface drainage. The shaded area is that portion of the larger Basin and Range Province that is characterized by internal drainage.

topographic (or hydrographic) Great Basin is an irregularly shaped area centered on the modern state of Nevada, but it also includes most of the western one-third of Utah, a portion of northeast California and southeast Oregon, and much of the Mojave Desert–Owens Valley area of eastern California (Figure 1.4). Everywhere in this region, surface streams fail to reach the sea; the water they carry is either absorbed, evaporated, or drains into isolated reservoirs such as Sevier, Pyramid, and Great Salt Lakes, among many others. While this Great Basin is accurately characterized as a region of interior drainage, it is not a single bowl that receives water from the surrounding terrain. In reality, it consists of many small sumps that collectively prevent the development of through-flowing drainage. The borders of this region are, in some places, well defined by the pattern of gullies and creek beds that draw water into it. In other places, the boundaries are more obscure, especially in the more arid tracts, where the surface drainage pattern is not well defined. Under the dry climate that prevails throughout much of the Great Basin, surface water tends to be ephemeral and limited. For this reason, the drainage pattern is

sometimes so weakly expressed on the land surface that it is difficult to trace with absolute certainty.

The eastern and western margins of the Great Basin are relatively easy to locate and are coincident with the crests of the Wasatch Range and Sierra Nevada, respectively. Elsewhere the boundaries are more obscure. The northern boundary is a meandering line that separates the tributaries of the Snake-Columbia river system from streams that either flow into the Humboldt River or vanish into small basins before reaching it (Figure 1.4). This northern boundary swings west from the extension of the Wasatch crest in southwest Wyoming through the southeast corner of Idaho, and weaves along the divides of the Raft River, Grouse Creek, and Jarbidge Mountains in the area where Idaho, Nevada, and Utah meet. In central Nevada, the boundary curves south around the watershed of the north-flowing Owyhee River, and then swings into southeast Oregon, where it forms a broad loop enclosing many isolated lakes, including Harney, Malheur, and Summer. After looping through southeast Oregon, the boundary follows the crest of the Warner Range in northeast California to the south, swings west to enclose the Eagle Lake and Honey Lake basins, before merging with the Sierra crest in the area north of Lake Tahoe. After following the Sierra crest south for more than 400 miles, the boundary bends west through the Techachapi Mountains to meet the San Gabriel–San Bernardino Mountains, enclosing the western Mojave Desert region. Eastward from the crest of the San Bernardino Mountains, the boundary extends into the area around Joshua Tree National Monument and meanders across the southern Mojave Desert separating the Mojave River drainage from washes that connect with the Colorado River system to the southeast. This portion of the boundary is rather indistinct, as no prominent mountain crest or other topographic feature is present to clearly mark its course. Approaching the Colorado River, the boundary abruptly swings north along the divide of the New York and Spring Mountains, which are separated from each other by the California-Nevada state line. In this area, the boundary of the Great Basin comes within a few miles of the Colorado River. In southern Nevada, the boundary curves to the northwest around Las Vegas Wash, which drains into the Colorado River (or at least it used to; it now flows into Lake Mead, backed up behind Hoover Dam). Technically, the city of Las Vegas does not reside within the Great Basin; this bit of topographic trivia brings some comfort to those who prefer to think of the Great Basin deserts as unspoiled and pristine. From the Las Vegas region, the boundary extends north into south-central Nevada to form a fingerlike embayment that omits the Virgin River–Muddy River drainage, which is part of the Colorado River system. This north-trending indentation into the area of internal drainage was created less than about 2 million years ago, when the modern Meadow Valley Wash began to flow into the Muddy-Virgin drainage system. After circling the Virgin-Muddy River watershed, the boundary turns east along the crest of the Pine Valley Mountains of southwest Utah, extends northward into the High Plateaus of southwest Utah, and eventually rejoins the Wasatch crest near Mount Nebo, which rises above the small town of Nephi, Utah.

THE GREAT BASIN AND THE BASIN RANGE

The topographic Great Basin is part of a larger natural province recognized by geologists and physiographers as the Basin and Range Province. The use of these two terms in older literature has been anything but consistent, creating some confusion about their exact meanings. The terms *Great Basin* and *Basin and Range* are not synonymous, though sometimes writers have used them that way. The Basin and Range Province was established primarily on the basis of **physiography** (essentially, the shape and arrangement of landforms), not by topography or drainage. As such, the Basin and Range consists of a expansive tract of land characterized by isolated, short mountain ranges aligned nearly parallel with one another, separated by intervening valleys. The term valley can be misleading because it evokes the image of a linear depression through which a river runs from its

Figure 1.5 Generalized Anatomy of a Fault-Block Mountain. The mountain ranges of the Great Basin are large blocks of rock uplifted and tilted by normal faults. The basins have developed where the blocks have rotated downward, creating low areas where sediment derived from the adjacent highlands accumulates.

source to its mouth. In fact, no such thing exists in the Basin and Range Province. The term basin is used because it is a much better description of the small lowlands that separate the mountain ranges, and into which the washes and gullies plunge from the nearby highlands. Thus, the term "Basin and Range" is an apt description of much of the American southwest where such a repetitive pattern of mountains and lowlands produces a highly corrugated, or wrinkled, landscape. The mountain ranges, of which there are more than 500 in the entire province, are usually less than 100 miles long and 10–25 miles wide. The flat-floored basins that separate the linear mountains are generally 50–100 miles long but may be substantially wider than the mountain blocks they separate. In the Great Basin portion of the Basin and Range, the difference in the elevation between the basins and the adjacent mountain peaks (known as the maximum relief) ranges from 4,000 feet to more than 7,600 feet. Unlike valleys in other regions, the basins of the Basin and Range, for the most part, are not carved by rivers. Instead, their flat surfaces consist of loose rock fragments eroded from the bedrock exposed along the confining mountain faces. The valleys thus are not erosional features; they are elongated depressions partly filled with varying amounts of sand, gravel, and silt litter. This "geological sawdust" extends deep beneath the surface. In places, more than 15,000 feet of it has accumulated in the gaps separating the mountain blocks.

The uplifted mountain blocks of the Basin and Range are almost everywhere bounded by normal faults, great fractures in the earth's crust that extend down under the basins. Immense blocks of rock have been displaced down these ramps to form the low valley areas, while the blocks on the opposite side of the fracture have either moved upward or have remained high relative to the descending block (Figure 1.5). Geologists are justifiably fond of the term "fault-block mountains" in describing the general structure of the Basin and Range highlands, as distinct from the folded types of mountain systems, such as the Appalachians, Alps, and California coast ranges.

In much of the Basin and Range Province there is a strong uniformity in the orientation of the mountains and valleys. In general, and particularly in the northern part of the Basin and Range Province across much of Nevada and western Utah, the trend of the mountains and valleys is mostly north-south. To the south, in southern Arizona and New Mexico, the directional fabric is a little less obvious, but it is still evident (see Plate 2). This alignment is what led the poetic geologist Clarence E. Dutton, upon viewing the first accurate terrain maps of the region in the late 1800s, to describe the province as "an army of caterpillars crawling toward Mexico (or coming "north out of Mexico," depending on who "quotes" Dutton!) In Dutton's fanciful description is an important seed of distinction between the Basin and Range and the Great Basin: the "caterpillars" were heading south to *Mexico*! In other words, the Basin and Range province extends well beyond the area of internal drainage that we have called the Great Basin. In fact, the southern portion of the Basin and Range province is not a *basin* at all: it is drained by the Colorado and Rio Grande river systems. This is the essential (but not the only) distinction between the Great Basin and the Basin and Range: the Great Basin is that portion of the Basin and Range, approximately the northern half of it, characterized by internal drainage. Thus the Basin and Range Province is a much larger physiographic entity than is the Great Basin, encompassing more than 300,000 square miles and stretching from Oregon well into Mexico. The landscape in the Basin and Range Province is a jumbled array of a block-faulted mountains, as is the Great Basin, but rivers in the southern portion of it eventually find their way to the sea.

The Great Basin is a land full of paradoxes. Consider, for example, that the region is generally very arid and water is scarce throughout this vast expanse of desert land. Yet the most commonly used definition of the Great Basin,

the hydrographic one, is based on the way the almost nonexistent water drains within it. Geologist Charles Hunt has pointed out that forests are minimal in this desert region, but on some of the highest peaks are found dense stands of bristlecone pine (*Pinus longaeva*), the oldest trees in North America. The oldest bristlecones of the White-Inyo Range in eastern California are at least 4,500 years old. On summer days temperatures can rise well over 100°F, while dust devils spin beneath the white-hot sky. But, on one February night in Elko, Nevada, I recall walking through eighteen inches of snow driven by a fierce wind, and shivering in the −25°F air. The aridity of the Great Basin is withering (over 70 percent of it receives less than ten inches of precipitation annually), making water a rare commodity in this parched land. And yet the region has extensive lake beds exposed almost everywhere on the surface of the low basins. The wild-and-woolly Great Basin was one of the last frontiers in North America to be explored and settled, and yet adventurous travelers and California-bound immigrants swarmed along the Humboldt River, slashing across the northern sector of the region, from the time of Jedediah Smith (1827) to the days of the great gold rush of the 1850s. Humans have always been scarce in the Great Basin. Even today it is one of the most sparsely populated regions in the United States. In 1990, Nevada had a population density of only eleven people per square mile; if the urban tumors of Las Vegas and Reno (containing 88 percent of the total population) are ig-nored, only a handful of residents are scattered across the sagebrush desert. But, in 1996, 9,400-year-old human remains, representing what might have been one of the earliest humans to live in North America, were discovered in Spirit Cave near Fallon, Nevada. With so little water and such barren lands, life for humans in the Great Basin has always been a struggle to sur-vive with minimal resources. Nonetheless, legends and myths concerning fabulous wealth and opulent cities have persisted since the seven-teenth century: Cibola, Sierra Azul, Tequayo, and Copala are just a few examples of mythical empires that were thought, at one time or another, to exist in the Great Basin. None of these places were ever found. Even in modern times, the vast majority of people who have sought wealth in the Great Basin, in either the mining camps or the casinos, have found nothing more than an illusion, proving that there's more than one type of mirage in this desert. For geologists, though, there is a grand paradox that surpasses all others: it is the contradiction between the elevation and the drainage in the Great Basin, and the conundrum it poses concerning the origin and development of this perplexing province.

THE FALLEN LAND

Great Basin: big hole, deep depression, immense indentation. The name elicits visions of bottomless voids and a sensation of vertigo, as if we were perched on the brink of an abyss, about to fall in, gravity leading us downward to the same place where all the rivers of this land flow. Upward thoughts of loftiness, of high places, of altitude don't come to mind easily when such a thing as a *Great Basin* is contemplated. After all, in order for anything to become a basin, it must be lower than adjacent areas: low, depressed, foundered, submerged, sunken.

Maps are magical. They are graphic images of places that can create a powerful connection between humans and terrain. Maps ignite our imagination about exotic places, they inspire our curiosity about the surrounding world, and they help us dream. Good maps, those compiled with cartographic skill and with accuracy of measurement, also represent fundamental truths of the land. They proclaim the realities of distance, space, elevation, and slope, free, or nearly so, from the weaknesses and obscuring effects of human perception. Study a good map of the Great Basin region, and note, in particular, the elevations of the peaks rising from the many mountain ranges: Ruby Dome in the Ruby Mountains, 11,387'; Wheeler Peak in the Snake Range, 13,063'; Deseret Peak in the Stansbury Range, 11,031'; Ibapah Peak in the Deep Creek Range, 12,087'; and Mount Jefferson in the Toquima Range, 11,941'. Now, find the low, level floors of the basins that separate the mountains. Look at the elevations of the towns situated in the valleys: Tooele in the Tooele Valley, 4,240'; Delta, in the flat Sevier Desert, 4,650'; Fallon in the Carson Sink, 3,962'; Eureka at the south end of Diamond Valley, 6481'; Elko along the Humboldt River, 5,067'; Cedarville in Surprise Valley, 4,675'. The Great Basin contains some of the highest peaks in the West, separated by nearly mile-high valley floors. Yet rivers still gather in its interior. We might expect rivers flowing on such an elevated tract as the Great Basin to find a route to lower ground, but they don't. The Great Basin is a basin, to be sure, but it is also a highland. Basin: highland—the juxtaposition seems absurd, anywhere except the paradoxical Great Basin. How is this seeming contradiction possible?

The Great Collapse:
Origin of the Great Basin

Dan DeQuille (William Wright), in his 1876 publication *The Big Bonanza*, recounted an explanation for the interior drainage of Nevada that was offered by an old-time prospector. The Almighty, the prospector said, had just about finished "creatin' and fashionin'… this here yearth" when he got to Nevada. It was late on Saturday evening as the Creator was winding rivers across the land. The intent was to make "one big boss river" that would outclass the Mississippi, Ohio, or Missouri, all of which had been created earlier. But, darkness fell on the last day of creation before the Almighty could wrap the Humboldt, Truckee, Carson, Walker, and other streams together. So, he simply "tucked the lower ends of the several streams into the ground, whar they have remained from that day to this." In the 1870s, this explanation was as good as any for the paradox of a highland basin. But, remember that geology reveals deep history. Maybe the prospector was right; however, as geologists learned more and more about the geological history of the Great Basin, the actual methods by which the Almighty might have accomplished the task of tucking rivers into the ground have become apparent.

The widest part of the Great Basin is girdled by two great mountain ranges: the Sierra Nevada on the west and the Wasatch on the

Figure 1.6 East-facing escarpment of the Sierra Nevada near Big Pine, California.

east. Rising like a great stone wall from the desert below, the eastern escarpment of the Sierra looks eastward over Owens Valley, and over range after range toward the Wasatch Front. West of the Sierra crest, the slope descends gently to the San Joaquin and Sacramento Valleys, with an average slope of only about 3 degrees. Where this gentle ramp feathers out onto the flat floor of the great Central Valley of California, the merger is so subtle as to be barely noticed by travelers on the trans-Sierra highways. The Wasatch Mountains are equally asymmetrical, with a sheer west face that drops almost vertically into the Ogden, Salt Lake and Utah Valleys. East of the narrow Wasatch crest, behind the sheer western front, the slopes descend gently into the Kamas Prairie, Morgan Valley, and Heber Valley. These so-called "back valleys" of the Wasatch hinterland, all roughly 6,000 feet above sea level, are "low" only in comparison to the highest peaks of the mountains, which tower a mile above them. Unlike the western slope of the Sierra Nevada, the eastern borderlands of the Wasatch extend as a high undulating plain all the way to the Rocky Mountains in Colorado. The Wasatch and the Sierra, despite their profound geological dissimilarities and slight differences in symmetry, are essentially physiographic mirror images of each other. The precipitous walls of the two ranges face each other across more than 500 miles of corrugated desert terrain (Figure 1.7). They are the bookends of the Great Basin.

Between the two bordering ranges lies the basin-range-basin-range staccato. However, superimposed on the repetitive rise and fall of the land is a subtle, but revealing, pattern of elevation. The valley floors are lowest adjacent to the east and west margins of the Great Basin. The two largest cities in the region—Salt Lake City and Reno—are located in these low peripheral basins; on opposite sides of the Great Basin, both have downtown elevations of about 4,500 feet. In the middle of the Great Basin along a transect joining these cities—for example, in the Ruby Mountains region—the valley floors are 500 to 1,500 feet higher than at the periphery of the province. In other words, superimposed across the rugged terrain of the Great Basin is an almost imperceptible arch, a swelling of the land surface nearly lost amid the heaving ascent of individual mountains and fearsome drops into the basins around them. The Great Basin is a dome-shaped region, broken like a pile of bricks that have fallen from an ancient arch that formerly extended across the region.

The domed shape of the basin is perhaps more evident from the distribution of lakes within it than it is from topographic maps. The two largest lakes in the region today, Pyramid Lake (Plate 3) and Great Salt Lake, are ponded against the mountains in the lower ground on the west and east sides, respectively. About 15,000 years ago, during the last of the great ice ages, water was much more abundant in the Great Basin than it is today. But the Ice Age topography was similar to that of our time. Consequently, both Pyramid and Great Salt Lakes were preceded by enormous Ice Age predecessors: Lake Bonneville in Utah and Lake Lahontan in Nevada (more details about these lakes are presented in Chapter 9). Then, as now, the water in the Great Basin tended to drain toward the lower margins of the arched region and away from its high axis in east-central Nevada. The arch-like form may be subtle, but it is strong enough to direct the flow of the Humboldt River to the west from the axis, toward the Humboldt Sink. What does this arch-like form of the Great Basin tell us of its origin? Plenty! ≈

Remember the asymmetry of the Sierra Nevada and the Wasatch Range, the western and eastern walls that confine the Great Basin? When the Wasatch block was elevated, it was also tilted down to the east, while the Sierra block was rotated down to the west as it ascended. Although the Sierra seem to be the younger of the two mountains, both began to rise along fractures known as normal faults during the past 10–20 million years (mid-Miocene time by the geological clock). These great rifts in the earth's crust were the product of stretching, or extensional, forces that developed as the foundation of the modern landscape began to evolve. We will explore this phase of Great Basin prehistory more closely in Chapter 8, but the important point here is that the province has been distended, not crushed or torn, into its present form. Geologists know this from the geometry of the fractures in the Wasatch Fault system, the Sierra Nevada Fault Zone, and virtually every other range-forming fault in the Great Basin. For the Wasatch and Sierra, the faulting involved rotation and tilting of the upthrown blocks, producing the strong topographic asymmetry of these mountains. As a general rule (and, of course, there are some exceptions), all of the mountain ranges of the Great Basin reflect this kind of lurching displacement; the rotation of mountains in the eastern Great Basin tilted most of them to the east, while in western Nevada and eastern California the mountain blocks tend to list toward the west (Figure 1.7). If we project the gentle slopes of the Sierra and Wasatch back into the Great Basin from the west and east, they meet high above the sagebrush desert in north-central Nevada, defining a vanished dome that, at one time, encompassed the Great Basin before it was a great basin. This dome has collapsed, but vestiges of it are still preserved in jostled terrain and in the back slopes of the mountains that border the fallen portion. The Great Basin unfolded as the ancient dome was sprained beyond the breaking point by extensional forces. The dome cleaved into hundreds of blocks that tilted in opposite directions on either side of the axis. Some blocks foundered lower than others to become the foundations of the basins, while the high-

standing blocks gave rise to the mountains. The collapse of the Great Basin seems to have occurred primarily between 20 and 10 million years ago. A variety of evidence (plant fossils, modern structure of the crust, etc.) suggests that the most profound collapse occurred between 16 and 13 million years ago. Prior to that phase of collapse, the region had an average elevation of more than 9,800 feet. Since that time, the basin-forming blocks have plunged several thousand feet, while the mountain blocks have dropped, for the most part, by less than a few hundred feet. Thus, the entire landscape between the Wasatch Range and the Sierra Nevada has slumped from its former dome-like profile, but by variable amounts. Recent geological studies have indicated that the collapse of the Great Basin was not quite as simple an event as it may seem. It is now clear that some areas collapsed earlier than others, by different amounts, at varying rates, and through a variety of specific mechanisms. Nonetheless, the resemblance of the modern landscape to a slumped dome is not a coincidence: the Great Basin is a fallen land.

WATER IN THE GREAT BASIN

The collapse of the Great Basin to form an area of internal drainage began millions of years ago, and it continues today. There was never any magic moment in time when, all of a sudden, the streams in this region lost their connection to the sea. The Great Basin fell in increments, as the individual basins evolved. The entire province, as it is recognized today, originated in a piecemeal fashion as the isolated basins collapsed to their present elevations. As the collapse progressed, one stream at a time adjusted its course to the changing land surface, each seeking the shortest path to the lowest ground.

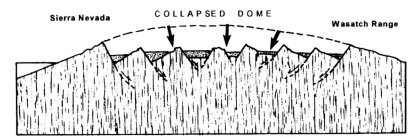

Figure 1.7 The Dome-like Profile of the Great Basin. On a regional scale, the Great Basin can be considered a collapsed dome between the Sierra Nevada on the west and the Wasatch Mountains on the east.

Many of the mountain-basin faults are still active and, as a consequence, the Great Basin continues to evolve. The edges of the modern Great Basin are its most geologically active portion. Within the Great Basin, earthquakes are most frequent (and powerful) along the base of the Sierra and Wasatch Ranges. The most recent volcanic activity is also clustered along the periphery of the province. This geological clamor signifies the continuing growth of the Great Basin, a subject to which we will return in the last chapter of this book.

Water reaches the Great Basin primarily from the Pacific Ocean, the great incubator of cold-front storms that pound western North America, generally from October to April. Although most of the moisture is wrenched from such weather systems by the Sierra Nevada, the residual amount that falls in the Great Basin constitutes the single largest source of water in the region. Smaller, but significant, amounts of rainfall arrive from the Gulf of Mexico and Baja regions, usually during the late summer "monsoon" season. Spectacular July and August thunderstorms, triggered by tropical moisture creeping north during those months, are especially common in the southeastern part of the region. Sometimes, rainfall is produced by low-pressure systems that develop *within* the Great Basin, usually in the spring and fall months, and most commonly in the central part of the province. Though this "internal" rainfall is of minimal significance on a regional scale, it can amount to nearly half of the annual rainfall total in the interior of the Great Basin. In the distant geological past, water reached the Great Basin the same way it does today, though the proportions between the three sources might have been different under varying climatic regimes. No matter how the water got to the Great Basin, more and more of it was retained as the internal drainage systems began to evolve in response to the falling land. At the beginning of the collapse, water was much more abundant in the region than it is today. The earliest sediments washed into the basins record numerous large lakes and rivers, as we will learn later. For now, the important point is that the desertlike climate that exists today throughout most of the Great Basin is

a relatively recent development. The physiographic evolution of the Great Basin preceded the development of its modern aridity; the origin of the mountains and basins and the birth of the desert are two different things.

Few people traveling through the modern Great Basin dispute its desertlike character. The parched appearance of the land is no illusion; 190,000 square miles of the Great Basin, almost all of it, receives less than 20 inches of precipitation each year. About 75 percent of this region gets less than 10 inches of annual precipitation. The driest part of the Great Basin is along its western margin, adjacent to the Sierra Nevada, where rainfall averages less than 5 inches each year. The highest mountains receive much more precipitation; each year 30–40 inches of water falls as rain and snow on the lofty peaks of such ranges as the Deep Creek Mountains, the Ruby Mountains, the Snake Range, and the White-Inyo Range. But these places are really just small islands of moisture surrounded by a vast expanse of aridity. The water that falls on the isolated high peaks drains into the bordering valleys but doesn't normally travel very far before it either is lost by evaporation into the dry air or is swallowed by the thirsty soil. The Humboldt River, the only permanent stream to flow through the Great Basin, carries a meager 500,000 acre-feet of water into the Humboldt Sink in western Nevada in an average year. For comparison, the Colorado River, which also flows through desert terrain (at least in its lower reaches) carries between 10 and 15 million acre-feet of water per year. In view of the limited precipitation in the region, it is not surprising to learn that a substantial fraction of the water in the Great Basin actually originates beyond its borders in the high Sierra Nevada and Wasatch Ranges. For example, the Great Salt Lake, huddled against the eastern edge of the Great Basin, receives two-thirds of its water from the Wasatch and Uinta Mountains to the east. In a similar manner, Walker Lake, located in western Nevada, gets more than 80 percent of its water from the Walker River, which carries moisture from the Sierra Nevada. Imagine how scarce water would be in the Great Basin if it was not surrounded by these high watersheds.

It is not only the limited supply of water than makes the Great Basin arid. This region is also relatively warm, which results in generally high rates of evaporation. Evaporation of moisture from the Great Basin averages about 50 inches per year. In the southern portion, where elevations are lower and temperatures higher, the evaporation rate may exceed 70 inches per year, while in the cooler northern reaches of the province, as little as 40 inches of moisture vaporizes annually. A simple comparison of these regional statistics, less than 20 inches of water received versus an average of 50 inches of water lost, reveals that much of the Great Basin is a region of net water loss. In fact, most of the Great Basin qualifies as a true **desert**, using the definition for that term that climatologists have developed: a region which receives less than half as much moisture as is lost by evaporation. However, not all the Great Basin is true desert; in fact, only about 40 percent of the region can be accurately classified as such. Regions that lose more moisture to evaporation than they receive, but less than twice as much, are technically known as **steppes**. Steppes are still dry areas, of course, but their aridity is less extreme than that of true deserts. More than 85 percent of the Great Basin is a mosaic of desert and steppe, with the remaining fraction representing the higher mountain slopes and peaks where water supply and water loss are less out of kilter. Nonetheless, the Great Basin, whether we classify most of its land as desert or steppe, is undeniably dry.

The aridity of the Great Basin results from a coincidence between its location and relatively recent geological events. Centered on a latitude of 38° North, the Great Basin is positioned along the northern fringe of the Tropic of Capricorn. In this zone, the climate is strongly influenced by one of the subtropical high-pressure belts that circle the earth at latitudes of about 30° north and south of the equator. Warm air rises from the equator, diverges aloft, moves both north and south, and descends back to the earth's surface in these two zones. As the air falls, the atmospheric pressure is increased, which, in turn, reduces rainfall. Thus, two belts of relatively dry climate girdle the earth on either side of the equator: the Tropic of Capricorn (in the Northern Hemisphere) and the Tropic of Cancer (in the Southern Hemisphere). Many, but not all, of the world's deserts and steppes are located within these belts. The Great Basin is not directly under the descending air in the Tropic of Capricorn, but it is close enough to be influenced by it. The limited rainfall in the Great Basin reflects, in part, its location close to this belt of prevailing aridity. But, there is another factor to consider.

The western edge of the Great Basin is defined by the crest of two great mountain systems: the Sierra Nevada and the southern Cascade Range. The Sierra Nevada, nearly 400 miles long and 60 miles wide, reaches a maximum elevation of 14,495 feet at Mount Whitney. Along the crest of the Sierra Nevada, there are twelve other peaks higher than 14,000 feet. The Sierra Nevada extends northward where it merges, almost imperceptibly, into the southern Cascade Range in the vicinity of California's Lassen Peak. Though the geology of the Cascade Range is utterly dissimilar to that of the Sierra, it continues the topographic spine of the Pacific mountains north through northern California, Oregon, and Washington. The Cascade Range is generally lower than is the Sierra Nevada, but its crest is dotted with high volcanic cones that rise, in some places, to elevations exceeding 14,000 feet above sea level. Together, the Sierra–southern Cascade mountain system forms the dramatic western edge of the Great Basin. As we have already discussed, most of the winter precipitation that falls in the western portion of North America originates from the northwestern Pacific Ocean in, or near, the Gulf of Alaska. As the cold-front storms race inland from the ocean, they tend to approach the Great Basin from the northwest. Less frequently, in summer and early fall, tropical storms migrate inland from the southwest. In either case, the moisture-laden air masses must rise over the Sierra-Cascade barrier before they reach the Great Basin. As the storms ascend the western slopes of the mountains, the air is cooled, water vapor condenses, and heavy rain and/or snow falls on the ascending slopes. By the time the air masses reach the Sierra-Cascade crest, they have been wrung like a

Figure 1.8 Sand Mountain near Fallon, Nevada.

sponge, and the mountain slopes have been blanketed in snow or pounded by rain. Moving east, over the mountain barrier into the Great Basin, the air masses descend the abrupt eastern escarpment. As the air spills into the Great Basin, it is compressed and warmed by virtue of its descent to lower elevations. Rain is rarely produced from such dehydrated warm air. The sapping influence of the mountains on passing storms is known as the **orographic effect**, popularly referred to as the "rain shadow." The Great Basin is located directly in the extensive rain shadow cast by the Sierra-Cascade system. Because of this, it is the great graveyard of many Pacific storms.

The prevailing dryness of the Great Basin affects more than just the flora, fauna, lakes, and rivers of the region. Many of the landforms that make the desert so stunning are directly related to the arid conditions. For example, across the Great Basin, there are many spectacular dune fields, including the "Little Sahara" area of western Utah, Sand Mountain near Fallon, Nevada (Figure 1.8), the Eureka and Stovepipe Wells dunes in the Death Valley region, and the Kelso Dune Field in the Mojave Desert. Though sand dunes also develop in other (e.g., coastal) environments, it is in deserts that they become such magnificent elements of the landscape. The correlations between deserts and dunes stem from two primary factors. First, the lack of moisture restricts natural vegetation that might serve as windbreaks, and allows the wind to howl full force over the desert floor, transporting great clouds of sand and dust as it goes. Second, dry

sand is less cohesive, and therefore more susceptible to wind agitation, than it would be in a wetter environment with more abundant soil moisture. All of the Great Basin dune fields exist in areas where the winds tend to lighten due to some interfering pattern of hills, ridges, mountains, or gullies. In such protected pockets, the weakened winds lose the ability to sustain sand transport, and the dunes result from grains accumulating in the same place over a long period of time. When the wind picks up sand grains from the granular desert soils, the larger pebbles and gravel tend to remain in place. This process of selectively removing the fine soil particles via wind is known as **deflation**. Over time, deflation results in **desert pavement**, an armor of closely spaced pebbles that mantles the surface after the smaller grains have been swept away by the wind. Desert pavement almost never develops on the surface in wetter regions, where the soil is more cohesive and where dense vegetation covers much of the ground. Some of the larger stones in the desert pavement, after centuries of sandblasting, develop smooth triangular facets oriented toward the prevailing winds. Such rocks are known as **ventifacts,** another landscape feature unique to dry environments.

Also related to the aridity of the Great Basin are the many salt-encrusted temporary lake beds known as **playas**. Flat playa "lakes" exist in the center of almost every valley in the Great Basin. Some of them, such as the Black Rock Desert of northwest Nevada, are so large that it is difficult to see across them. Occasionally, when rare floodwater gathers from the adjacent hillsides, the beds of playas can be covered by an inch or two of water. In a few days at most, the water will have vanished into the thirsty air, leaving behind an encrusting mixture of silt and salt on the incredibly flat floor of the playa. Dunes, playas, slat flats, desert pavement, and ventifacts—all testaments to the parching dryness of the Great Basin—each contribute a distinctive touch to the striking landscape.

Everything about the modern landscapes of the Great Basin—the mountains, the valleys, the dunes, the playas, the internal drainage, the aridity, and even the flora and fauna—is of relatively recent origin. The physiographic collapse of the region began less than 20 million years ago (and continues today). The internal drainage is a reflection of the depression of individual basins between the mountains and thus can be no older than the collapse of the region. The Sierra-Cascade mountain system, the primary cause of the pervasive aridity, has a complex history of ascent, but most of it occurred over the past 5 million years. Before that uplift of these bordering ranges, in the absence of the rain shadows they cast to the east, water was much more plentiful in the Great Basin region. Thus, the deserts (and steppes) of the Great Basin are also relatively recent, reflecting climatic shifts induced by geological events that began only a few million years ago. Not surprisingly, plant fossils that provide a record of the prehistoric vegetation in the Great Basin reveal a lush pre-desert flora hardly comparable to the sparse xerophytic verdure that covers the terrain today. So, the Great Basin desert is a fallen land, a dry land, and an evolving land. On a planet with a history encompassing more than 4.5 billion years, any landscape less than 20 million years old is virtually inchoate. The bleakness of the Great Basin, and the luring austerity of its harsh emptiness, are embellishments added to the land during the latest instant of geological time. If we scratch through that coat of desolate paint, the geological foundation of the land

Figure 1.9 Desert pavement and ventifacts on the floor of Panamint Valley in the southern Great Basin.

reveals other, completely different, worlds that preceded the modern desert. The rocks in the Great Basin, hundreds of millions of years old, are the documents in which the existence of these former worlds is recorded. Grab a rock—any rock, anywhere in the Great Basin—and you hold a fragment of an incredible autobiography. Look closely at the rock—the fossils, the mineral grains, the color, and the texture. Each of these means something and provides clues about the unimaginably long history of an incomparable land. Let the rock carry you back to the time of its origin: before the mountains, before the basins, before the dryness. As you grasp the history preserved in the rocks, and stand surrounded by the awesome desolation of the modern Great Basin, let yourself drift into the past to wander through those former worlds. Be prepared for surprises!

CHAPTER TWO

The Breaking of a Continent

It was on a Friday afternoon more than twenty years ago that Dave and I raced west on I-80 from Salt Lake City, watching the blanket of smog that enveloped it shrink in the rear-view mirror. As we traveled toward Wendover, Dave and I talked about rocks and fish, the dual objectives of our weekend adventure to the Ruby Mountains region of northeast Nevada. We mostly talked about fish because neither of us knew very much about the rocks of the Ruby–East Humboldt Ranges. Dave was a bright kid, one of the best of a talented crop of undergraduates then enrolled in the Geology department. But, at this early stage in his academic career, he hadn't gained much experience with the regional geology of the Great Basin, or anywhere else for that matter. Unlike Dave, I could not claim youth as an excuse for my ignorance of Ruby Mountain geology. Instead, it was the geological complexity of the region that had fostered my nescience. I knew that the Ruby Mountain–East Humboldt Range region was what geologists referred to as a "metamorphic core complex," a newly invented term (in the early 1980s) for a dome-like mass of highly altered rocks, partially buried under a cover of relatively undeformed rocks that had shuffled about along a shear zone over the underlying dome. There were, at the time, about twenty metamorphic core complexes known in the Great Basin, and their nature was a hot issue in regional geology. From reading a few descriptions of these features, I envisioned them as incredibly complex structures, with wild patterns of folding and metamorphism that would alter the appearance and orientation of rock masses. Also, these areas contained numerous igneous bodies that cut across older rock units and were disrupted by many young, normal faults and

volcanic units that would complicate the whole mess even more. Even though many Great Basin geologists were working on the core complexes, and they were the subject of much-heated debate at professional meetings, I avoided them like the plague. Metamorphic core complexes scared me. The complexity of these structures struck me as virtually incomprehensible and I often wondered how anyone could make sense of them. I could not fathom their bewildering three-dimensional structure, and I had no clue how they developed, or when. I was fascinated by every other aspect of Great Basin geology, and considered myself something of an expert on the geological history of the region. But, the core complexes left me cold … and bedeviled. So, in response to my inability to understand the metamorphic core complexes, I did the logical thing: I ignored them, or at least tried to.

But, now that Dave and I were headed into the heart of the Ruby Mountains metamorphic core complex, I had decided to bite the bullet and try to make some sense of the rocks. I had recently learned of some new radiometric dates that would help simplify the geologic imbroglio by providing some temporal reference points amidst the rocky garble. Some of the new dates indicated that certain rocks near the Ruby Mountains were older than 2 billion years, and I was greatly intrigued by those ancient rocks. They were, perhaps, the oldest rocks in the Great Basin, and I wanted to know how they fit into the story of the Ruby and East Humboldt Ranges as well as the province as a whole. All I had to do, it seemed, was find the various rocks that had been dated, assess their relationship with surrounding rock units, and use these clues to unscramble the history of the entire complex. From what I had read, it seemed like a feasible thing to attempt. In fact, several geologists had already been engaged in such an effort and had published their preliminary conclusions. I simply needed to follow their footsteps, using the published information as signposts on the highway to geological enlightenment. I was beginning to think, as we passed through Wendover en route to Wells, that metamorphic core complexes might not be so bad after all. And even if they were, in fact, the geological nightmare that I envisioned them to be, there was always the fish to pursue!

We rolled into Elko just as the sun began to touch the western horizon. After refueling the truck, we headed south over the low pass through the Elko Hills and down into Lamoille Valley. The west-facing slope of the Ruby Mountains loomed before us like an enormous gray wall, touched with rosy highlights from the sinking sun. The gaping, U-shaped glacial canyons were already deep in shadow, and the bare rock of the skyline was dusted by early autumn snow. Lamoille Valley in those days was a ten-mile wide expanse of sagebrush, with irrigated parklike pastures around the little village of Lamoille, near the mouth of Lamoille Canyon. As we passed the ranch houses of Lamoille, situated along gurgling creeks flowing from the rose-gray mountains, I couldn't imagine a more beautiful setting. The drive from Elko to Lamoille was for Dave and me a journey through paradise. We felt like we had discovered our own personal Shangri-La, a place so hidden and remote that it would never be discovered, let alone despoiled, by the outside world. We had no inkling then that one day the booming suburb of Spring Creek would spread like a pox across Lamoille Valley, complete with high-density housing, four-lane paved roads, traffic lights, schools, fast-food outlets, and West Coast–style strip malls. That Elko would ever become a city large enough to have a suburb was the last thing we would have predicted during that serene drive across Lamoille Valley twenty years ago.

We began to ascend the road up Lamoille Canyon as the sunset dissolved into dusk. There was still enough dim ruddy light, though, to reveal the canyon's wide, U-shaped gape. Neither Dave nor I had anticipated such a sweeping

glacial spectacle as Lamoille Canyon displayed at twilight. The scale of the yawning chasm, its ghostly illumination, and maybe the rapid gain in elevation, all made me a little dizzy as the truck labored up the grade, past fluttering aspens and beaver ponds. Lamoille Creek paralleled the road and, in most places, ran swiftly down the steep incline of its boulder-strewn bed. In some areas, where the gradient of the creek flattened out a bit, or where beavers had managed to tame the torrent, the small stream slowed to a languid roll. Such slack-water stretches thrilled us; we imagined the hungry fish that must be waiting under the cutbanks along the edge of the calm water. We couldn't see the rocks exposed in the walls of the canyon very well, but it didn't matter. We had discovered what appeared to be a fisherman's paradise and, at the moment, we could not have had less interest in the rocks.

Eventually we came to a deserted campground along the stream and decided to pull in for the night. From the camping area, we walked to Lamoille Creek to examine its course and flow. Although the little stream was swift and its bed rocky, we thought it had fishing potential. But it was too late, and too dark, to fish, so we unrolled our bedding, built a small fire, and heated up some beans for our supper. The moonless night was so dark that it was all but impenetrable to our eyes. After supper, we lay awake for awhile, straining to look through the black-light veil surrounding our camp. I had a sense that we were surrounded by very old rocks, but in the darkness it was impossible to verify this notion. Dave noticed some subtle movement a few yards into the darkness, and raised up on his elbows for a better view. It turned out to be an everyday Great Basin porcupine, *Erethizon dorsatum*, out for a midnight stroll. I wanted a closer look at the little pincushion, so I stumbled after it with a plate of leftover beans, hoping to tempt the creature to hang around for a while. By the time I caught up to it, the prickly rodent had already vanished into the undergrowth along the creek. I put the plate of cold beans down, yearning to lure our friend back. However, it never reemerged from the brush. When I checked the plate early the following morning, there was no sign of any nocturnal visitors. I guess porcupines have little interest in cold beans.

We spent the morning hours of the next day exploring Lamoille Creek, looking for stretches where the fishing might be good. We didn't have much success: in most places, the creek was either too turbulent and swift for easy fishing, or the banks were choked by a tangle of thick vegetation that prevented us from getting to the water. We found a few intact beaver dams where we could sneak upstream to the impounded ponds and cast our flies over the wall of logs and branches. We caught several small brook trout this way, but otherwise the fishing was not very exciting at this point in the canyon. As noon approached, my interest began to shift to the geology of the canyon. While Dave continued fishing, I began to direct my attention to the rocks exposed along the banks of the creek. There were outcrops of hard gray limestone, partially silicified, with numerous cross-cutting fractures filled by white calcite. I noticed also boulders of silvery schist, a high-grade metamorphic rock, scattered near, or in, the creek bed. I also noticed lots of hard, nondescript brown rock that most geologists would call argillite. Occasionally, a chunk of speckled granite, a slab of fine-textured monzonite, or a block of reddish-gray gneiss with wildly contorted bands would appear in the rubble adjacent to the creek. I was intrigued by the variety of rocks, but felt my dread of metamorphic core complexes creeping back with each new type that appeared along the stream. As I continued walking, Lamoille Canyon was looking more like a geological hodgepodge with each step. How old were these rocks? From what formations were they derived? Which were exposed in the immediate vicinity? Which had been transported from somewhere else? Which of the rocks produced the new radiometric dates I had read about? I had no answers for any of these questions.

I returned to the truck after my hike to find Dave sitting on the tailgate eating a sandwich. We decided to leave the creek after lunch and seek better fishing spots elsewhere in the area. While we ate, we gazed upward at the steep canyon walls, assessing the geology of Lamoille Canyon as best we could from that vantage point. We saw the massive gray-brown layers twisted upon themselves in the cliffs above us. We noticed how abruptly these rocks terminated farther down the canyon: a sure sign of faulting. It was impossible to make sense of the convoluted strata from where we sat, but it was clear to both of us that the rocks of the central Ruby Range were highly deformed. I told Dave about the rocks I had seen along the creek and described the general geology of the metamorphic core complexes, at least to the extent that my minimal understanding of them allowed. I told Dave that we were in the midst of a mauled and mangled mountain range, butchered by more than two billion years of geologic upheavals. Dave listened silently as I described the recumbent folds, the thrust faults, and the penetrating bodies of igneous rock that I thought I could see in the walls of Lamoille Canyon. When my arm-waving harangue was finished, Dave maintained his steady gaze down the canyon for a few seconds. Without breaking his stare, he finally spoke: "You think there's any brown trout down there?"

As we prepared to leave Lamoille Canyon, I dug out the maps we had brought along. We didn't have detailed maps of the entire region, but we had enough USGS quads to get a general feel for the lay of the land around us. Our interest was still more directed toward fishing than to geology, so Dave and I scanned the maps for blue: lines for creeks, splotches for lakes. There were lots of small lakes and streams in the high country of the Ruby Mountains. We could have continued up the canyon to trails that led to alpine ponds such as Verdi Lake, Island Lake, and Lamoille Lake. South of Lamoille Canyon, we noticed that the headwaters of the South Fork of the Humboldt River included many small streams that might offer good fishing. In fact, there were so many places to hunt fish that making a decision about where to go next was not easy. As my eyes drifted over the map, I noticed a small blue oval southwest of the town of Wells. It was labeled "Angel Lake" and looked as though it could be reached via a short, but steep and winding, road that ascended the east face of the East Humboldt Range from the town of Wells. Since we had only one day left, and Angel Lake was more or less on the way back to Salt Lake City, I received little resistance from Dave when I suggested that we make that locale our last camp. Dave agreed that since both big fish and geological erudition had eluded us in Lamoille Canyon, we had little to lose by gambling on Angel Lake. We packed up so quickly that I almost forgot the plate of beans that I had set out for the porcupine. When I ran to retrieve it, I found that there was not a single bean left on it. It had been licked clean by something that left no trace other than the spotless plate. I paused for a second, staring into the undergrowth for some sign of life, which never came.

En route to Angel Lake, we stopped in Wells to refuel the truck again and then headed south, following a dusty dirt road. We rumbled up the rocky road and eventually began to ascend the sage-covered slope rising toward the northern end of the East Humboldt Range. The road climbed steadily as it swung over several ridges and hills. After about fifteen miles or so, I noticed the orange needle on the temperature gauge creeping toward the "H" end. I wasn't greatly concerned with the rising temperature gauge (it was *modus operandi* for my old truck), but mentioned to Dave that we might have to stop for a short pause on the way to the lake. He just nodded as he lifted his eyes to scan the impressive wall of stone that was beginning to rise in front of us. A few miles ahead and high above us, we could see that the road etched a zig-zag into the face of the mountain. From our perspective, the road looked impossibly steep and precariously carved into the almost vertical mountain face. As we continued along the road, the needle on the gauge kept rising: I noticed that it now covered the last "e" in the word T-E-M-P-E-R-A-T-U-R-E.

As we passed the entrance to a campground, the road swung sharply into the first of several switchbacks. The grade steepened considerably. I could feel myself sink into the back of the seat as the truck nosed up the road. I shifted into third gear, then into second. The roar of the truck changed into a painful whine as it continued up the grade. Dave, in the passenger seat,

squirmed as we rounded a corner carved into the steep mountain slope. The land dropped away to the right of truck with a terrifying plunge toward the flat sagebrush plain below. Looking past Dave, through the open passenger-side window, I saw nothing but transparent blue air floating beside us. I felt a little disoriented, like I was flying, as we continued our ascent toward Angel Lake. The spell was broken, however, when the truck's whine deepened into an ominous growl. An instant later, it seemed, I began to smell smoking oil and hot rubber. The orange needle of the temperature gauge was banging against the right-hand peg on the dial. It was time to stop.

We eased to the side of the steep road, getting as close to the fearsome edge as we dared, which wasn't very close at all. The truck was essentially parked *in* the road, albeit off to the right side a little. The Angel Lake road appeared to be so lightly traveled that I had little concern about oncoming traffic. On the other hand, I had grave concerns about the precipitous, unguarded edge of the road: just looking over it was enough to make me reel a little. While we waited for the truck to cool down, we walked up the road to survey the scenery around the East Humboldt Range. The horizon seemed to expand without limit, fading like a ghost into the infinite distance that completely engulfed us. The sky was arched overhead like a endless blue canopy, streaked with feathery bands of white clouds. So grand was the scale of the desert that the boundary between the heavens and the horizon was blurred by its spaciousness. The world above and the world below just blended together, like forever merging with eternity. Surrounded by such enormity, I felt less than small: I was downright minuscule, a submicroscopic nothing in the midst of a boundless macrocosm of rock and sky.

Walking back to the truck, I avoided the menacing edge of the road by hugging the inside of the curves. I noticed the contorted and banded rocks in the road cuts: gneiss again, but this time of a somewhat darker gray color than the gneiss in Lamoille Canyon. The distinct light and dark bands of the gneiss, along with their wild convolutions, suggested a rock with a long and agonizing history of geological distress. I picked up a large chunk of the rock to examine more closely. It had almond-shaped clots of mineral grains that tapered into light-colored bands that, in turn, rippled through the rock: "augen structures," as geologists call them. I wondered about the age of these rocks, if maybe they were the two-billion-year-old materials that I was seeking. They certainly looked old, older than anything I had ever seen before anywhere in Nevada. I had no way of knowing, at the time, how old the rocks really were, but I convinced myself that these must be the oldest rocks in the East Humboldt Range. With such thoughts in my head, my sense of smallness took on a new dimension as I approached the cooled-down truck. I not only was dwarfed by the physical aspects of nature here but was lilliputian in a temporal sense as well. Compared to these rocks, our entire lives—for that matter, the entire history of our species—represents nothing more than a fleeting instant in time. All our failures and accomplishments, all our recollections of the past, and all our dreams of the future take place in nano-time. Nothing that occurs on such a limited temporal scale can have any enduring significance, except to nugatory little specks of cytoplasm such as ourselves, locked in that same infinitesimal timespan. Against the backdrop of such limitless space and time, the irony and arrogance of human self-adoration is both absurd and comical.

Such are the emotions that old rocks can foster. Struggling to comprehend the age and significance of the oldest rocks in the Great Basin leads to more than just an understanding of deep history; it also provides a stabilizing perspective on our own existence in the natural universe. However, our main concern here is to establish a historical foundation for the geologic evolution of the Great Basin. For that, we must turn our attention to places like the Ruby Mountains, where geologists have identified the oldest rocks in the region. Accurately interpreting the stories in these rocks, and even recognizing them when they are encountered, requires some novel approaches. Let's consider first what makes truly ancient rocks special.

THE NATURE OF OLD ROCKS

By human standards, most rocks are old. Take a random hike through any Great Basin mountain range and you're likely to be stepping over 20-million-year-old solidified lava or maybe even 400-million-year-old limestone that originated as gray mud on the floor of some long-vanished sea. Finding such ancient rocks is, literally, a matter of hitting the ground with your hat. But, what about the *really* old rocks? I mean those that are several *billion* years old. How often would you drop your boot on rock with a ten-digit age? In the Great Basin, and anywhere else for that matter, the answer is: almost never!

Really old rocks are hard to find for several reasons. Approximately 75 percent of the surface of the earth is covered by sedimentary rocks, the type that consists of mineral grains produced by the relentless weathering of bedrock exposures. Once they form from decomposing and crumbling bedrock, these grains are washed away in streams, blown about by the wind (if they are small enough), or simply dissolved by water. Eventually, this geological refuse arrives at some location where its movement ceases and it begins to settle out in layers of mud, silt, or sand. Since the production of these mineral grains, along with their transportation and deposition, all occur at the surface, it is not particulary surprising that three-fourths of the planet's exterior is covered by a veneer of such material. The sediment grains may remain loose and unconsolidated, or they may become hardened into solid stone. Either way, old rocks tend to bury themselves under their own debris. The older a rock is, the more likely it has become concealed beneath layers of younger rock and detritus. When it comes to primeval rocks, the earth is no exhibitionist; it tends to hide the elderly.

But, even in those places where for one reason or another sediment layers have not accumulated, it is still not easy to find truly primordial rock. Our planet is a convulsive orb, with an active plate tectonic system. The dynamic interactions between slabs of the lithosphere (the rigid exterior of the planet, above the deeper

Figure 2.1 Banded gneiss from Farmington Canyon in the Wasatch Mountains of Utah.

and softer zones), result in a global-scale rock recycling system that has been in operation for at least 4 billion years. In places, the lithospheric slabs are jammed under other plates, in a process known as subduction. The descending slab (along with some of the rock around it) eventually melts as it falls deeper into the hot interior of the earth. The liquid magma oozes upward from its source to erupt as volcanoes or to cool below the surface as a mass of granitic rock. The subduction process thus reforms old rock by melting it into magma, from which new igneous rocks develop once the fluid has migrated upward into cooler domains and solidified. In the various subduction zones around the world, the average rate of lithospheric consumption via subduction is, roughly, an inch or so per year. Against the backdrop of Earth's immense history, this seemingly slow process of rock destruction could easily have reformed all the original rocks of the planet seven times over! That we have any multi-billion-year-old rocks at all, anywhere on this writhing planet, is a minor miracle in itself.

Finding old rocks is therefore not easy, but there are even other problems to contend with. When really ancient rocks are found, they are commonly so disfigured by metamorphism (the application of heat, pressure, and chemically active fluids) that scientists can determine very little about their original nature. As is the case with people, old rocks have richer histories than young ones. A 2-billion-year-old rock has likely been subject to many more volcanic rampages, wrenching mountain-building events, and crushing earthquakes than a youngster 50 million years old. The process of metamorphism is related to such earth upheavals, and results in changes in the composition, texture, and internal structure of rocks. All of the oldest rocks known are, to one degree or another, metamorphosed. Often, geologists can only approximate the original nature of such ancient rocks, because their primary features have been masked by the veil of metamorphism. The challenge for the geologist investigating such rocks is similar to trying to understand the nature of wheat kernels by studying a loaf of bread.

Another uncertainty that besets geologists interested in extremely old rocks is the very thing that makes them intriguing: their age. The standard methods of measuring rock ages from the ratios between radioactive elements and their accumulated stable daughter products do not always work well with very old rocks. The process of metamorphism can modify the ratios in such rocks, producing imprecise or erroneous calculations of the absolute age. In extreme cases, metamorphism can reset the internal atomic clocks so that the calculated date represents the time of metamorphism, not the age of the rock. To make matters even worse, the spotty exposures and variable appearance of most very old rocks in the Great Basin make it difficult to correlate, or match up, the small and isolated outcrops where they are exposed across this large region. The actual distribution and patterns of these old rocks—crudely dated, variously metamorphosed, and mostly concealed under younger material—is thus very difficult to reconstruct. Geologists who specialize in the study of primeval rocks are to be commended for their valor, admired for the difficulty of their task, and, perhaps, consoled for their lunacy.

THE PRECAMBRIAN INTERVAL

Traditionally, all rocks older than about half a billion years have been lumped by geologists into an immense abyss of earth history known as the Precambrian. Although the Precambrian interval encompasses about 87 percent of Earth's history, it has not been as precisely subdivided into smaller increments as have the more recent chapters of the geological time scale. The relatively crude partitioning of

Precambrian time reflects the uncertainties and difficulties inherent in studying such ancient rocks. When scientists attempt to fathom Precambrian time, they simply don't have much to go on, and what they do have is not always certain. Nonetheless, geologists are able to recognize some very broad subdivisions of the immense Precambrian interval (Figure 2.2). The definitions of each component, and the boundaries that separate them, are relatively imprecise, but there is general agreement on the basic framework of the earliest stages of Earth history.

The first half-billion years of our planet's history is currently referred to as the Hadean Eon, the earliest portion of Precambrian time. The Hadean Eon begins with the initial accretion of the earth from a great cosmic cloud of dust and gas, an event that appears to have occurred about 4.6 billion years ago. This proto-earth was nothing like our modern planet: it had no continents or oceans, there were no volcanoes, no atmosphere other than the residual gases (such as hydrogen, methane, and others), and no interior activity. This was the embryonic earth: a black-sky world with a homogenous interior and a barren surface blistered by the radiation from the newborn sun. Once it formed, this lifeless dirt clod of a planet began to evolve into the active haven of life that we know today. But that transition took some time. Specifically, it took the entire span of the Hadean Eon (4.6 to about 4.0 billion years ago). It was during the Hadean Eon that the interior structure of the earth began to develop, setting into motion the internal dynamics that sustain the geological commotion of the modern world. Earth breathed its first geological breath. An inner core of solid iron slowly developed, with a zone of swirling fluid, the outer core, around it. The multilayered mantle evolved above the molten iron of the outer core and, eventually, heat energy was transferred through it to the rocky exterior shell, known as the lithosphere. Heat-driven currents beneath the rigid lithosphere caused it to break apart into mobile plates. Some of the plates separated from each other and molten rock oozed upward through the cracks to erupt as the world's first volcanoes. Other plates collided, forming the first mountain chains as earth-

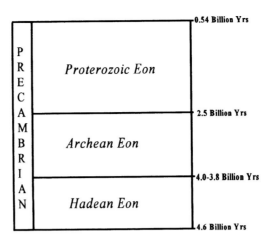

Figure 2.2 The Precambrian is an informal age designation that encompasses the earliest 87 percent of earth history. It has been subdivided into three formal eons: the Hadean, Archean, and Proterozoic.

quakes rocked the land. As the volcanoes erupted, they belched great clouds of steam and other gases from their summits. Eventually, as the young earth cooled, the steam condensed into liquid water and rain pounded the surface for the first time. Some of the volcanic gases, notably nitrogen and carbon dioxide, accumulated in the primeval atmosphere to replace the lighter gases that had escaped into space. Gradually, liquid water collected in the low places and the first oceans emerged on the Hadean earth.

Thus, the Hadean Eon was the time of the geological awakening of the planet Earth. Its transformation into a vibrant, active sphere was driven by heat. The originally solid and homogenous proto-earth was at least partially liquified by the internal heat from gravitational compression and the decay of radioactive elements. In addition, scientists believe that at this early stage in the history of the solar system, the earth orbited the sun through a cosmic junkyard littered with rocky and/or icy debris left over from the accretion of the larger proto-planets. These smaller objects would have been gravitationally drawn into the earth, striking its original surface with an intensity that we can only imagine. The impacts would have generated additional heat that might have led to further melting. No one knows for sure if the entire proto-earth was transformed into a large molten droplet or if only a portion of the embryonic planet was liquified. But, it is fairly certain that the early earth was heated and melded to some degree during its first half-billion years of existence. So, the Hadean Eon is well named: it was a hellish

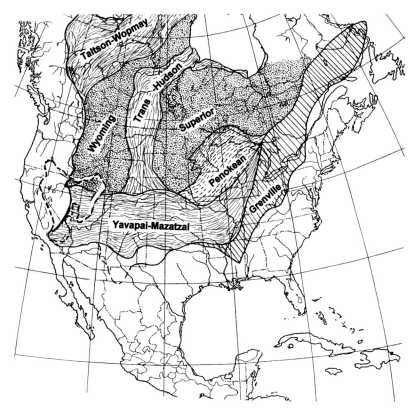

Figure 2.3 Precambrian Basement Provinces of North America. The Wyoming and Superior Provinces (shaded) consist primarily of Archean rocks, while rocks in the other provinces are mostly of Proterozoic age. This pattern reflects the piecemeal construction of the core of North America during the Archean and Proterozoic Eons.

epoch everywhere on (and in) the young planet Earth!

So few rocks of the Hadean Eon are known that it is sometimes referred to as the "rockless eon." It is easy to understand why. For most of the Hadean, there were no rocks at all, only the original dust grains (microscopic specks of mostly metallic grit) compacted into a spherical mass. Only near the end of the Hadean Eon were the normal rock-forming processes operating on a broad scale. Even then, the rocks that formed would now have a minimum age of nearly 4 billion years. Given the geological chaos that has been going on since the close of the Hadean Eon, it is very unlikely that such ancient rocks would have survived to the present time, or be recognized as such even if they did. No one has ever found a trace of Hadean rocks in the Great Basin. With a couple of notable exceptions, the same could be said of the entire planet.

The end of the Hadean Eon is marked by the complete evolution of the inner structure of the earth and the maturation of the plate tectonic system that it drives. The close of the Hadean, estimated to have occurred between 4.0 and 3.8 billion years ago, coincides with the beginning

of the next phase of Precambrian time, the Archean Eon. During the Archean Eon, the earliest small continents began to materialize as the active plate tectonic system surged along. Where plates were subducted, partial melting produced fluid rock (magma) that was relatively deficient in metals and rich in silica (SiO_2). As this primitive magma ascended toward the surface, some of it cooled to form masses of relatively low-density rocks such as granite. These granite masses were buoyant with respect to the denser materials around and beneath them, and, in essence, they "floated" higher above the surface than the primordial rock that surrounded them. The first continents on the earth emerged as small chunks of granitic rock. Elsewhere, especially along the boundaries between lithospheric plates, the magma reached the surface to erupt in volcanic events. Linear chains of volcanic mountains, known as volcanic arcs, created elongated landmasses that rose as islands above the primal seas.

Mobilized by the shifting plates to which they were attached, the Archean microcontinents and volcanic arcs occasionally collided with one another, building larger continental masses. Gradually, throughout the Archean Eon, the continents grew larger via the accretion of small chunks of subduction-related granitic rock or volcanic-island arcs. This process of Archean (and somewhat later) continent building is clearly expressed on a map of the ancient basement rocks in North America (Figure 2.3).

The oldest rocks of North America, exposed primarily in the Canadian Shield region, are arranged in an intricate mosaic of various kinds of metamorphic and igneous rocks. Beyond the Great Lakes–Hudson's Bay region, these old materials are mostly buried under younger rock but can be glimpsed in a few places. Recall that it is very difficult to find, interpret, and date rocks of Archean age. There is enough information, however, to allow geologists to recognize the basic patterns in the geological foundation of North America. In fact, the embryonic core of our continent is now recognized to consist of "provinces" plastered together like a gigantic geological collage. The provinces consist of complex associations of metamorphic rocks (such as banded gneiss), and bodies of igneous rocks (such as granite), and metamorphosed volcanic rocks known as greenstones. Each province of the Precambrian basement contains rock assemblages of more or less uniform character and structure. This is not to suggest that the provinces are geologically simple; in fact, they commonly consist of rock suites of bewildering complexity. But, each province can be delineated on the basis of rock types and ages, degree and pattern of metamorphism, and the orientation of geological structures such as faults and folds. The Precambrian provinces are reasonably discrete packages of rock that formed together during a specific event, or series of events. Geologists have named the provinces for the regions where they are best exposed and/or are best developed.

The patchwork pattern revealed by the rocks that comprise the ancient core of North America reflects the process of sequential continental growth during the Archean and later phases of the Precambrian. The oldest rocks in North America, some nearly 4 billion years old, are found in the Superior and Wyoming Provinces. Not surprisingly, these provinces are composed primarily of oceanic sediments and other materials, now intensely metamorphosed, associated with bodies of granite. These bundles of rock seem to represent small masses of crustal material—Archean "microcontinents," or "microplates"—that became welded together via collisional tectonic events during the time that the ancient core of North America was being constructed. Other Precambrian provinces—the Penokean, Trans-Hudson, Yavapai, Mazatzal, and Grenville—represent continued continental growth in the later Proterozoic Eon (2.5 billion years to 0.54 billion years ago).

The Proterozoic Eon is the last major subdivision of Precambrian time. In general, rocks of Proterozoic age are usually less severely metamorphosed than are Archean materials, and some are hardly altered at all. Proterozoic rocks are also much more widespread than Archean sequences and are known from many areas in North America. As the name suggests, it was during the Proterozoic Eon that primitive life became abundant on the earth, though it

probably originated through the gradual processes of chemical and biological evolution in the preceding Archean Eon. Partly as a consequence of the growth of living systems, the atmosphere and oceans of the ancient earth became "modernized" during the Proterozoic Eon as well. The atmosphere was cleansed of carbon dioxide and began to accumulate free oxygen for the first time near the end of Proterozoic time. The oceans continued to evolve as well during this time; some metals (such as iron) were oxidized and removed from the primordial seawater as oceanic sediment. The salinity of the oceans increased as minerals were washed into the sea from the growing continents or vented from undersea volcanoes. Continental growth continued during the Proterozoic Eon as well. By the end of the eon, ancient North America was approximately half its present size.

The rapid growth of continents was one the most profound global phenomena during the late Archean and Proterozoic Eons. The ancient cores of all the modern continents were steadily enlarged as smaller landmasses were carried toward their margins by lithospheric plate movements. Each microcontinent or volcanic arc that collided with a larger proto-continental mass was crushed against the edge of the larger block, as it was detached from the lithospheric plate it was riding on (Figure 2.4). Near the end of the Proterozoic Eon, this process of microcontinetal accretion had produced at least three continents comparable in size to the major landmasses of the modern world. *Ur*, from the German term for "original," consisted of the older cores of modern India, Australia, and Antarctica. *Atlantica* was comprised of the old rocks of modern South America and Africa. *Nena*, whose name is an acronym for **n**orthern **E**urope and **N**orth **A**merica, was the proto-continent on which the terrain that would later become the Great Basin was located.

The construction of Nena spanned an interval of more than one billion years. It began in the late Archean, when the Wyoming, Superior, and other provinces to the north accreted to form what has been called *Arctica*. Later, in the early and middle Proterozoic, additional land

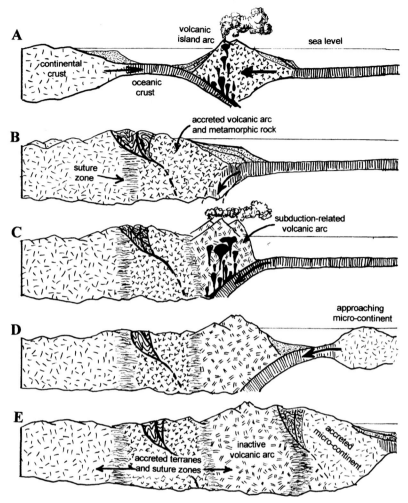

Figure 2.4 Continental Accretion. A, a volcanic island arc approaches a small continent; B, the island arc collides with, and is sutured to, the continental block. A zone of deformed and metamophosed rocks marks the suture between the two masses; C, a subduction zone gives rise to a volcanic arc along the leading edge of the continent; D, a microcontinent approaches the growing continent; E, accretion of the microcontinent adds material to the larger continent. Events of this type produced the mosaic of Precambrian basement in North America. In the Great Basin, the boundary between the Archean Wyoming Province and younger rocks to the south represents a tectonic suture.

In the diagram: volcanic island arc, sea level, continental crust, oceanic crust (A); accreted volcanic arc and metamorphic rock, suture zone (B); subduction-related volcanic arc (C); approaching micro-continent (D); accreted terranes and suture zones, inactive volcanic arc, accreted micro-continent (E).

masses were accreted to Arctica to form the larger, and younger, Nena. In southwest North America, the growth of Arctica into Nena during the first two-thirds of Proterozoic time is thought to have involved the addition of several individual microcontinents, now recognized as Proterozoic provinces: the Mojave (or Mojavia), the Yavapai, and the Mazatzal (Figure 2.5). The progressive suturing of smaller provinces onto the southwest margin of Nena is important in understanding the age, type, and distribution of the oldest rocks in the Great Basin.

Even when Nena emerged as a fully developed land mass, the great era of continent building had yet to reach its dramatic climax. In the late Proterozoic, about 1 billion years ago, Nena, Ur, and Atlantica merged to form *Rodinia*, the first genuine supercontinent on Earth. Because the suturing of Nena to the other late Proterozoic continents occurred either on the eastern side or far beyond the present west-

ern extent of North America, that event left no trace in the geological record of the Great Basin. But the Grenville belt in the southeastern United States is thought to have been formed in the last great Precambrian collision. The name Rodinia is derived from the Russian word *rodina*, or "motherland." Once this supercontinent developed in the late Proterozoic, it contained virtually all the world's landmasses and was surrounded by a global ocean.

As impressive as it was, Rodinia was short-lived. It was only 100–200 million years after Rodina formed that it began to break apart. Continental fragmentation, or rifting, begins when upward-moving currents develop in the semisolid portion of the upper mantle, that zone of the interior earth beneath the rigid and rocky lithospheric plates. The material rising from the mantle is very hot. It reaches the base of the lithosphere, where it causes the rocks to thermally expand. Eventually, the heat from the rising plume causes the partial melting of rocks near the base of the lithosphere, producing bodies of magma. The early stages of rifting are expressed on the surface of a continent by a dome-like uplift, a consequence of the thermal inflation of the rocks above the hot zone. As rocks are stretched across the top of the swelling dome, large fractures (faults) open across the highest portion. Volcanic activity begins as magma generated far below the surface oozes upward along the opening fractures (Figure 2.6). The faults that form across the top of the domed surface widen to produce deep rift valleys as the stretching forces (tension) intensify. Typically, these faults form in a tri-radiate pattern, in which three great fractures diverge from a common point at angles of about 120° to each other. Several such three-arm fracture systems may develop on the bulging surface of a continent being severed from below.

Even though volcanic activity in the domed region may be intense, only a small fraction of the magma produced beneath the continent ever gets to the surface; most of it is deflected as underflow, moving in opposite directions deep under the continental mass. As rifting proceeds, the next stage in the process involves the initial separation of the portions of the continent on

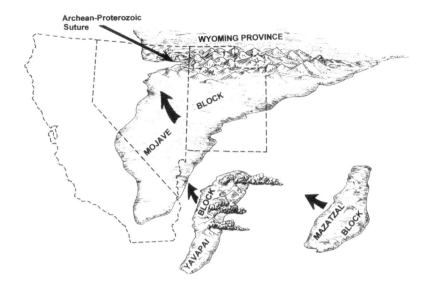

Figure 2.5 The Assembly of Nena during the Proterozoic Eon. The Mojave block has already collided against the southern edge of the Wyoming Province, and a major mountain system marks the suture. The Yavapai block (shown as an island arc) and the Mazatzal block (depicted as a microcontinent) approach from the southeast. Later in Proterozoic time, these two isolated blocks would collide with the Mojave-Wyoming portion of Nena.

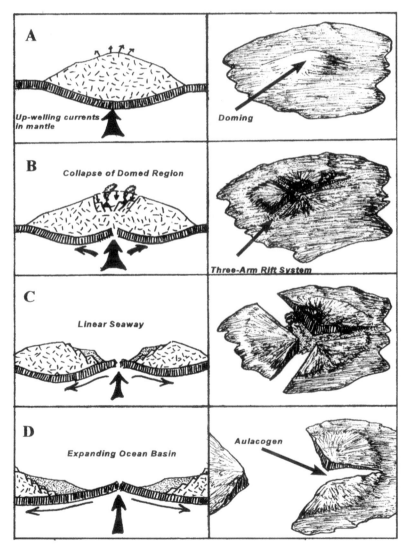

Figure 2.6 The process of continental rifting, seen in cross-section (left) and from above (right). A, rifting begins with the development of a zone of ascending hot material and surface doming; B, domed region collapses along fractures while underflow in the mantle stretches the continent in opposite directions; C, after initial rifting, a linear seaway separates the two continental fragments; D, the ocean basin widens through time and an aulacogen represents the remnant of the three-arm rift system.

either side of the zone of subsurface heating. The rigid portions of the continent not affected by the rising mantle currents are tugged in opposite directions by the deep currents. The rock below the domed region is thermally weakened and offers little resistance to the stretching forces produced by the underflow. As the opposite ends of the continent move apart, the domed region begins to collapse into the space created between the diverging blocks. The three-arm rift systems widen and deepen, while volcanic activity continues. Eventually, the floors of the rift valleys drop below sea level, and a linear seaway develops as seawater rushes into the low trough. At this point the continent has been severed into two (or more) large portions, separated by a shallow, narrow sea. The linear seaway represents two arms of the three-arm rift systems that have either penetrated the margins of the continent, or joined with rifts from adjacent centers. Normally, the narrow seaway is bent or kinked in places that represent the union of rift valleys from adjacent three-arm centers, or the angles that separated the individual rifts diverging from a common center. In the modern world, this stage is represented by the Red Sea and Gulf of Aden, which isolate the Arabian Plate from the African Plate. The rifting of Arabia from Africa began only a few million years ago and is still in progress. The Red Sea and the Gulf of Aden intersect at an angle of about 120°; they represent two flooded rift valleys that formed from the same center. The modern Arabian Peninsula represents a small lithospheric plate that has been broken free from the African Plate in the recent geologic past and has yet to move far from its source.

Once the continental blocks have been separated and the linear seaway has been established, continued volcanic activity above the upwelling mantle currents results in an undersea chain of volcanoes known as a **mid-oceanic ridge**. Here, the phenomena of sea-floor spreading produces new oceanic lithosphere as the undersea eruptions continue. Slowly, large plates creep away from the axis of the mid-oceanic ridge in opposite directions, forcing the rifted continental blocks farther apart. In the modern world, rates of sea-floor spreading vary between about 2 to 10 centimeters per year (cm/yr), and average around 5 cm/yr. For comparison, the average rate of sea-floor spreading is about the same as the rate of human fingernail growth. Though sea-floor spreading is a slow process by human standards, it eventually leads to the widening of the linear seaway into an expansive ocean basin. As the broad ocean basin develops, it is continually flanked by continental blocks that were formerly united. Thus, continental rifting not only produces smaller continents but also results in a new ocean basin. If a supercontinent such as Rodinia is rifted into several smaller fragments, several new ocean basins could be created between the smaller continental masses.

The complete assembly of Rodina required nearly two billion years, beginning in the early Archean. Its fragmentation seems to have occurred in only a few hundred million years during the latest Proterozoic time. Both Precambrian events were significant in understanding the distribution and types of Precambrian rocks in the Great Basin, but to unequal degrees. There are some hints of the assembly of Rodina, especially the Nena portion of it, in the oldest rocks of the Great Basin. The rifting of Rodinia is much more clearly expressed in the Great Basin, and that continental breach influenced virtually every other geologic event since the late Proterozoic. To understand the ways in which these geologic milestones are recorded in the Great Basin, we must shift our attention to the actual record: those elusive and secluded rocks of Precambrian age.

THE OLDEST ROCKS OF THE GREAT BASIN

The oldest rocks in the Great Basin are exposed primarily in the northeastern portion of the region (Figure 2.7). Archean metamorphic rocks have been identified in the northern Wasatch Mountains (Farmington Canyon Complex), in the Raft River, Grouse Creek, and Albion Ranges near the common intersection of the Idaho, Nevada, and Utah state lines, and in the East Humboldt Range near Wells, Nevada. Additional exposures of rocks that are probably Archean in age have been mapped on Antelope Island in the Great Salt Lake and near the mouth of Little Cottonwood Canyon in the Wasatch Mountains, southeast of Salt Lake City. As is the case with all Archean rocks, the ages, origins, and patterns of distribution of these rocks are not known with absolute certainty, but from them geologists have developed some general ideas about the earliest history of the Great Basin.

How old are the oldest rocks in the Great Basin and what is their nature? Almost certainly, the oldest rocks in the province occur in the Farmington Canyon Complex and coeval units. The term "complex" is a good choice for this sequence of rocks: it consists mostly of a complex assemblage of high-grade metamorphic rocks, including beautifully banded gneiss and platy schists of several different types (see Figure 2.1). Such rocks are particularly well exposed in Farmington Canyon in the northern Wasatch Mountains, from which the name of the sequence originates. The metamorphic rocks are cut of numerous masses of igneous rock such as granite and pegmatite that were emplaced into the metamorphic rocks as fluid magma injected along fractures or other planes of weakness. The pegmatites of the Farmington Canyon Complex are similar to granite in composition, but consist of large crystals (up to six inches on a side) of tan feldspar, clear to white quartz, and splintery black hornblende. Since the granitic masses cut through the metamorphic rocks, they must be younger than the Farmington Canyon Complex as a whole. The age of the metamorphic rocks in the Farmington Canyon Complex is at least 2.6 billion years; portions of it may be as old as 3 billion years. These beautiful rocks are highly prized for building and ornamental stone: a good way to observe the variety of rocks in the Farmington Canyon Complex is to stroll through the older neighborhoods in Farmington, Utah, where many of the original homes were constructed with native rock. I wonder how many of the residents in those old homes know that they live within walls built from some of the oldest rocks in North America.

In the Raft River–Grouse Creek Range–Albion Mountains area, rocks nearly as old as the Farmington Canyon Complex are known as the Green Greek Complex. The Green Creek

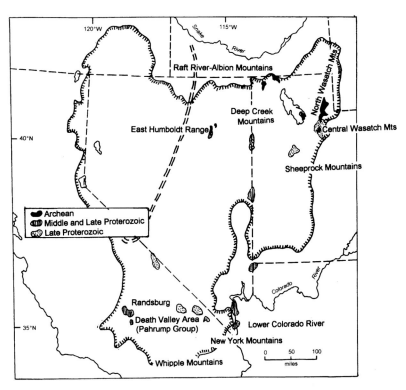

Complex is comprised mostly of light-colored granitic gneiss that is mingled with dark, platy schist and amphibolite, a rock consisting almost entirely of intergrown crystals of the mineral hornblende. The minimum age of the Green Creek Complex is about 2.5 billion years, but it has been affected by numerous episodes of metamorphism that make its age determination somewhat uncertain. In the East Humboldt Range, high-grade metamorphic rocks, again mostly banded gneiss, also appear to be at least 2.5 billion years old. Finally, near the mouth of Little Cottonwood Canyon east of Sandy, Utah, geologists have mapped the Little Willow Formation, a sequence of gneissic rocks that might be as old as 2.5 billion years but may also include younger metamorphic units. All of these rocks—the Farmington Canyon Complex, the Green Creek Complex, and the Little Willow Formation—represent complicated sequences of originally sedimentary and igneous rocks that have been affected by extreme temperature and pressure conditions. For the Farmington Canyon Complex, and probably for the other rock sequences as well, there apparently have been several episodes of metamorphism since these rocks originally formed. The multiple events of metamorphism occurred at temperatures as

Figure 2.7 Important Precambrian rock outcrops of the Great Basin. The oldest rocks of the Great Basin are exposed in the Raft River, Albion, and Grouse Creek Mountains near the junction of the Utah, Nevada, and Idaho borders. Younger Precambrian rocks, of Proterozoic age, are much more widespread to the south. The 0.706 Line (double dashed line), indicating the edge of the rifted Precambrian continent, slants diagonally through the center of modern Nevada.

Figure 2.8 Gneissic rocks of the Little Willow Complex in north-central Utah.

high as 500–800°C and pressures nearly 10,000 times the surface pressure of the earth. In many of these ancient rock sequences, numerous zones of crushed and sheared rock provide additional evidence of the extreme conditions to which they have been subjected.

The oldest rocks in the Great Basin, such as the Farmington Canyon Complex, are thus tortured rocks. All of these rock units are of Archean age, with the exception of the Little Willow Formation, a portion of which might be early Proterozoic. In areas of the Great Basin south of the Utah-Nevada-Idaho corner, Precambrian rocks occur, but most of these are clearly younger than the Archean. Parts of the Mojave block in the southern Great Basin have recently been dated as late Archean, but this is an exception to the general rule. Elsewhere in the Great Basin, Archean age rocks only exist north of a east-west line that extends from just south of Salt Lake City westward toward the East Humboldt Range. South of that line, most Precambrian rocks are of Proterozoic age, substantially younger than the severely mangled Archean materials to the north. The relatively small area of the Great Basin where Archean rocks occur is regarded by most geologists as a portion of the Wyoming Province that extends into the Great Basin. The Farmington Canyon Complex and coeval units therefore represent some of the oldest rocks in North America: they originated at a time when the entire continent was in its infancy.

What is the nature of the Archean-Proterozoic boundary in the northeastern Great Basin? Recall that the oldest Precambrian provinces in North America, such as the Wyoming Province,

were the embryonic seeds of the continent. Later in Precambrian time, microcontinents and volcanic arcs collided around the edges of these proto-continents to build a larger landmass. One by one, the microcontinents crashed into the periphery of the older masses, raising mountains in the process and crushing the rocks that were caught between the impacting blocks. The Archean-Proterozoic boundary in the northeastern Great Basin represents one such suture, specifically the collision zone between the southwest edge of the Wyoming Province and a microcontinent that approached from the south. When did this collision occur and what was the nature of the impacting microcontinent? Clues to those mysteries can be found south of the Archean-Proterozoic boundary, in those places in the Great Basin where younger Precambrian basement rocks are found.

Protozoic Rocks

The Proterozoic Eon is the longest formal unit of the geological time scale. It encompasses almost two billion years, from 2.5 billion years ago to about 0.54 billion (540 million) years ago. Not surprisingly, geologists have subdivided this immense interval into early, middle, and late eras. Once again, the uncertainties inherent in unraveling Precambrian history prevent absolute precision in establishing the boundaries between the three phases of the Proterozoic Eon. Currently, the early Proterozoic (sometimes referred to as the "Paleoproterozoic") is thought to encompass the interval from the close of the Archean, some 2.5 billion years ago to about 1.6 billion years ago. The middle portion of the Proterozoic (or "Mesoproterozoic") spans the interval from 1.6 to about 1 billion years ago. The final subdivision of Proterozoic time (the "Neoproterozoic") extends from about 1 billion years ago to the beginning of the Paleozoic Era, about 540 million years ago. Terms such as "neoproterozoic" and "paleoproterozoic" are more or less arbitrary units of Proterozoic time, and, because their use has not yet become universal, we will simplify our discussion by using the terms early, middle, and late as subdivisions of Proterozoic time.

In the Great Basin, there are many more exposures of Proterozoic rocks than of Archean materials. Proterozoic rocks are exposed primarily in the eastern and southern Great Basin. Some of the rocks associated with the Archean complexes already discussed are Proterozoic in age, so we can learn something about this eon from places such as the East Humboldt Range and the northern Wasatch Mountains. However, even better exposures of Proterozoic rocks occur in the Death Valley region, in many localities across the Mojave Desert region, in southeastern Idaho, in the Sheeprock and Deep Creek Mountains of western Utah, and along the lower Colorado River near the southern boundary of the Great Basin. Studies of the relatively abundant Proterozoic rock exposures have resulted in a more detailed understanding of the history of the Great Basin near the end of Precambrian time than was the case for the older Archean Eon.

The Proterozoic rock record in the Great Basin can generally be divided into two strikingly different rock sequences: 1) the early and middle Proterozoic rocks are mostly high to moderate grade metamorphic rocks, such as schist and gneiss, and 2) the overlying, and relatively unmetamorphosed, late Proterozoic package is dominated by sedimentary rocks such as sandstone, shale, and conglomerate, sometimes intercalated with dark-colored lava flows. A profound gap, or unconformity, normally separates the two Proterozoic rock assemblages wherever they are both exposed. The twofold nature of the Great Basin Proterozoic rock assemblages suggests that some significant changes in the rock-forming processes must have occurred during Proterozoic time. To understand what those changes might have been, it is necessary to take a closer look at the rock sequences that record the events.

Early and Middle Proterozoic Rocks

Rocks of early to middle Proterozoic age (2.5 to 1.0 billion years) in the Great Basin consist mainly of platy schist rich in mica minerals, banded gneiss, dark-colored volcanic materials, and coarsely crystalline igneous rocks such as granite or monzonite. In general, these rocks are similar to the older Archean complexes but have experienced somewhat less severe metamorphic effects and are associated with more voluminous granitic rocks. Proterozoic metamorphic-igneous complexes are particularly well exposed along the lower Colorado River, in the hills surrounding the old Mojave Desert mining camps of Randsburg, and in many eastern Mojave desert mountain ranges, including the Avawatz, Owlshead, Whipple, Turtle, and New York Mountains. Early and middle Proterozoic rocks are thus concentrated in the southern and eastern portions of the Great Basin. Elsewhere, such rocks are either unknown or have yet to be adequately studied by geologists.

The early and middle Proterozoic rock sequences in the southern Great Basin represent a mixed package of metamorphic and igneous materials that contains rocks as old as about 2.5 billion years. Thus, some of these rocks are actually Archean in age, but most are Proterozoic. These rocks include the layered gneiss of the Ivanpah, New York, and Turtle Mountains, along with the oldest basement gneiss in the Death Valley region. These ancient rocks, along with some younger material, were deformed and metamorphosed about 1.7 billion years ago, during what geologists have called the Ivanpah Orogeny. The cause of the Ivanpah Orogeny is still somewhat conjectural, but it seems to be related to the collision, or suturing, of the younger Yavapai block to the Mojave block. The Mojave block itself was probably a microcontinent that had collided with the southern margin of Nena sometime prior to the Ivanpah Orogeny. Evidently, this microcontinent (called Mojavia by some geologists) included some Archean rocks, which explains the presence of such ancient materials in the southern Great Basin, far south of the Wyoming Province. The zone of collision is difficult to define from the spotty exposures of early and middle Proterozoic rocks in the Great Basin, but seems to be aligned in a north-south direction and located in western Arizona, slightly east of the modern Colorado River. After the collision between the Mojave and Yavapai blocks, great volumes of magma rose into the deformed rocks of the southern Great

Basin to form many bodies of granite ranging in age from 1.6 to 1.4 billion years old. These granites are widespread in the southeast California–Nevada–Arizona region and can be observed in the Piute and Homer Mountains of the Mojave Desert, the Lucy Grey Mountains of southern Nevada, and in many other localities along the lower Colorado River corridor. The formation of so many bodies of granite, coupled with the slightly earlier accretion of the Yavapai block, added more rock to the growing core of North America.

Thus it appears that continent-building was still in progress throughout most of the early and middle Proterozoic Eon. The southwest edge of North America was steadily enlarged either through the collision and suturing of microcontinents or the development of volcanic arcs along its margin. The last major Proterozoic continent-building event in the Southwest was apparently the accretion of the Mazatzal block. This block, which was probably a chain of subduction-related volcanoes, was either built along, or jammed into, the southern edge of ancient North America. The Mazatzal Orogeny is thought to have occurred about 1.65 billion years ago in what is now the central Arizona–New Mexico region. Consequently, its effects are not very widespread or striking in most of the Great Basin. By the close of middle Proterozoic time, after the accretion of the Mazatzal block, nearly 800 miles of new land had be added to the southern margin of ancient North America since the Archean Eon. The foundation of the Great Basin was in place as part of the giant supercontinent of Rodinia.

Late Proterozoic:
Destruction of a Supercontinent

Rocks formed during the latest portion of the Proterozoic Eon are much more widespread in the Great Basin than are those from any other Precambrian interval. In fact, late Proterozoic rocks occur in so many of the mountain ranges of the eastern and southern Great Basin that merely describing them all would require a considerable number of pages. However, most of what is known about the late Proterozoic

history of the Great Basin has been gleaned from the rock sequences exposed in the Death Valley region, the White-Inyo Mountains, the northern Wasatch Mountains, and in a few isolated ranges such as the Snake, Deep Creek, and Sheeprock Mountains. From these well-studied rock assemblages, along with less-dramatic exposures elsewhere, geologists have discovered evidence from some profound changes during latest Precambrian time in the Great Basin: these rocks tell the story of a dramatic shift in the plate tectonic setting of North America, and the beginning of a new geological framework for North America.

Recall that south of the Wyoming Province throughout the Great Basin the ancient metamorphic and igneous basement rocks are of early and middle Proterozoic age. With a few exceptions, the youngest granites of the Yavapai and Mojave blocks are about 1.4 billion years old. The late Proterozoic sedimentary rocks in the Great Basin are difficult to date accurately, but the oldest of them are probably around 800 or 900 million years old, and most are much younger. Thus, in most of the Great Basin, a time gap of approximately 500 million years normally separates the youngest late Proterozoic sedimentary rocks from the metamorphic-igneous basement on which they rest. Such a hiatus is known as an **unconformity**, a horizon within a rock sequence that signifies a missing interval of time. We have no way of knowing what events occurred in the Great Basin during the time represented by the unconformity: it is a half-billion years of unknown history. Perhaps there were other collisions between Rodinia and island arcs or microcontinents. Perhaps volcanoes erupted, or maybe the seas invaded onto the dry land of coastal Rodinia. Whatever happened during the hiatus remains a mystery because the rocks that might have recorded the events were later eroded, leaving geologists with nothing but a wavy line, an old erosion surface, beneath the late Proterozoic rock layers. However, when the geological record resumes again, things in the Great Basin appear to have changed in a dramatic way during the final phases of Precambrian time.

Of the many places in the Great Basin where

late Proterzoic rocks can be examined, one of the best is the region around Death Valley. Here, in mountains such as the Panamint, Kingston, and Amargosa Ranges, rocks of late Proterozoic age may be as much as 12,000 feet thick. So varied is this thick stack of rock layers that geologists have split it into several components: the lowermost Pahrump Group (consisting of the Crystal Spring, Beck Spring, and Kingston Peak Formations), overlain in succession by the Noonday Dolomite, the Johnnie Formation, and the Sterling Quartzite. Because of relatively recent faulting in the region, the exposures of late Proterozoic rocks are discontinuous in the Death Valley region, as are most rock outcrops anywhere in the Great Basin. Nonetheless, geologists can trace these strata from mountain to mountain over a distance of more than one hundred miles. This thick and widespread package of rock layers exhibits both vertical and lateral variations that tell us a great deal about the history of the Great Basin and the conditions under which the rocks formed.

The oldest portion of the Proterozoic package in Death Valley is the Pahrump Group, which consists of three different rock formations and averages around 7,000 feet in thickness (Figure 2.9). The Crystal Spring Formation is the lowest unit of the Pahrump Group and is a heterogeneous assemblage of mostly sedimentary rocks, for the most part little affected by metamorphism. Included in the Crystal Spring Formation are such rocks as feldspar-rich gritty sandstone and conglomerate, fine-grained mudstone and shale, and thicker layers of limestone and dolomite composed of carbonate minerals precipitated in a shallow, warm primordial sea. In addition to these sedimentary rock types, the Crystal Spring Formation also includes sheets and vein-like bodies of diabase, a dark-colored igneous rock that represents magma that squirted upward along fractures or zones of weakness in the layered sedimentary rocks and hardened in place. In places where the molten magma came into contact with the carbonate rocks, metamorphism produced talc-rich marbles, a rock that is mined in several localities near Death Valley. The diabase makes up a relatively

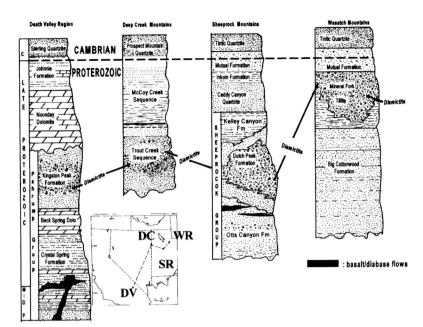

small portion of the Crystal Spring Formation, but it produces radiometric dates suggesting that the oldest part of the formation may be as old as 1.2 billion years. Younger rocks higher in the Crystal Spring Formation are probably around a billion years old. Thus, the Crystal Spring Formation is not entirely of late Proterozoic age; it began to accumulate some 200 million years before the last phase of that eon commenced. The Crystal Spring Formation rests directly on much older (1.4–1.8 billion years) metamorphic rocks of the Mojave Province.

The heterogeneity of the Crystal Spring Formation documents a late Proterozoic world of rapidly changing conditions in the southwestern Great Basin. The conglomerate and coarse sandstone that occurs near the base of the formation represents gravel eroded from the metamorphic rocks of the Mojave Province and washed into local basins by swift rivers. The finer-grained mudstones probably record low coastal plains or mudflats along the primordial coast. The carbonate rocks (limestone and dolomite) almost certainly indicate deposition under shallow oceanic conditions. Finally, the diabase demonstrates that, while these profound changes were occurring, the earth's crust was fractured and molten fluid from below was leaking toward the surface. With only a few datable rocks, it is very difficult to determine exactly how much time is represented by the Crystal Spring Formation.

Figure 2.9 Late Proterozoic sedimentary rocks of the Great Basin. Glaciomarine diamictites can be traced across the region, from Death Valley to southern Idaho.

Figure 2.10 Proterozoic diamictite of the Kingston Peak Formation, with pebbles stretched by subsequent metamorphism.

Was it 50 million years? 200 million years? However long the interval may have been, it clearly was a tumultuous time. The land rose, then fell. The sea crept inland, drowning the low areas. Molten rock oozed upward through rock fractured by the wrenching geological forces.

Above the Crystal Springs Formation, two additional formations complete the Pahrump Group: the Beck Spring Dolomite and the Kingston Peak Formation. The Beck Spring Dolomite averages only about 500 feet thick, making it the thinnest formation of the Pahrump Group. Dolomite is a mineral composed of calcium-magnesium carbonate [$CaMg(CO_3)_2$] that forms on the sea floor beneath shallow warm seas. The Beck Spring Dolomite (some geologists prefer the term "dolostone" for this rock) records a prolonged period of submersion under shallow seas. The Kingston Peak Formation, the youngest of the Pahrump Group, is also the thickest; in many places, it is more than 3,000 feet thick. Unlike the Beck Spring Dolomite beneath it, the Kingston Peak Formation is an incredibly mixed assemblage of granular sedimentary rocks with lesser amounts of limestone and volcanic rocks. The dramatic shift from monotonous marine deposits of the Beck Spring to the wildy variable rubble in the Kingston Peak Formation signifies some dramatic upheavals during the late Proterozoic Eon. Because the Kingston Peak strata tell us a great deal about the portentous events of the time, the formation may be one of the most geologically intriguing and significant rock units in the Great Basin. It is definitely worth a closer look because, as we shall see, it reveals the grand finale of Precambrian history in the Great Basin.

Diamictites and Diabase: Clues to Chaos

In the middle and upper portions of the Kingston Peak Formation a very distinctive type of sedimentary rock makes up a sequence of beds more than 3,000 feet thick. These rocks consist of a hodgepodge of large rounded stones, several inches in size, embedded in a matrix of dark-colored sand and silt. The fine-grained sand-silt matrix is made up mostly of tiny rock particles, from about 1/50 to 1/1000 of an inch in size. Such microscopic rock particles represent finely ground and pulverized rock, while the larger chunks suspended in the matrix obviously experienced less grinding and crushing. The term **diamictite** is used by geologists to describe rocks comprised of such a wide variety of particle sizes. Diamictites, with their large cobbles and pebbles randomly scattered through the fine matrix, have a very distinctive appearance: they look a little like a mass of dark-brown or black concrete, containing large light-colored stones. Unlike most other sedimentary rocks, diamictite deposits are usually not well-layered, suggesting that the material was dumped rapidly where it accumulated, rather than piling up gradually in discrete layers. With their characteristic lack of well-defined stratification and "polka dot" appearance, diamictites are almost impossible to mistake for anything else. They are strange rocks in many ways.

The most intriguing aspect of diamictite is not so much its appearance but its origin. Diamictite is obviously a sedimentary rock, consisting of rock particles that accumulated in some basin of deposition. What mechanism could have delivered the rock grains to the place where they accumulated without sorting them by size while at the same time dumping them in a non-layered heap? Certainly not rivers or the wind: these two mechanisms of sediment transport are both selective in the size of particles that can be carried and normally produce well-stratified accumulations of sediment. The only plausible explanation for the unusual features of the Kingston Peak diamictite is that it was dropped from the melting ice of a glacier. We know from direct observations of modern glaciers that as glacial ice melts it releases rock particles of all sizes, from giant boulders (called dropstones) to microscopic fragments, at the same time and in the same place. Unless the mixed rubble is redistributed by meltwater streams, the glacial dross accumulates in nonlayered heaps along the edge of the melting ice. Such sediment dumped along the margins of a melting glacier is known as till. Till, since it accumulates only where glacial ice is melting, is generally not deposited as a sheet over a broad area. Instead, it normally occurs on land, piled into ridges or mounds that mark the terminus of some vanished glacier. The diamictite of the Kingston Peak formation is almost certainly of glacial origin; so confident are some geologists of the glacial ancestry of the diamictite that it is sometimes referred to as **tillite**, a rock made of lithified till. The distinction between a tillite and a diamictite is that the former is a genetic term, describing the source of the sediment (till, deposited from a glacier), while the latter is a textural term denoting the lack of sorting that is characteristic of such deposits. The late Proterozoic materials under consideration here are definitely diamictites, and they are probably also tillites.

There is one important qualification, however, that must be added to that interpretation of the Great Basin diamictites as tillites. Unlike the till that accumulates on land in the vicinity of melting glaciers today, the late Proterozoic diamictite in the Great Basin is not distributed in localized mounds, ridges, or heaps. In fact, diamictite can be found in late Proterozoic strata across the entire province, from the Mojave region to southeastern Idaho! In addition to the glacial rubble in the Kingston Peak Formation, diamictite also occurs in the Deep Creek Mountains of western Utah (Trout Creek Group), the Sheeprock Mountains of west-central Utah (Dutch Peak Formation), the Wasatch Range (Mineral Fork Tillite), and in extreme northern Utah and southern Idaho (parts of the Pocatello Formation and other units). The diamictite, or tillite, exposed in these localities appears to have once been part of an extensive sheet that covered the eastern Great Basin, rather than the localized accumulations that we might expect if the sediments were strictly glacial in origin. Geologists studying the

Figure 2.11 Dropstone in the Mineral Fork Tillite, Little Cottonwood Canyon, northern Utah.

distribution and types of stones embedded in the diamictite, along with variations in the thickness and type of sediments associated with them, have come to the conclusion that the sediments composing the diamictites actually accumulated in a shallow ocean basin, not on land. There is still no doubt about the glacial nature of these deposits, but it appears that the ice was melting at sea. This is probably how the tillite became so widely distributed across the Great Basin during the late Proterozoic. As glaciers slid off land into the ancient sea, they probably broke apart into large plates of ice that could drift offshore for thousands of miles. We know that the glaciers had to originate on some nearby landmass, rather than as sea ice. If they didn't, it would impossible for them to contain the stones and rock debris that was deposited as they melted in the ocean. As the fragments of glacial ice drifted along at sea, they would melt steadily, dropping rock particles to the sea floor. Eventually, the floating icebergs would vanish entirely, while the ocean floor was buried under piles of glacially transported residue. In recognition of the part-glacial, part-oceanic character of the late Proterozoic diamictite-tillite sequences in the Great Basin, most of them are now referred to as "glacio-marine" sediments.

One very intriguing aspect of the late Proterozoic glacio-marine deposits of the Great Basin region is their apparent synchroneity across this vast region. From Idaho to Death Valley, the tillites appear to have been deposited during the same time interval: from about 800 to 700 million years ago. The approximate nature of these age estimates reflects the sad reality that geologists have not developed dating techniques that can establish the precise ages of ancient sedimentary rocks such as the Great Basin tillites. How then can the age of these rocks, regardless of the precision, be determined at all? Fortunately, in many places in the Great Basin where the tillites are exposed, they are associated with rocks that can be dated with the various techniques that rely on the progressive decay of radioactive elements. Known generally as radiometric dating, such methods can be applied best to igneous rocks, those that form from the cooling and solidification of the molten fluid known as magma. Volcanic rocks, along with rocks like granite (representing magma that solidified slowly, deep underground), are therefore most amenable to radiometric dating techniques. The late Proterozoic tillites of the Great Basin are commonly intermingled with several different types of volcanic rocks that provide the basis for age estimates. The most common igneous material associated with the tillites is a dark-colored, dense volcanic rock known as **diabase**. These rocks not only are important as temporal benchmarks in the late Proterozoic rock record but they also add some fascinating details to the Proterozoic history of the region.

The late Proterozoic diabase of the Great Basin is a rock similar to the relatively young black basalt that covers the surface of the modern region in places like the Black Rock Desert of Utah or the Amboy Crater region of the Mojave Desert. The ancient diabase and the much younger basalt both consist predominantly of dark-colored minerals such as plagioclase and augite, and both were formed when magma reached the surface to pour from an erupting volcano as an incandescent lava flow. The young basalt flows have resulted from the movement of magma upward along the fractures known as normal faults that form as the Great Basin region is stretched in an east-west direction. This extension produces the faults and, as they continually open, magma oozes upward and eventually erupts on the surface (more details on this phase of Great Basin geology are presented in later chapters). Similarities between the recent basalt and the Proterozoic diabase suggest comparable origins. In fact, the diabase associated with the tillites probably does represent magma that moved up along faults that formed when the earth's crust was stretched to the breaking point. But, there are some important, and revealing, differences between the ancient diabase and the younger basalt.

The diabase (and similar Proterozoic volcanic units) is usually composed of small intergrown mineral crystals, as opposed to the noncrystalline appearance of the more recent basalt. This finely crystalline appearance probably reflects the great age of the Proterozoic volcanic units; the radiometric dates they produce cluster around 850 million years. During their immense history these rocks have experienced many episodes of geological havoc; most have been heated, crushed, stretched, and stewed in chemically active vapors and fluids many times. Consequently, the diabase has been modified, to one degree or another, from its original condition. This may be why the diabase has developed a fine-grained crystalline appearance, in contrast to the comparatively dull complexion of the younger, and much less modified, basalts. While they can still be recognized as volcanic rocks, some the Proterozoic volcanic rocks are better thought of as metamorphic ("changed form") rocks. In fact, one geologist has referred to the volcanic rocks associated with the Kingston Peak Formation as "metabasalt." In addition to this (mostly) age-related difference in texture, the Proterozoic diabase units in the Great Basin commonly exhibit crude rounded shapes known as "pillow structures." Pillow structures almost always signify the eruption and cooling of lava under deep water, where the cold temperatures and high pressures cause the magma to harden in the form of rounded blobs, no more than a foot or two across. Each blob, or pillow, is surrounded by a glassy rim where the lava solidified instantly as it came into contact with the cold seawater. The interior of each pillow may remain fluid after the solid rim has hardened. Often, the rim cracks open and viscous lava squirts out of the pillow to form another rounded blob adjacent to the earlier-formed pillow. As the undersea eruption continues, the pillow-shaped masses of hardened lava form one at a time on the sea floor. When the submarine eruption is over, the resulting lava flow resembles a heap or sheet of coalesced pillows. This interpretation of pillow lava has been verified by direct observation of modern undersea erup-tions in places where the sea floor is being ripped open by stretching forces. The pillow structures in some of the Proterozoic volcanic rocks of the Great Basin is a certain indication of their submarine origin.

The diamictite-diabase association in the late Proterozoic rock sequences of the Great Basin is widely distributed across the region. In the Panamint Mountains near Death Valley, volcanic rocks are intercalated with diamictites in the Surprise Member of the Kingston Peak Formation. In extreme southeast Idaho, the diabasic rocks are known as the Bannock Volcanic Member of the Pocatello Formation, which is sandwiched between the diamictites of the Scout Mountain Member. In the Sheeprock Mountains of west-central Utah, the diabase occurs within the upper part of the Otts Canyon Formation, just below the massive diamictites of the Dutch Peak Formation. The volcanic rocks are absent from some exposures of the late Proterozoic strata in the eastern Great Basin, most notably in the central Wasatch Mountains, where there is no trace of them in the Big Cottonwood Formation or the overlying Mineral Fork Tillite. Nonetheless, the diamictite-diabase couplet appears to be a widespread feature of the Proterozoic rocks throughout most of the Great Basin. Local absence of the volcanic component may reflect erosional gaps (or **unconformities**) in the rock record, or it may simply indicate those places where, for one reason or another, no lava was erupted on the sea floor during the time that the glaciers were melting.

Radiometric dating indicates that the Proterozoic diabase units are generally around 850 million years old. The diamictites appear to be related to a global glacial event known as the Sturtian glaciation, which occurred about 750 million years ago. Thus, the diamictites and volcanic units suggest that from between about 850 and 750 million years ago, an ocean basin developed in the Great Basin region. The land near the ocean was covered by glaciers that slid into the sea, breaking up into ice rafts or icebergs. At the same time, volcanic eruptions were in

Figure 2.12 Tilted layers of light-colored quartzite in the Big Cottonwood Formation at Storm Mountain, Utah.

progress on the sea floor, far beneath the floating and melting chunks of glacial ice. Pillow lavas formed across most of the sea floor, while sediment grains of all sizes fell from the melting ice above to accumulate under, over, and around volcanic rocks. But is there a more fundamental link between the diabase and the diamictite? Does the association of such dissimilar rocks evince some pivotal geological event? Or is it just a coincidence that these two disjunctive types of rocks should appear together, at virtually the same time, in the Proterozoic rock sequences of the Great Basin?

Continental Rifting: The Great Break-up
Because tillite-volcanic units are so widespread and distinctive throughout the Great Basin, this rocky twosome has been used to subdivide the entire late Proterozoic succession into pre-tillite and post-tillite portions. The significance of the tillite-volcanic horizon becomes apparent when the patterns of deposition and type of sediment in rocks deposited before the tillite are compared to the same characteristics for the post-tillite rocks. Prior to the tillites, sediment was washed, primarily by rivers, into several different isolated basins, producing a complicated pattern of distribution across the Great Basin (Figure 2.13). Some of these basins were greatly elongated, like troughs, and subsided enough to accommodate a pile of sediment nearly 20,000 feet (4 miles!) thick. Sediment filling these basins was generally coarse, river-washed sand

Figure 2.13 Patterns of sediment deposition for Late Proterozoic strata in western North America deposited before (vertical ruling) and after (stippled pattern) the tillite-diabase sequence. Note that post-tillite sediments accumulated in a linear north-south zone, whereas the pre-tillite sediment accumulated in irregular isolated basins that formed as Rodinia was beginning to break apart.

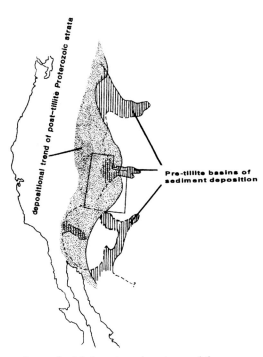

and gravel, rich in mineral grains and fragments derived from pre-Proterozoic granite, gneiss, and schist. Rocks representing such sediment are found in the Uinta Mountain Group of northern Utah, the lower part of the Crystal Spring Formation of the Death Valley Region, and portions of the Otts Canyon Formation of the Sheeprock Mountains. After the glacial deposits and volcanic rocks were formed, the pattern of sediment accumulation shifted dramatically: post-tillite units almost all thicken to the west and are uniformly aligned along a north-south zone that extends parallel to the eastern boundary of the modern Great Basin. Nearly all post-tillite rock units in the Great Basin follow this north-south, west-thickening trend. In addition, the sediment that accumulated in this zone is generally fine-grained silt and sand that appears to have been reworked by the currents and waves of the sea. Thus, the sediment that buried the tillites after they had accumulated on the sea floor during the Sturtian glaciation appears to have accumulated along some linear basin of oceanic character that became deeper to the west. The uniform pattern of sediment accumulation, along with the type of sediment, is what might be expected along a linear coast, aligned north-south. Sand would accumulate along the beaches of this coast, while the finer sediment would be carried farther offshore by

waves and currents. Scouring and reworking of the beach sands would carry some of this sediment to the deeper ocean floor west of the coast. Over time, a greater volume of sand and mud would accumulate under the deeper water than along the beach. Thus, a west (seaward)-thickening wedge of sediment would form along the coastal region.

This prominent shift in sediment accumulation patterns, following the diamictite-diabase interval, has been interpreted as evidence of continental rifting, or the break-up of Rodinia. In this scenario, the irregular pre-tillite basins represent the early stages of continental separation, when great rift valleys develop on land while volcanoes begin to erupt. Late Proterozoic rivers of the Great Basin region would have shifted their courses to flow through the linear troughs. These rivers would have washed great amounts of coarse sand and gravel into the deep rift valleys while the earliest of the diabase flows were pouring from the erupting volcanoes. Recall that fracture systems that form during periods of continental rifting usually exhibit three arms, separated from each other by angles of about 120°. There is evidence that perhaps as many as four of these three-arm rift centers developed early in the late Proterozoic in western North America; possible remnants of them have be identified in central Montana, in the Uinta Mountains of Utah, in the Death Valley region, and in central Arizona. After the inception of rifting, the volcanically active rift systems grew longer as their floors continued to sink under the weight of accumulated sediment. In each of the rift centers, two of the three lengthened arms eventually reached the edge of Rodinia, or connected with the rift valleys extending from adjacent rift centers. Thus, the small rift centers merged into one large, more or less linear zone by the union of rift valleys growing outward from each center. This coupling of rift valleys created a single, north-south-trending zone of continental separation. As the floor of the large unified rift valley continued to subside, it eventually fell below sea level. At this point, seawater rushed into the low trough. In this manner, a new ocean basin was created between the main mass of Rodinia and a relatively small sliver of it that had been severed from its western edge. This narrow seaway ran north-south through what is now the Great Basin. As continental rifting continued, subsequent volcanic eruptions took place underwater, giving rise to the pillow structures observed in many of the late Proterozoic diabase units. Along the newly formed coast, glaciers slid into the sea, breaking up into numerous icebergs that drifted along in the narrow ocean. After the glaciers had disappeared, and after the rift had fully developed, sediment was washed from the land into the embryonic seaway by rivers that drained the adjacent landmasses. The post-rifting sediment that buried the diamictite and diabase accumulated as a west-thickening mass, aligned in a north-south direction, in a widening ocean basin.

The narrow seaway that opened as Rodinia was rifted in the late Proterozoic appears to have been centered in east-central Nevada. To the west, toward Utah and the eastern margin of the modern Great Basin, the broken edge of Rodinia passed into unfractured rock in those portions of the ancient supercontinent unaffected by fragmentation. However, there are some "scars" of the rifting event preserved in the deeper crust of regions outside the zone of separation in central Nevada. Remember that rifting is achieved by the growth and coalescence of two of the three arms of the rift centers. What becomes of the third rift valley extending from each center? These valleys are sometimes called "failed rifts," because they neither reach the edge of the continent nor join with rifts from adjacent centers. Instead, they form deep, fault-bounded troughs that intersect the newly rifted margin of the continent at a high angle. Inland from the coast, these troughs gradually disappear as their floors rise to merge with the unfractured rock of the continental interior. After rifting is complete, the failed rifts become inactive and eventually become filled with sediment deposited by rivers that drain through them. Geologists call such linear basins, bounded by fractures and filled with sediment, **aulacogens**. Four late Proterozoic aulacogens appear to have formed in western North America during the rifting of Rodinia. Two of them, the Uinta and Amargosa aulacogens, are located in or near the Great

Basin. The aulacogens are the "scabs" of continental rifting: great fractures that have been filled by, and buried under, younger rocks. The postulated rifting of Rodinia in central Nevada is supported by the presence of several aulacogens nearby.

In addition to the aulacogens, there is an interesting, though subtle, difference in the chemistry of much younger granitic rocks in the Great Basin that confirms the notion of continental rifting during the Proterozoic. If the rift opened in central Nevada, as is suggested by the distribution and type of late Proterozoic rocks, then western Nevada and all of California should be underlain by relatively young oceanic crust, composed of volcanic rocks such as basalt or diabase, that were erupted underwater. East of central Nevada, the deeper rocks should be composed of the thicker and older crust of Rodinia, which was similar, in terms of bulk composition, to the rock granite. To make a long story short, the deeper levels of the crust west of the zone of rifting should consist mostly of younger oceanic rock, while the crust east of the zone of rifting should be more "continental" in character, and older. These differences in age and type of ancient crust east and west of the zone of rifting should be reflected in the chemistry of magma that moved upward through the dissimilar materials much later in time. As we will learn in Chapters 7 and 8, many bodies of magma worked their way upward through these ancient rocks in the Great Basin between about 200 and 30 million years ago.

As the molten masses rose through the surrounding rock, they assimilated some of the older material and took on certain chemical signatures of the rocks through which they passed. In particular, the magma bodies absorbed different ratios of two isotopes of the relatively rare element Strontium (Sr), depending on what type of rock they penetrated en route to the surface. Geologists have found that relatively young oceanic rocks typically have a ratio of Strontium-87 to Strontium-86 ($^{87}Sr/^{86}Sr$) of less than 0.706. On the other hand, older "continental" rock tends to have a $^{87}Sr/^{86}Sr$ ratio greater than 0.706. Thus, if we analyze the young granite bodies in the Great Basin that have pierced the older and deeper crust and find $^{87}Sr/^{86}Sr$ ratios greater than 0.706, we may assume that the magma that formed the granite passed through thick continental rock, not thin oceanic (and younger) oceanic crust. If the ratio is less than 0.706, then the magma probably rose through thin, and younger, oceanic rock. Such analyses have been performed on hundreds of igneous rock masses in the Great Basin. The high $^{87}Sr/^{86}Sr$ ratios (greater than 0.706) tend to be concentrated in the eastern half of the Great Basin, while lower ratios (less than 0.706) typify the western portion of the province. A line can be drawn through the middle of the Great Basin that separates places where the $^{87}Sr/^{86}Sr$ ratios are greater than 0.706 from locations where they are less. This line is referred to as the "0.706 Line" (see Figure 2.7). The 0.706 Line passes diagonally through central Nevada from southwestern Idaho to the Owens Valley region of eastern California. This line signifies the edge of the fractured pre-Proterozoic continental crust. West of the 0.706 Line, no ancient continental crust exists: all of the rocks comprising the Precambrian basement in that area appear to be of oceanic character. The location of the 0.706 Line, and the fact that it can be drawn at all, is a reflection of the continental rifting event that occurred in late Proterozoic time.

What became of the western fragment of Rodinia that was severed along the zone of continental separation? If Rodinia was rifted along the 0.706 Line, then a piece of the ancient basement must have moved away toward the west as the new-formed ocean basin expanded between the fragments of the rifted supercontinent. The rocks in this vanished block of material must have been similar to rocks comprising the Farmington Canyon Complex, or the Little Willow Complex, or the early Proterozoic basement rock in the Mojave region. After all, the severed piece of Rodinia was continuous with such Great Basin rock assemblages prior to the rifting event. We suspect that the other fragment of Rodinia moved west, but exactly how far? Where is it now?

Geologists have identified rocks that are remarkably similar to the Archean-Proterozoic materials in the Great Basin in two very unlikely places: Siberian Asia and Antarctica! Portions of the Precambrian basement exposed in the interiors of both continents are comprised of rock similar in age, composition, degree of metamorphism, and geological structure to coeval materials in the Great Basin. Which of these blocks of "exotic" rocks, now firmly locked into place as part of the ancient mosaic of Asia and Antarctica, represents *the* western fragment of Rodinia? No one knows for sure, and geologists have debated the issue for years without any final resolution. It may be that the late Proterozoic rifting event in the Great Basin resulted in two separate fragments that drifted away, later colliding with the ancient cores of Asia and Antarctica, respectively. If so, then rock sequences now wedged into the interior of these distant continents represent fragments removed from the western edge of Rodinia and transported thousands of miles from their source.

The timing of the rifting of western Rodinia cannot be established with any great precision. Radiometric dates from the rifting-related volcanic rocks suggest that the process began around 800 million years ago. It may have taken tens of millions of years before the western edge of the ancient supercontinent was completely removed from the eastern portion. Some geologists have suggested that rifting may have continued to near the end of the Proterozoic Eon, currently estimated to have occurred about 545 million years ago. Based on studies of regions such as the Arabian Peninsula, where continental rifting is in progress today, scientists have recognized that continental masses can be rifted in only a few million years. The process does not require the immense span of time, some 250 million years, that separates the geological estimates for the beginning and ending of the Proterozoic rifting in the Great Basin. Perhaps there was more than a single rifting event during this time. The western edge of Rodinia may have been removed in several pieces, one at a time, over a long interval. On the other hand, there may have been only a single rifting episode involved in the break-up of western Rodinia. If so, then the extreme range of estimated dates bracketing that event more likely reflects the lack of data and the imprecision of geological dating methods than it indicates a prolonged process. There is little doubt, however, that by the time the Proterozoic Eon drew to a close, a new rifted edge of western Rodinia had been formed in the central Great Basin.

POST-RIFTING ROCKS AND FOSSILS IN THE GREAT BASIN

As we have already learned, the break-up of Rodinia produced a new continental margin running north-south through central Nevada. Sea floor spreading in the embryonic ocean basin forced the severed fragments of Rodinia farther apart as time progressed. In the eastern Great Basin, where the fractured eastern remnant of ancient North America met the oceanic crust of the newly formed seaway, a broad coastal plain developed along the new continental margin. Geologists refer to such continental borderlands as "passive margins," because there is no interaction of lithosphere plates that could produce violent earthquakes and intense volcanic activity. Passive margins are geologically serene because they are merely places where two types of crust (oceanic and continental) meet within the *same* plate. Without some violent interaction with another lithospheric plate, there is no mechanism to ignite volcanoes, shatter rock to produce earthquakes, or build mountains. The eastern seaboard of North America is a modern example of such as passive continental margin. In contrast to the rowdy western brow, the modern eastern edge of our continent is geologically tranquil, owing to the absence of any encounters between lithospheric plates. The current residents of Boston, New York, and Philadelphia may be subject to many hazardous conditions, but they can rest assured that their cities are not likely to be engulfed by hot lava or leveled by devastating earthquakes anytime soon.

After the rifting of Rodinia, the geological setting of the Great Basin was similar to that of the modern East Coast. Rodinia no longer existed; the supercontinent had been rifted into several smaller fragments representing the cores of the modern continents. For the final 100 million years or so of the Proterozoic Eon,

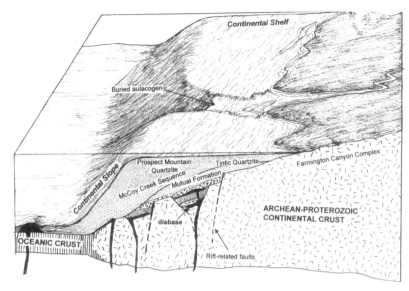

sediment was gently washed west from the exposed portion of the rifted margin to the coastal plain along the narrow sea that covered the central and western Great Basin region (Figure 2.14). The north-south-trending coast line was initially indented by at least two elongated inlets: the Uinta and Amargosa aulacogens. These two depressions, relicts of the rifting process, eventually became completely filled with sediment and were buried by sheets of sand and silt deposited along the coastal plain and just offshore in the shallow seaway. The aulacogen-filling sediment includes such rock units as parts of the Big Cottonwood Formation and Uinta Mountain Group of Utah and the younger parts of the Kingston Peak Formation and Noonday Dolomite in the Death Valley region. The sheets of sediment that covered the filled aulacogens are represented by such formations as the Johnnie and Sterling Formations in Death Valley, the McCoy Creek Group of eastern Nevada and western Utah, the Mutual Formation near Salt Lake City, and the Brigham Group of northern Utah and southern Idaho. Farther offshore, in the White-Inyo Mountains region of eastern California, fine silt and carbonate mud accumulated in deeper basins at the foot of the continental slope. Such sediments are now represented by the shale, siltstone, and dolomite of the Wyman and Reed Formations of the White Mountain area.

The easternmost strata of this terminal Proterozoic rock sequence are dominated by

Figure 2.14 The late Proterozoic passive margin of North America. Post-rifting sediments, such as the Mutual Formation, McCoy Creek sequence, and Prospect Mountain Quartzite were deposited over the broken western edge of former Rodinia. The continental shelf extended from the shoreline in central Utah westward into central Nevada. Beyond the 0.706 Line (the boundary between continental and oceanic crust), the continental slope descended toward the deep ocean floor in what is now western Nevada and California.

Figure 2.15 Laminated Noonday Dolomite in Mosaic Canyon, Death Valley National Park

sandstone and siltstone, consisting of grains that were deposited by sediment-choked rivers flowing across the coastal plain. Finer-grained sediment, such as silt and clay, accumulated farther offshore beneath the shallow turbid water of the proto-Pacific Ocean. In places on the late Proterozoic sea floor, carbonate mud collected, later hardening into limestone and dolomite. Thus, the west-thickening wedge of sediment that was draped over the broken edge of North America in latest Proterozoic time was a mixed assemblage of materials that gradually built a gently sloping continental shelf extending from central Utah to eastern Nevada. West of the shelf margin, and farther from the ancient continent, the sea floor plunged down a steep continental slope as it passed into the deep abyssal plains in what is now western Nevada and California. By the end of the Proterozoic Eon, the outer edge of the continental shelf extended well into central Nevada and was located not far from the 0.706 Line.

The late Proterozoic sea floor of the Great Basin region was thus shallow and sandy to the east and deeper and muddy to the west. There were probably several small islands of land dispersed in the sea offshore of the coast. These small landmasses represented microcontinents produced during the rifting episode, but they contributed little sediment to the widening ocean basin. The shallow water that submerged the continental shelf was probably warm and clear as it lapped around the edges of the microcontinents, or washed back and forth along the beaches to the east. Some of the sediment that accumulated beneath the waves bears evidence of another fascinating aspect of this late Proterozoic seaway: it was far from lifeless!

Late Proterozoic Life in the Great Basin

In the mountain ranges that surround Death Valley, good exposures of the late Proterozoic Noonday Dolomite can be found in the southern Panamint Range, in the southern Black Mountains, in the Alexander Hills, and at the south end of the Resting Spring Range. In these places, the outcrops of the normally buff-colored dolomite often reveal faint, wispy laminations running through the rock. The fine layers, only a millimeter or so thick, are commonly crinkled and wavy, making the rock surface look a little like a head of cabbage cut in half. Sometimes, these undulating laminations are arched upward into a dome-like shape. Such features in carbonate rocks such as the Noonday Dolomite almost certainly were produced by the successive growth of sheets, or "mats," of photosynthetic algae (or similar primitive creatures) on the sea floor. In shallow marine basins today, the growth of algal colonies on the sea floor results in laminated structures and mats nearly identical to those found in the Noonday. In the modern seas, marine algae and cyanobacteria secrete a coat of slime, the glycocalyx, in which millions of cells are contained. The slimy secretion protects the microorganisms but also traps sediment particles suspended in seawater. If too much sediment collects on the scummy surface of the mat, the microbiological colony is extinguished. The muddy encrusted surface of the glycocalyx is then repopulated by a new algal/cyanobacterial colony and the process starts over again. Wherever there is sufficient sunlight to support photosynthesis, the algal mass grows outward across the shallow sea floor as an expanding mat. If the sea floor is uneven, such that high and low spots exist, the mat may be develop bumps or pillars over the high spots, where the photosynthetic algae grow more profusely under the influence of stronger sunlight. Sometimes, the pillars rising from the algal mat may be as much as 5–6 feet tall and nearly 3 feet in diameter. Such hemispherical structures are called **stromatolites** by geologists, and they are often found in association with the laminated sediments in many late Proterozoic rock successions in the Great Basin. So, it appears that the bottom of the shallow Proterozoic sea floor was carpeted by a slimy green coat of algae, at least during the time that the Noonday sediments were being deposited in the Death Valley region. In places, large knobs or pillars rose from this mat to within a few inches of the surface of the shallow sea.

Algal structures similar to those of the Noonday Dolomite have been observed in other late Proterozoic rock units across the southern Great Basin. In addition, geologists have recovered some actual microfossils of the primitive single-celled organisms that might have been constructing the mats and mounds on the sea floor. Even the Beck Spring Dolomite, about 100 million years older than the Noonday Dolomite, has produced microscopic filaments, spheroidal structures, and vase-shaped "cells" that probably represent the oldest fossils in the Great Basin. Beyond the modern Great Basin, but still close enough to be related to its early history, Proterozoic microfossils have been found in the Grand Canyon region (Kwagunt and Galeros Formations), Uinta Mountains (Mount Watson Formation and Red Pine Shale), and central Arizona (Mescal Limestone). Although these tiny organisms may seem primitive by today's standards, they were extremely important residents of the late Proterozoic seaway that extended through the Great Basin. They were producing oxygen as a by-product of their photosynthetic metabolism. These simple organisms jettisoned oxygen into the sea that eventually escaped into the atmosphere. Ultimately, it was these crude unicellular creatures that oxygenated the oceans and the atmosphere, creating the lush haven of life that we enjoy today. Without them, our world may never have become as habitable as it now is. Without the abundant supply of oxygen the algae and cyanobacteria produced, complex life (such as ourselves) might never have been possible. In fact, we do not have to look far from rock units such as the Noonday Dolomite to find evidence of the earliest complex life in the Great Basin. The evidence for more sophisticated forms of life appears in rocks only slightly younger than the Noonday Dolomite.

The youngest Proterozoic rock units in the Great Basin include such sequences of strata as the Wyman Formation and Reed Dolomite of the White-Inyo region, and the Sterling Quartzite and Johnnie Formation of the Death Valley Area. These rocks were deposited about 600 million years ago, just before the beginning of the Cambrian Period (545–505 million years ago). In them, geologists have found some rather odd structures that suggest that life in the Great Basin sea had crossed the multicellular threshold. Horizontal, tubelike structures in the Wyman Formation seem to indicate the presence of "worms" plowing through the soft mud of the sea floor. No actual body parts are preserved as fossils in the Wyman Formation, so geologists cannot be certain about what types of ancient "worms" might have left the markings. The Sterling Quartzite has also produced some trace fossils similar to those of the Wyman Formation but including some impressions of what may be more complex soft-bodied creatures, similar in a general way to modern jellyfish. Thus, it appears that the microbial organisms that produced the laminated structures in the Noonday Dolomite (and other rock units) were joined by larger, and much more complex, creatures near the end of the Proterozoic Eon. These fossils suggest that the pace of evolution of marine organisms was accelerating just before the beginning of Cambrian time. In a way, the advent of multicellular life in the Great Basin sea was similar to lighting the fuse to a biotic bomb. Once these complex creatures evolved, the abundance and diversity of marine life soon expanded dramatically on a global scale. This great "Cambrian Explosion" transformed the late Proterozoic marine ecosystem from a relatively simple microbial-based architecture to an intricate scheme of fascinating complexity, involving hundreds of different kinds of very complex organisms.

The rifting of Rodinia created new ocean basins all over the world, including the one in the Great Basin. It was in those new oceans that some extraordinary organisms and amazingly rich biotic communities developed during the Cambrian explosion. This great burst of evolution was a unique event in the history of life on our planet: nothing like it has occurred in the 545 million years since the beginning of the

Cambrian Period. Without the rifting of Rodinia, there would have been less shallow marine habitat available for this great evolutionary drama to be performed. The fragmentation of the ancient supercontinent also created the template for the later geological history in the Great Basin. On a global scale, this event—the breaking of a supercontinent—heralded an entirely new biological and geological age. It is to that age that we travel next.

The wind whistled through the windows of the old Bronco as it raced east on Interstate 40, past Daggett, across the creosote plains of the Mojave Desert, and into the blinding October sunrise. The brilliant light from the rising sun, smeared across the windshield by the splattered bugs and grime, made it difficult for the occupants to see anything directly ahead. A young man, smallish in stature, wearing a sweat-stained baseball cap over his shaggy, untrimmed locks sat in the back seat. He stared out of the window, watching the desert fly past at what seemed to be about 90 miles per hour. Ned was one of four students in the truck, all young geology undergraduates out for one of their impulsive romps in the desert in the late 1970s.

They called them "field trips," these extemporaneous escapades; but they were a far cry from the formal, organized outings that normally accompany college classes. The purpose of these outings was primarily therapeutic: a treatment for their "classroom fever," a malaise closely related to the cabin fever of cold regions. As classmates, the crew in the Bronco had all suffered through the same lectures, same exams, same laboratory exercises, and the same professors. They also all shared the same boundless zeal for desert road trips, especially when midterm exams were over, or when the pressures of a term paper deadline became burdensome. No excuse was really needed, however, and such weekend defections from responsibility could occur at any time, without notice.

Often, such trips germinated when one of the four heard something in a lecture or read something somewhere that sounded intriguing. Their current trip to the desert was stimulated by a lecture on the correlations of Cambrian strata between the Grand Canyon, Mojave Desert, and Death Valley regions. Mark, who was perhaps the brightest of the crew, had observed after the lecture that some of the localities mentioned by the speaker were only a few hours drive from the southern California campus they all attended. Even as Mark spoke, all eight eyes in the group widened simultaneously, the grins broke out, and they all knew it was time for another expedition. And, that was about all it *ever* took to set the wheels of a field trip in motion. Preparing for such a trip was a drill they all knew well, a natural process in which the individual roles never required discussion or rehearsal.

First, they would determine whose old car or truck was least likely to break down at the time. The Bronco belonging to Gary was the choice that particular week, due to leaking brake lines, bad fuel pumps, or slipping clutches on every other available alternative. Then, each of the comrades would arrange some excuse for avoiding whatever part-time jobs they happened to be holding. If all else failed, a case of the weekend flu, or a plea of academic urgency could usually facilitate an escape. Next, the boys would determine how much fuel would be needed, and pass the hat to pay for gasoline at the outrageous price of 55 cents per gallon. Food was never a problem: Pat lived in the dorms and could scrounge leftovers from the kitchen, where he had some inside connections. Such provisions as surplus oranges, stale bread, day-old strips of cold bacon, and congealed spaghetti were usually available in quantity. The menu varied constantly, but it was only necessary that the food be edible, even if only marginally so. A much more serious concern was beer, an essential staple on every trip. A little informal research would usually identify a

CHAPTER THREE

WATER BUGS IN STONE HOUSES

liquor store somewhere that was selling six-packs for 99 cents, and the boys would always stock up according to the time-tested formula: a six-pack per day per person, more if they could afford it. The last element in the planning was to determine when to leave and where to rendezvous. They always left early.

The chatter in the truck that October morning constantly shifted, as it always did, from one topic to another. Sometimes, geological matters such as Cambrian stratigraphy, the theme of their current jaunt, were debated. Later, the boys might commiserate with each other about some difficult assignment or perhaps discuss strategies for dealing with implacable professors. Occasionally, they would exchange their generally radical political views or share perspectives on the two or three female majors in the Geology department. Women were still an underrepresented minority in geology programs in those days, so it was natural that the fellows took notice of those brave pioneers in a male-dominated field. That, of course, was not the only reason why they found them intriguing.

Ned, the runty lad in the back seat of the Bronco, was thumbing through some photocopied publications on Cambrian rocks when the conversation shifted to graduate schools. This always made him uneasy because he considered himself the dimmest light in this otherwise luminous array. Ned knew, even at that young age, that all of his cohorts were destined for distinguished careers. And he was right; they all are now distinguished professors or researchers at major universities. Ned's own academic record was adequate for graduation but far from distinguished. He always wondered why his brighter compatriots tolerated someone like him, who contributed so little to the intellectual atmosphere of the trips. He secretly felt very lucky to be included in the group. He always learned a great deal on these adventures with his gifted mates; more, perhaps, than from any of his professors on campus. On the other hand, Ned was sure that no one in the group ever learned anything substantial from him. For some reason, perhaps his small stature and timid nature, he was always treated like a little brother by his friends. They offered assistance

when he struggled with school work, and politely answered his naive questions. Still, Ned was usually left out of the discussions about which university offered the best doctoral program, or which professors had what grants for graduate students. Ned feared, at the time, that he might be the only one in the group who had no professional future in geology at all. Sometimes, he would even wonder how he ever wound up in such a rigorous science in the first place, and his chronic self-doubt led him to continuously second-guess the decisions he had made. Would he have fared better as a history major? What about English? Perhaps philosophy? In spite of all his anxiety, it was only when the topic of advanced study came up that Ned had difficulty concealing his inferiority complex from his buddies. Otherwise, the four youngsters blended well into a compatible group of brash roosters out to conquer the scientific world. They meandered all over the Mojave Desert during those years with a swaggering self-assurance that must have been as amusing to their professors as it was exhilarating to them. Even Ned, a master at disguising his feelings of inadequacy, had just a touch of arrogance about him, however ill-founded it might have been.

The destination of the crew on that fall trip long ago was the southern Marble Mountains in the Mojave Desert (Plate 4), a site that was famous for trilobite fossils. It also afforded excellent exposures of several different rock units of Cambrian age, hence the emphasis on the site in the lecture they had all attended earlier in the week. Several hours after they left the university, the Marble Mountains came into view under a sun that was beginning to shed considerable warmth across the desert. The Bronco pulled off the highway onto a rough dirt road that led south toward the outcrops of Cambrian rocks, still a few miles away. The desert panorama, splashed across the horizon in browns and grays and framed in bright green yucca, was stunning in the warm autumn light. The group paused for a photograph and, since it was approaching midday, their first beer. The area around the Marble Mountains is now included in the Mojave National Preserve, but in those years it was still a spacious and vacant desert terrain with no government designation as anything else. The boys were free to go anywhere they chose, any time, by any means. So, after a brief moment of silent admiration of the desert panorama, the Bronco roared to life again and charged down the road, throwing dirt and rocks from the spinning rear wheels. As they careened down the road from the highway, they gave a cheer every time the Bronco became airborne over high spots in the rough gravel road. The wild ride was a mirror of their lives in those carefree days: full-throttle, rollicking, and a little reckless. Even Ned found it easy to put aside all of his self-doubts in the exhilaration of the moment.

Trailing a cloud of dust, the Bronco swung in and out around the curves in the road as it approached a ridge composed of tilted layers of brown and gray rock. Checking the detailed geological map he had brought, Mark's eyes danced from the ridge to the paper and back to the ridge. Pointing northeast, toward the lower end of the ridge, he directed Gary, the driver, to a faint track through the creosote and yucca. The Bronco slowed down a little to maneuver over some rocks, and continued to bounce along the barely distinguishable road. Soon the vehicle rolled to a stop just a few yards from an outcrop of hard, brown rock. The doors of the truck flew open, and the youngsters stumbled over empty beer bottles as they clambered out. As soon as they shouldered their packs, the lads assaulted the ridge, flailing at the outcrops with rock hammers and examining the details of the stone with hand lenses. They paced off the approximate thickness of the layers and found a few fossils, unraveling the sequence of strata on the fly: the Prospect Mountain Quartzite, the Latham Shale, the Chambless Limestone, the Cadiz Formation. The rock succession made sense; it was just as it had been described in the lecture the previous week and in the articles they had passed around in the truck. As the crew scampered around the rugged hills, they had little difficulty relating each rock unit to the model for Cambrian stratigraphy that had been described in the classroom. As usual, they had all the answers. Except for Ned, that is, who didn't even know what the questions were.

After a laborious climb to the apex of the ridge, the four sat down to rest and chew on some of the leftover dorm bacon, now so shriveled in the desert air that it resembled beef jerky. Sitting on layers of 530-million-year-old limestone, they talked briefly about shallow tropical seas, the positioning of North America along the equator in Cambrian time, and the paleoecology of trilobites. They were pleased with their success in deciphering the sequence of strata exposed along the ridge. Ned began to feel a little better now that the focus was back on rocks and the debate about graduate programs was finished. After a brief rest, the kids rose, slung their packs, and marched upward along the ridge in single file. Ned brought up the rear, straining to hear the conversations going on ahead of him.

As they marched upward, more rock layers passed under their boots. Each feature of the rock was identified and analyzed without pause: cryptalgal laminations, oncolites, cross-bedding, trilobite fragments, inarticulate brachiopods. The words streamed from their mouths, one at a time, as the various features were observed during the ascent. Even Ned correctly identified a rock here and there, though his fear of saying something dumb kept him mostly mute during the ascent. The next halt was at a saddle nearly a thousand feet above the desert floor. There they caught their breaths, drank some water, and surveyed the surrounding terrain. There were still three hours or so of good daylight left, and the boys, as was their habit, intended to make the most of it before setting up camp. The group decided to split up in order to cover the most ground, and to maximize their chances of finding something … well, extraordinary. Of course, no one in the group knew exactly what that extraordinary thing might be, but the prospect of finding it thrilled them nonetheless. Gary, nursing a painful ankle sprained in some sort of athletic sparring with his girlfriend, chose to angle off the ridge, and loop back to the Bronco. Mark and Pat continued up the ridge to what appeared to be the summit of a hill, then disappeared over the top of it. Ned, undecided as always, was left sitting alone on a boulder of silty limestone at the saddle. Eventually he rose and slowly started walking the sidehill in the general direction of the Bronco and the descending sun.

Ned always felt some relief whenever he became separated from his friends. Walking alone, he didn't have to respond to any questions, or offer any interpretations of geological phenomena. In solitude, he could talk freely, without fear of embarrassing himself with naive questions or erroneous perceptions. In fact, though Ned enjoyed being part of the group, he was most at ease when he was isolated from them. He was much more afraid of being humiliated than he was of being alone, and he had never understood why most people seem to fear loneliness. To Ned, the inner peace that came from being alone, far removed from any possible ridicule, made loneliness feel good. He was not antisocial in a pathological sense; it was just that being around people was most often an exertion for him. Not surprisingly, he loved the gloriously vacant deserts of the Mojave.

The sun floated just above the western horizon when Ned happened to see a small trilobite fossil on a slab of greenish-gray shale lying on the slope. The fossil was only about an inch long, but it was nearly compete, with all three major body lobes preserved as a thin brownish-black plate protruding from the soft rock that enclosed it. The angled rays of light from the setting sun highlighted each irregularity, every bump and furrow, on the fossil as Ned tilted the slab. Though the fossil was of good quality, it was not particularly impressive for the Marble Mountains; but Ned was mesmerized by the object as he picked up the slab, laid it across his lap, and seated himself on a large boulder. Inspecting the fossil carapace, Ned identified as many of the anatomical features as he could recall: the pygidium, the facial suture, the glabella, the genal spines. He chanted each name aloud, one by one, to no one except the lonely Mojave yuccas on the slope. In silence that was broken only by a lonesome whisper of wind, he sat enchanted by the half-billion-year-old corpse entombed in the rock resting in his lap.

Then, without any conscious effort or advance warning, a vision crystallized from Ned's imagination. As Ned sat on the mountain slope, he lifted his eyes from the fossil and looked out over the creosote-dotted Mojave plains. He imagined the desert floor undulating up and down, forward and back. The sand and rocks far below melted into crystal clear water, shimmering in the slanted light of the sun. Near the base of the mountain, he envisioned small waves in the sea breaking against the rocks, creating foamy swirls that ran a short distance up the shore before falling back. The cool maritime breeze began to lift Ned's baseball cap. He could smell the salt. As he looked toward the horizon, he imagined rippled waters receded into the distant haze, which, in turn, merged imperceptibly with the darkening sky. Below him, near the water's edge, Ned pictured rippled sand, submerged beneath maybe two feet of water, over which many small dark objects groveled about. He imagined the dark objects were trilobites, the same type and size as the fossil preserved on the slab, wriggling over the rippled sandy sediment, their brownish segmented bodies mottled with camouflage patterns of purple and maroon. The vision continued even as Ned rose, dropped the slab of rock, and descended the slope.

At the base of ridge, Ned dreamed of looking into clear water to see a trilobite flex its body into an arch, nose into the sand, and, with a shivering motion, disappear from view. Elsewhere, particularly where swirling water scoured some of the soft sand away, Ned pictured trilobites emerging from the sediment, using the same shivering motion to shake sand off their bodies as they crawled away. Under the shallow water, Ned could perceive round, green-gray objects, about a half-inch in diameter, rolling back and forth in the agitated water. He could almost feel the mass of soft, slimy algae on its shell of crusty material as he fancied reaching into the water to grab one of the small spheres. In his dream world, there were also mats of greenish brown material that he presumed to be algae washed up on the slope descending to the water. Ned imagined the sliminess of the mats, similar to the feel of the round objects. Ambling slowly along the imagined shore, Ned contemplated a low spit that extended offshore into the distance. He dreamt of walking along that spit, and seeing rounded, biscuit-shaped stromatolites beneath the clear, surging water. The mountain he had just descended vanished from Ned's dream world, replaced by a nearly flat plain, rising ever so gently from the water's edge and imperceptibly fading into the infinite distance. The low land exposed beyond the water was gullied here and there, barren, and naked: no shrubs, no trees, no insects, no life at all. Everything in the imaginary landscape had a misty, dreamlike quality, making Ned feel as if he had suddenly been transported to an exotic planet in some other solar system. Delighted by the strangeness of the unearthly world he was in, Ned's eyes widened, as he continued to survey the illusory scene around him. Then, on the etheral salt breeze, he heard his name, called from some mysterious source farther along the shore. As his dream world began to melt away, Ned turned toward the sound. He strained to discern the large, dark object, obscured by the growing darkness, directly ahead of him.

The tailgate of the Bronco was down, and the Coleman stove resting on it was hissing softly beneath a large pot, as Ned walked into camp. The boys teased him about getting lost, not hearing their calls, and almost missing his share of the chili beans. Ned just smiled his quiet smile, saying nothing of his apparition on the mountain. Later that night, around the fire, they finished all the beer and jabbered about all sorts of geological matters until well after midnight. Eventually, the kids all retired to their sleeping bags as the coyotes filled the blackness with their yelping calls. Ned laid awake for awhile, staring at the stars in the Mojave night, reflecting on the vision of the Cambrian world he had experienced. He couldn't quit thinking about the imaginary scenes and how magical the feeling was of being somewhere else, in another time. Even though hours had passed, he still felt the exhilaration of the dream, the rapture of walking in a vanished world. He had never experienced such a powerful sensation, and he knew, even before it began, that some sort of spiritual transformation awaited him. From his sleeping bag, he reached out to grab a small piece of shale from the rocks the boys had used to line their firepit. Holding this chunk of the ancient sea floor in his hand, he drifted off to sleep, still floating in the euphoria of his vision.

The boys left early the next morning and spent most of the day enduring the long drive back to the university. They talked about future outings, to other places, to explore other geological wonders. The Marble Mountains trip had been fun, but no one in the group seemed to regard it as anything special. No one, that is, except Ned. As usual, he said very little as he gazed out at the desert from the back seat of the Bronco. But, inside, his spirit was still aflame with the elation of his time transcendence. Ned's epiphany in the Marble Mountains changed him, and he knew things were never going to be the same. From that moment on, his passion for geology, and for science in general, never faded. The Cambrian vision had somehow purged him of all his insecurity and self-doubt. In time, he came to realize that his imaginary visit to the prehistoric world may have radiated from his soul, but it materialized only because of his scientific insights. He understood that his dream had been built from many small pieces of knowledge, assembled into a conceptual pattern by abilities he never suspected he possessed. Ned had learned that mysticism and science, often regarded as opposite poles of the human intellect, sometimes complement each other. Scientists can be dreamers too, and Ned had just experienced one hell of a dream. Somehow the cold rock and lifeless fossils combined to ignite a fire in his soul that was never extinguished.

The details of Ned's story may be unique, but he is clearly not the only person to become captivated by the Cambrian rocks of the Great Basin. Since the mid-1800s, some of the world's foremost geologists have been lured to the magnificent exposures of Cambrian rocks in the region. The rapture that Ned felt must also have swelled in the hearts of Charles Walcott and Fielding Meek as they stalked Cambrian fossils in the 1880s. In the following century, scores of illustrious geologists including Charles Resser, Allison Palmer, Richard Robison, Lehi Hintze, John Stewart, Clemens Nelson, and many others followed the same charm into the Great Basin to help establish the modern framework of knowledge about the Cambrian history of the region. And still they come, by the hundreds, every field season to seek additional clues about a

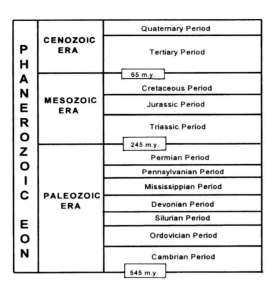

Figure 3.1 The Phanerozoic Eon. The last 545 million years of Earth history is subdivided into three eras, each of which consists of numerous periods. Despite a duration that exceeds a half-billion years, the Phanerozoic Eon encompasses only the most recent 13 percent of the history of our planet.

P H A N E R O Z O I C E O N	**CENOZOIC ERA**	Quaternary Period
		Tertiary Period
		65 m.y.
	MESOZOIC ERA	Cretaceous Period
		Jurassic Period
		Triassic Period
		245 m.y.
	PALEOZOIC ERA	Permian Period
		Pennsylvanian Period
		Mississippian Period
		Devonian Period
		Silurian Period
		Ordovician Period
		Cambrian Period
		545 m.y.

captivating time in a remarkable place. Such is the magic of the Great Basin Cambrian!

CAMBRIAN SEAS OF THE GREAT BASIN

The Cambrian Period (545–505 million years ago) is the first period in what geologists refer to as the Phanerozoic Eon: the eon of "visible life." The period was named for that portion of Wales traditionally known as Cambria, where geologists first collected fossils from rocks of this age. The Phanerozoic Eon is the last major division of geologic time, embracing the most recent 13 percent of earth history. Thus, during the first 87 percent of the history of our planet—the combined durations of the Hadean, Archean, and Proterozoic Eons—life on the earth was not "visible." Perhaps "obvious" is a better word than "visible" because, as we have already seen, life certainly existed in these earlier eons, and, based on the sizes of the Proterozoic fossils described in the preceding chapter, at least some of it was "visible." But, compared to the complexity, abundance, and diversity of life in the Phanerozoic Eon, Precambrian life was rare and primitive. In rocks of Phanerozoic age, fossils become dramatically more abundant and diverse. Using the rich fossil assemblages and their nonrandom occurrence through time, geologists have subdivided the Phanerozoic Eon with much greater precision than is possible for the sparsely fossiliferous pre-Phanerozoic eons. Thus, the Phanerozoic Eon consists of three

eras: the Paleozoic, Mesozoic, and Cenozoic. Each era is further subdivided into numerous **periods**. The Cambrian Period, for example, is the first period of the Paleozoic Era. Six later periods of this era have been established by geologists (Figure 3.1).

As the Cambrian Period began in the Great Basin, no major shifts in the geologic or tectonic setting seem to have occurred. The patterns that had been established by the Proterozoic rifting of Rodinia continued with little change into the Phanerozoic Eon. Most of the Great Basin was sea floor, and sediment continued to accumulate along the passive margin of western North America. However, it seems that just after the beginning of the Cambrian Period, the sea level dropped for some reason. The lowering of the sea level may have been due to climatic events (a cooling trend would cause such an oceanic event) or perhaps by the reduction in the overall rate of sea-floor spreading. Whatever the cause may have been, the decline of the sea level forced the ocean to recede toward the west in the Great Basin, as the falling water drained off the exposed land to the east. Such a withdrawal of seas from land is known as a **regression**. During the early Cambrian regression, the shoreline migrated west to a position near the modern Nevada-Utah border, while the depth of the ocean was reduced in those areas farther west that remained submerged. This moderate regression produced an unconformity, or gap, in the Cambrian rock record of eastern Nevada and western Utah: the earliest Cambrian sediments were exposed to the elements after the regression and were eventually removed by erosion. Offshore of the post-regression shoreline, in the western Great Basin, the falling sea level resulted in the deposition of thick sequences of limestone and dolomite, carbonate rocks that form most readily in shallow seawater. In the White-Inyo Mountains, for example, early Cambrian rock units such as the Reed Dolomite and Deep Spring Formation consist dominantly of carbonate rocks that record the shallowing of the seas in that area. Because the western Great Basin region remained submerged after the regression, sediment continued to accumulate throughout Cambrian time, and no unconformities have been identified in that area. In fact, the early Cambrian succession in the White-Inyo Mountain region is more than 20,000 feet thick and is considered one of the most complete rock sequences of its age anywhere in the world. Within this sequence of rock, there are numerous indications of ex-tremely shallow water: small-scale cross-bedding, ripple marks, mud cracks, and algal features all suggest vigorous water movement and very shallow depth at the time the sediments were accumulating. Some geologists have speculated that the water in this part of the Great Basin may have been only three or four feet deep, even though the site was, at the time, a hundred miles or more offshore!

The regression of the sea in early Cambrian time was a short-lived phenomenon. Almost immediately after the sea level dropped, and the shoreline reached its westernmost position, the ocean began to creep eastward over the exposed portion of the older sea floor. Once again, the cause of the oceanic encroachment is obscure but probably reflects some combination of climatic, geological, and oceanic events. Geologists use the term **transgression** to describe the invasion of dry land by an advancing sea. Transgressions can result from several different causes. A rise in sea level, a drop in the land surface, or a change in the climate could all trigger a transgression. The transgression that followed the early Cambrian regression was rapid and dramatic; it is believed to have been accelerated by an actual rise in the level of the oceans, the simultaneous subsidence of the land surface, and, perhaps, by climatic changes. The subsidence of the land occurred as a result of the continued cooling of the broken edge of North America as it drifted farther from the "hot" oceanic ridge that was located well to the west by Cambrian time. Moving steadily away from the molten rock beneath the oceanic ridge, the rifted lithosphere beneath the Great Basin became thinner and more dense as it cooled, sinking deeper into the dense mantle material below. This type of subsidence is known as "thermal subsidence," because it is caused by the loss of heat and the subsequent effects of such

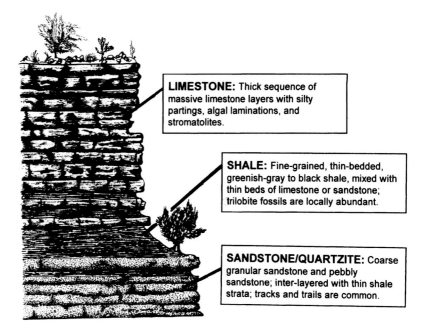

LIMESTONE: Thick sequence of massive limestone layers with silty partings, algal laminations, and stromatolites.

SHALE: Fine-grained, thin-bedded, greenish-gray to black shale, mixed with thin beds of limestone or sandstone; trilobite fossils are locally abundant.

SANDSTONE/QUARTZITE: Coarse granular sandstone and pebbly sandstone; inter-layered with thin shale strata; tracks and trails are common.

post-rifting cooling. Because thermal subsidence occurred at the same time as the climate-controlled sea level rise, the early Cambrian regression was abruptly reversed just before middle Cambrian time in the Great Basin. The shoreline raced east rapidly in one of the most dramatic oceanic assaults on land ever recorded: the great Cambrian Transgression.

CAMBRIAN TRANSGRESSIVE SEQUENCE

In the eastern Great Basin, the land exposed when the shoreline initially migrated west was reinundated by the advancing and rising seas near the end of early Cambrian time. As time progressed, the shoreline migrated farther to the east, into present-day Colorado, and the water deepened in areas in the Great Basin that were already submerged. As the shoreline pushed east through western Utah, the sandy beach migrated in that direction as well. Earlier sandy deposits, farther west, became buried under fine-grained muddy sediments that were deposited offshore from the east-marching shoreline. Eventually, when the Cambrian shoreline reached the interior of North America, the Great Basin region was so far offshore that little granular sediment reached the sea floor from the small amount of land that was still exposed hundreds of miles to the east. Limestone

Figure 3.2 The rock record of a transgression. The ascending succession of sandstone (beach and shallow offshore deposits), shale (mud deposited in deeper, less-agitated offshore basins), and limestone (open shallow marine deposits) records the submersion of the land beneath an advancing sea. This generalized pattern is typical of many basal Cambrian sequences in the southern and eastern Great Basin.

Figure 3.3 Unconformity between light-colored Cambrian strata and darker underlying Archean rocks in the northern Wasatch Mountains.

Figure 3.4 Early Cambrian rock units of the southern and eastern Great Basin. In the east, early Cambrian sediments were deposited on metamorphic basement materials. In the west, the Cambrian/Precambrian boundary is more difficult to locate because it occurs within a thick succession of sedimentary rock layers.

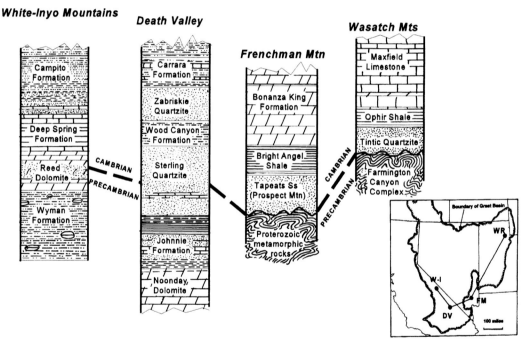

sediments formed in this shallow, open marine environment when calcium carbonate ($CaCO_3$) was precipitated directly from seawater. Thus, the record of the great Cambrian transgression in the eastern Great Basin is a threefold package of sedimentary rock: sandstone representing beach and near-shore sediments at the base, shale or siltstone signifying offshore mud in the middle, and thick layers of limestone (open ocean sediments) at the top (Figure 3.2). The entire sequence records deepening water and the development of more open and offshore marine environments, precisely as we might expect

from an oceanic invasion. This pattern is nearly universal in the lower (basal) portion of the Cambrian strata deposited in the eastern Great Basin. The lowermost sandstone is known by several different names, including the Prospect Mountain Quartzite (in eastern Nevada and western Utah), Tapeats Sandstone (in southernmost Nevada), and the Tintic Quartzite (in west-central Utah). The offshore muds are represented by such rock units as the Pioche Shale, the Ophir Shale, and the Bright Angel Shale. The open marine limestone sequences are normally the thickest portion of the transgressive sequence and are typified by such rock units as the Bonanza King Formation, the Howell Limestone, the Chambless Limestone, and many other units across the eastern Great Basin.

Immediately beneath this transgressive sequence, throughout the eastern Great Basin, a prominent unconformity separates the Cambrian rocks from the underlying Proterozoic and/or Archean basement. The older rocks beneath the Cambrian transgressive sequence were probably buried by very early Cambrian sediments, but these were removed by erosion when they became exposed following the early Cambrian regression. In places, the erosion that occurred between the regression and the subsequent transgression exposed metamorphic basement rocks of Proterozoic or Archean ages. During the great Cambrian transgression, this older erosion surface was buried under younger Cambrian rock layers, starting with the sandstone component of the transgressive sequence. Thus, in the eastern Great Basin, basal Cambrian sandstone, such as the Tintic Quartzite, commonly rests on metamorphic basement rocks such as the Farmington Canyon Complex (Figure 3.3). The surface between these two rock units, the great Cambrian/Precambrian unconformity, represents a gap in the rock record that may signify as much as 2 billion years of time (Figure 3.4). This unconformity vanishes as it is traced into the western Great Basin because, in that region, the sea floor remained submerged throughout the entire span of late Precambrian-Cambrian time.

POST-TRANSGRESSION PATTERNS: THE CORDILLERAN GEOSYNCLINE

Near the end of the Cambrian Period, the great transgression finally ended. By that time, however, most of North America had been submerged. Only the highest ridgelike spine of the Cambrian continent, known as the Transcontinental Arch, remained above sea level at the end of the Cambrian Period (Figure 3.5). The Transcontinental Arch extended from the Great Lakes region to New Mexico as a more or less linear tract of elevated land. No part of the Transcontinental Arch passed into the Great

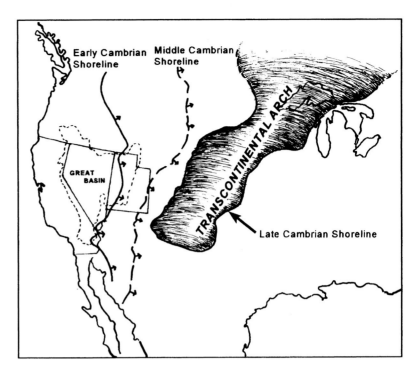

Basin, so by late Cambrian time the entire region was completely under water. Sediment of various types was accumulating everywhere on the ancient sea floor, from Colorado to California. This ancient basin of marine sedimentation has traditionally been known as the **Cordilleran Geosyncline.** Derived from the Spanish word for "mountain chain," the term "Cordilleran" refers to the mountainous portion of western North America, from the Rockies to the Pacific coast. "Geosyncline" is a geological term that originated in the 1870s for a subsiding basin or trough in which sedimentary and volcanic rocks accumulate. Until the plate tectonic revolution of the 1960s and 1970s, the development of geosynclines was thought to be the first step in the mountain-building process. These sinking troughs were envisioned to descend so deeply into the earth's interior that the materials they carried melted to form bodies of magma. The fluid magma then rose through the surrounding rock, driving the whole pile of sediment, now metamorphosed by the heat and pressure of its previous descent, upward in the form of a complex mountain range. Though this theory of mountain building has now been abandoned in favor of mechanisms involving the interactions between lithospheric plates, the basic concept of a geosyncline is still valid. When great amounts

Figure 3.5 The Transcontinental Arch and the great Cambrian Transgression. After an initial regression, the Cambrian shoreline migrated steadily east, as indicated by its position in early Cambrian (solid line with arrows) and in later Cambrian (dashed line with arrows) time. By the end of the Cambrian Period, the entire Great Basin was submerged, as was most of North America. Only the high Transcontinental Arch remained above sea level.

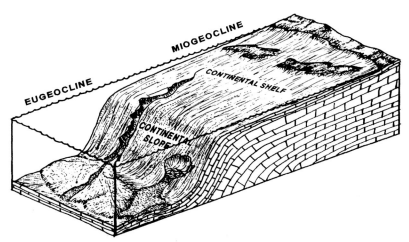

EUGEOCLINE MIOGEOCLINE CONTINENTAL SHELF CONTINENTAL SLOPE

Figure 3.6 The Cordilleran Geosyncline. By Cambrian time, the Cordilleran Geosyncline was differentiated into an eastern platform and ramp (miogeocline) and a western deep basin (eugeocline), joined by a relatively steep continental slope. Submarine landslides occasionally carried material down the slope and onto the deep sea floor in west-central Nevada.

of sediment accumulate on the sea floor, the crust below does, in fact, subside under the immense weight of the sediment load. The crust of our planet behaves as if it was floating on a dense fluid: anytime weight is added, the crust sinks; if weight is removed (through erosion, for example), the crust rises. This condition, in which a balance between weight and elevation is maintained by vertical movements of the crust, is known as **isostacy.** When the earth's crust sinks to accommodate a load placed on it, it is said to have experienced an "isostatic adjustment," or to have responded "isostatically." The crust under the Great Basin region was behaving in precisely this manner during the Cambrian, and for much of the remaining portion of the Paleozoic Era. As more sand, mud, and carbonate ooze accumulated on the sea floor, it sank steadily under the ever-increasing weight. So, the Cordilleran Geosyncline, though it is an old term which is considered antiquated by some geologists, is still an apt description of the Great Basin region during the early Paleozoic Era. The entire province was part of a north-south trending trough of sediment accumulation extending from the foothills of the Transcontinental Arch westward to the position of modern California. This trough was initiated by the late Proterozoic rifting event, but during the Cambrian Period it became a full-fledged ocean basin thousands of miles wide. The Paleozoic rocks of the Great Basin all accumulated along the eastern margin of this wide basin. The linear Cordilleran Geosyncline thus reflects only a portion (the eastern edge) of the broad ocean basin. Where was the western side of the basin? No one has ever

found the western counterpart to the Cordilleran Geosyncline with absolute certainty. During Cambrian time, the fragment(s) of Rodinia that were rifted to the west continued to drift in that direction, into Antarctica and/or Siberia, perhaps. As they moved away from North America, they were rafted along on oceanic plates that were also carrying sediments. Those sediments have long since vanished from North America; they exist today somewhere on the other side of the Pacific Ocean. What we know for sure is that all of the Paleozoic strata of western North America were deposited in the same great oceanic trough, the Cordilleran Geosyncline, which represented the eastern side of a much larger oceanic basin.

After the great Cambrian transgression, the Cordilleran Geosyncline developed an east-west dichotomy that would persist throughout the Paleozoic Era (Figure 3.6). The inner portion of the marine basin, that part closest to the shoreline, was submerged by very shallow water and sloped gently toward the deeper water to the west. By middle Cambrian time, this broad shelf was at least 300 miles wide, stretching from the coast in Colorado well into Nevada. This shallow-water portion of the Cordilleran Geosyncline is essentially the ancient continental shelf, and is known as the **miogeocline.** The water over the miogeocline was probably no more than about 500 feet deep in most places. In some places, it was probably less than 100 feet deep. The shallow seas surged continuously across the miogeocline, as waves, currents, and tides kept the water in constant motion. Western North America was located about 10 degrees north of the equator during most of the Cambrian Period, so the tropical sun would have sailed high above this shallow sea. In response to the strong sunlight, the warm and clear tropical water must have sparkled in pale blue radiance as it rose and fell with the passing swells.

Several hundred miles west of the Cambrian shoreline, the gentle slope of the miogeocline abruptly steepened, as the continental shelf passed into the continental slope. The continental slope marked the boundary between the rifted continental crust and the thin, dense oceanic crust, both buried beneath a blanket of sediment. This slope was located in central Nevada, about where the 0.706 Line, described in the preceding chapter, has been mapped. West of the continental slope was the deep, level floor of the Cordilleran Geosyncline. Here, conditions were much different from those on the miogeocline. The water west of the slope was deep, dark, and cold. The actual water depth was probably around 10,000 feet (two miles) or more. Water this deep circulates very slowly, is well below the depth to which light can penetrate, and is perpetually cold. This deep and murky portion of the Cordilleran Geosyncline is referred to as the **eugeocline**; it is essentially the abyssal counterpart of the shallow miogeocline. Sediment that accumulated in the gloomy depths of the eugeocline was much different from the material that was deposited on the gleaming floor of the miogeocline. Even before the distinctions between the miogeocline and eugeocline were fully appreciated by geologists, the contrasts between the Cambrian rocks of the western and eastern Great Basin were impossible to overlook: traditionally, geologists have used the terms "western siliceous assemblage" and "eastern carbonate assemblage" to describe the eugeoclinal and miogeoclinal deposits of Paleozoic age. Cambrian rocks are the oldest rocks in the Great Basin to exhibit the miogeocline-eugeocline dichotomy, but the pattern is also evident in many younger Paleozoic strata. What are the differences between these two packages of rock, and how do they result from the differences in water depth?

The Glittering Miogeocline: A Limestone Factory

The shallow water that submerged the Cambrian miogeocline was warm, clear, and constantly swirling under the influence of the tides, currents, and wind-generated waves. Under these conditions calcium carbonate ($CaCO_3$) is precipitated readily, in a process that is similar to the way sugar crystals form rock candy from a water solution. Other chemicals may also be precipitated from warm seawater, but calcium carbonate is by far the most abundant. The tiny crystals of $CaCO_3$, known as the mineral calcite, accumulate on the shallow sea floor as a muddy

Figure 3.7 Outcrop of Cambrian limestone in the House Range of western Utah.

ooze. Because it is so abundant in seawater, many marine organisms use calcium carbonate to build their shells. Coral, clams, snails, and even some encrusting algae all secrete calcium carbonate as a body covering. After death, this shell material may become pulverized by the waves and currents, and the fragments accumulate on the sea floor along with the calcite crystallizing directly from seawater. Gradually, this gritty mud hardens to form the sedimentary rock **limestone**: it consists almost entirely of the mineral calcite derived either directly from seawater (inorganic $CaCO_3$) or from granulated shell material (organic $CaCO_3$). Sometimes the calcium carbonate included some magnesium in the form of the mineral dolomite, $MgCa(CO_3)_2$; if so, then **dolostone**, a sedimentary rock made of the mineral dolomite, resulted. Such limestone and dolostone is usually light gray to almost black in color, thick-bedded to massive, with small fragments of fossils embedded in a solid calcite matrix. Collectively, limestone and dolostone are known as **carbonate** rocks, because they consist almost exclusively of carbonate minerals, either calcite or dolomite.

Though the limestone of the miogeocline is composed almost entirely of calcite, it may contain some silty impurities or some organic carbon, giving it, respectively, a light-colored "muddy" appearance or an earthy black color. Sometimes the limestone of the miogeocline exhibits platy rock fragments cemented together by the limestone matrix. Such fragmentary limestones probably represent pieces of stiff, semiconsolidated lime mud that were ripped from the sea floor by currents or waves, redeposited as "chips" or pebbles in deeper, calmer water, and subsequently buried under

Figure 3.8 Clastic Cambrian limestone of the Carrara Formation of the Mojave Desert region.

additional carbonate sediment. These granular limestones attest to the vigor of the water movement over the shallow miogeocline.

The Eugeocline: A Murky Abyss

At the outer edge of the miogeocline, or continental shelf, a steep ramp known as the continental slope existed. This slope descended abruptly into the deep eugeocline, which began at its base. In the Cordilleran Geosyncline during the Cambrian Period the slope was located in central Nevada and the water at mid-slope was probably more than 3,000 feet deep. Though the continental slope that connected the miogeocline to the eugeocline was not flat, sediment still accumulated on it. These slope sediments were primarily derived from the shallow miogeocline to the east, from which they were carried into deeper water by submarine landslides and turbid water currents moving down the inclined sea floor. Thus, limestone still formed on the continental slope of west-central Nevada, but these sediments consisted of fragments that were transported from the shallow-water miogeocline. Many of the Cambrian strata in the slope area exhibit a coarse granular or fragmental appearance that geologists describe with the term "clastic," from the Greek *klastos*, meaning "broken in pieces." Such rocks represent rubble that slid down the submarine slope and collected in layers on the deep sea floor tilted to the west. Cambrian deposits of this type occur in the Harmony and Hales Formations of southern and central Nevada.

To the west of the narrow continental slope, in western Nevada and California, the deep Cam-

brian sea floor apparently leveled out into a more or less flat plain, submerged under water that might have been more than 15,000 feet deep. This portion of the Cordilleran Geosyncline is known as the eugeocline: it was a gloomy pit of very deep, cold, and nearly motionless water. These conditions differed dramatically from those over the shallow miogeocline to the east. Not surprisingly, a distinctive suite of sedimentary rocks, much different from the miogeoclinal limestones, formed in the eugeocline. In the deep basin of the eugeocline, little calcite sediment was deposited. This is because calcite crystals precipitated in shallow, warm water will dissolve if they descend into water deeper than about 12,000 feet. This dissolution is due to the changes in the temperature, pressure, and chemistry of seawater that occur with increasing depth. Thus, most of calcite particles that drifted toward the sea floor of the eugeocline from the shallower water above were dissolved as they fell through the deep water. In addition, on the cold, dark sea floor, populations of calcite-secreting organisms such as clams, snails, and corals must have been minimal, if they existed at all. Because little $CaCO_3$ sediment could reach the eugeocline, or be produced there, limestone is a minor component of the eugeoclinal rock assemblage.

Silica (SiO_2), produced when minerals like quartz dissolve from rocks exposed on land, is another chemical present in seawater. Unlike calcite, silica crystals can be precipitated from the cold seawater on the deep sea floor. The crystals of quartz that form in this manner grow slowly and are usually very small. In addition, some marine organisms, such as sponges and microscopic diatoms and radiolarians, secrete tiny bits of silica to support their soft tissues. These microscopic particles of silica can accumulate in deep water because they are not as susceptible to dissolution in the deep sea as are carbonate materials. The tiny particles of silica commonly act as "seed" crystals, stimulating the precipitation of inorganic silica from the deep bottom water. Eventually, thin layers and masses of microcrystalline quartz develop on the deep ocean bottom where limestone sediments cannot accumulate. After they harden, such layers are known as **chert**, a very hard and smooth-textured sedimentary rock. Though chert can also occur as irregular lumps in shallow marine limestone, thick and evenly layered sequences form almost exclusively on the deep sea floor. In addition to chert, tiny particles of wind-blown silt accumulated on the floor of the eugeocline as a black ooze. These grains, derived from land hundreds of miles to the east, make up the principal component of **shale**, a thinly layered sedimentary rock. Planktonic microorganisms that floated in surface water contributed additional material to the sediment that blanketed the floor of the deep eugeocline. The organic matter that reached the deep sea floor included carbon residues that commonly give the shale and chert a black or gray color. Microscopic silica skeletons of single-celled plankton can sometimes be identified in the eugeoclinal deposits of the Cordilleran Geosyncline. Occasionally, geologists have discovered volcanic rocks (undersea lava flows or volcanic ash particles) associated with the eugeoclinal suite of rocks. These volcanic components probably represent lava erupted at mid-ocean ridges or in volcanic island settings. Altogether, the chert-shale-volcanic assemblage (sometimes with minor amounts of limestone) of the eugeocline is strikingly different from the limestone-dominated miogeoclinal rock sequence. In the Great Basin, eugeoclinal deposits of Cambrian age have been identified in several different places in west-central Nevada, including the Hot Creek Range, where such interbedded chert and shale comprises the Swarbrick Formation (Plate 5). The eugeocline-miogeocline (western siliceous–eastern carbonate) dichotomy persists into later periods of the Paleozoic Era, but the Cambrian rocks of the Great Basin are the oldest to clearly exhibit this pattern.

CAMBRIAN RHYTHMS AND CYCLES

Superimposed on both the transgressive sequence and the post-transgression strata are some very rhythmic or cyclic patterns in the Cambrian rocks of the Great Basin. Since the mid-1960s, geologists have been intrigued by these repetitive changes in the type of sediment that accumulated on the floor of the Cordilleran Geosyncline. On the miogeocline, a cycle typically begins with a sequence of silty shale that represents relatively deep water conditions. The shaly sediments are overlain by limestone layers that suggest relatively shallow water conditions. Above the sequence of limestone layers, shaly sediments commonly reappear to indicate the return of deeper and calmer water conditions. In many Cambrian rock successions in the Great Basin, some of which may be thousands of feet thick, several of these shale-limestone couplets may be identified. Even in the transgressive sequences of the eastern Great Basin, the eastward march of the sea was anything but steady. This great marine advance was a jerky, stutter-step sort of encroachment—two steps forward, one step back—for millions of years. The large-scale oscillations in the rock record were originally referred to as Grand Cycles, a term that is sometimes still used, although "sequences" is the currently favored buzzword. Whatever they are called, several (at least five) major cycles or sequences have been recognized by geologists studying the thick succession of Cambrian limestone and shale layers in the Great Basin. What might be the cause of such rhythmic or episodic shifts in the type of sediment that accumulated in the Cordilleran Geosyncline during Cambrian time?

Most often, geologists infer that water depth is the principal control on the type of sediment that accumulates on the sea floor. Other factors, such as water temperature, water chemistry, proximity to land, or agitation can also influence sediment type, but these are all related in some manner to water depth. Studies of modern marine sediments have provided a sound basis for relating water depth to specific types of sediments that were deposited in the ancient Cordilleran Geosyncline. The cyclic patterns observed in the Cambrian rocks of the Great Basin probably for the most part signify periodic changes in the depth of the sea. But, water depth can be influenced by numerous events. Subsidence of the sea floor will lead to an increase in depth, even if there is no actual change in sea level. Prevailing climatic trends are also important. Under prolonged warm conditions, glacial ice on land melts faster than it forms, while the water already in the sea expands in response to the rising temperatures. The net long-term effect of a warm climate is a lifting of sea level, and, therefore, an increase in water depth (unless, of course, the sea floor is simultaneously rising at the same rate). A cold climatic cycle generally has the opposite effect: a reduction, or drop, in sea level. However, the climate also affects the rate of erosion on land. If the rate of erosion increases (as it might under colder conditions), more sediment is washed into the seas, effectively raising the bottom of the sea floor temporarily, at least until isostatic adjustments to the weight of the sediments cause it to sink. In addition, the rate of erosion on land is not affected by just the climate but is also strongly influenced by topography and geologic events such as mountain-building episodes. If this all sounds a bit confusing, then you are beginning to understand why no one has identified the exact cause of the Cambrian cycles. The periodic shifts in relative sea level, clearly augured by the Cambrian strata of the Great Basin, most probably result from the complex interplay between climatic changes, fluctuations in sea level, mountain-building forces, and erosional cycles.

Despite geologists' inability to fully, or precisely, explain the cyclicity in Cambrian strata of the Great Basin, its very existence has significance. We have described the Cordilleran Geosyncline as an ocean basin (a half-basin, actually) that developed along a passive, rifted continental margin. *Passive*? The word is used routinely by geologists to describe a setting where there is no interaction between separate lithospheric plates. Certainly, that was true of the Great Basin region during the Cambrian Period: the ocean basin and the land to the east were both situated atop a single plate. But, was the Cordilleran Geosyncline really as geologically docile as the word *passive* might suggest? The cyclicity of the Cambrian strata reveal that conditions on the sea floor were far from static and uniform. The Great Basin region, submerged during most of the Cambrian Period, was not nearly as geologically active then as it is today, but it wasn't dead, either. At least some of

the cyclicity exhibited by the Cambrian strata is probably linked to geologically induced vertical movements of the miogeocline, which, in turn, might have been accompanied by other sources of instability. This vacillation in sea level, once it began, appears to have continued throughout much of the Cambrian Period. Though the vertical shifts in sea level were not dramatic, they still left clear indications of cyclic instability in the rock record. The constant, albeit subtle and cyclic, fluctuations in the conditions on the sea floor within the Cordilleran Geosyncline were just a hint of the rousing events that would come in the later periods of the Paleozoic Era. Nonetheless, the geologic record plainly reveals that the Cambrian sea floor of the Great Basin was not a static, invariable substrate for the organisms that lived and evolved there. As life proliferated in the ever-changing Cambrian seas of the Great Basin, the inconstancy of their habitat resulted in bottom-dwelling marine communities that shifted constantly through time. For tens of millions of years, Cambrian organisms came and went as they evolved, died out, adjusted their submarine ranges, or migrated from place to place across the ocean floor. In the shifty undersea world of the Cambrian, some very exotic sea creatures evolved and spread across the shallow sea floor. Some of them, in comparison to modern marine organisms, were so bizarre that they might be mistaken for alien life from another world!

CAMBRIAN LIFE OF THE CORDILLERAN GEOSYNCLINE

Worldwide, Cambrian rocks have attracted a great deal of attention from paleontologists for almost two hundred years. The sudden appearance of abundant, large, and diverse fossils in rocks of this age, previously described as the great "Cambrian explosion," is a worldwide phenomenon, and paleontologists have spared no outcrop of Cambrian strata from the relentless search for more information on this dramatic biotic event. Aside from its importance in the history of life on our planet, the Cambrian explosion is intriguing for another reason, as well. The creatures that emerged during this celebrated zoological detonation include some of the most baffling animals to evolve on the earth. It is not just the abundance of fossils in Cambrian strata that have lured generations of paleontologists to the Great Basin; it is the character—so primeval as to seem unearthly—of these extinct critters that we find equally irresistible and fascinating.

Cambrian fossils were first collected in the Great Basin by the government surveys that entered the region in the late 1800s. Since that time, thousands of Cambrian fossils have been collected from localities concentrated in the eastern and southwestern part of the region. In western Utah, Cambrian fossils occur in dazzling abundance in the House Range, Cricket Mountains, Wah Wah Mountains, Drum Mountains, and the northern Wasatch Mountains near Wellsville and Brigham City. In California, such localities as the White-Inyo Mountains, the Providence Mountains, the Nopah and Resting Springs Ranges, and the Marble Mountains are world-renowned for the Cambrian fossils that can be found there. Between these two regions, Cambrian rocks are exposed in many of the mountain ranges of south-central Nevada, especially the Toiyabe, Hot Creek, Schell Creek, Highland, and White Pine Ranges. Fossil localities in these central Great Basin ranges have produced many additional specimens of Cambrian fossils. Cambrian rocks are either absent, not exposed, or metamorphosed to extreme degrees in the northwestern part of the Great Basin. Consequently, no important fossil localities of Cambrian age have been reported from that area (Figure 3.9).

Trilobites: Kings of the Cordilleran Seas

The most common fossil found in Cambrian strata, not only in the Great Basin but all over the world, are the primitive arthropods known as **trilobites** (Plate 6). Trilobites are distant relatives of the modern crustaceans, but are utterly dissimilar from them. Aside from the jointed, paired legs and a few other anatomical similarities, trilobites have little in common with the lobsters, crabs, and shrimp that inhabit the sea floor today. The name reflects the three-part

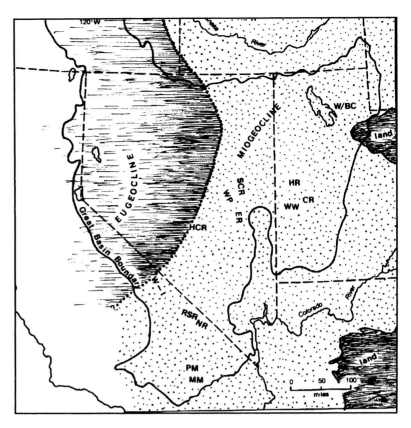

lengthwise subdivision of the segmented trilobite body: an axial (central) lobe, flanked by a pleural (side) lobe on either side (Figure 3.10). The three-lobed, segmented body of trilobites, usually an inch or two long, is their most distinctive feature; nothing like them exists in the seas today. Beneath the segmented body, trilobites possessed numerous pairs of jointed legs. Each of the paired appendages was, in turn, a double "leg": one large branch to perform the usually ambulatory function, and another smaller limb to support the external gills. Many of the details of the internal anatomy of trilobites are still uncertain, but it appears that most of these arthropods were bottom-living scavengers, algae grazers, or perhaps sediment feeders that found nutrients in the muddy ooze of the sea floor. However, some of the tiniest trilobites, only a few millimeters long, are also the most widely distributed, both in the Great Basin and on a global scale. This suggests that the smallest trilobites may have been free-swimming organisms that were capable of worldwide travel as they drifted with currents, or swam under their own power, in the Cambrian sea.

The upper (dorsal) surface of the trilobite

Figure 3.9 Cambrian paleogeography of the Great Basin. The edge of the shallow miogeocline was located in central Nevada, where a steep slope descended west to the deeper eugeocline. Good exposures of Cambrian strata occur in areas abbreviated as follows:
CR, Cricket Range, Utah;
ER, Egan Range, Nevada;
HCR, Hot Creek Range, Nevada;
HR, House Range, Utah;
MM, Marble Mountains, California;
NR, Nopah Range, California;
PM, Providence Mountains, California;
RSR, Resting Spring Range, California-Nevada;
SCR, Schell Creek Range, Nevada;
W/BC, N. Wasatch Mountains near Brigham City, Utah;
W-I, White-Inyo Mountains, California-Nevada;
WW, Wah Wah Range, Utah.

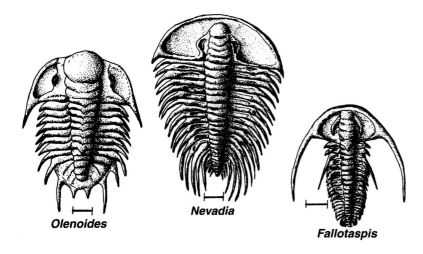

Olenoides　　**Nevadia**　　**Fallotaspis**

Figure 3.10 Cambrian trilobites from the Great Basin. While more than seventy different types of trilobites occur in the Cambrian strata of the Utah, Nevada, and eastern California, these three genera are among the most common. Scale bar = 1 cm in all drawings.Figure

body was covered by a thin, stiff carapace composed of chitino-phosphatic material (a mixture of proteins and mineral components), similar to the cuticle of modern crustaceans or the exoskeletons of insects. The soft underside of the trilobite was presumably naked, or covered by only a very thin calcified derma. Because trilobites were encased in a rigid (but thin) exoskeleton, each animal would have to periodically shed its carapace to accommodate growth. No one knows for sure how often trilobites shed their exoskeletons, but a good guess is that it must have occurred each time one of the transverse segments was added to the body. We might well envision a growing trilobite, straining against the confining carapace, eventually applying enough pressure against the shell to cause it to split open. In fact, there are thin creases, called facial sutures, in the head region of most trilobites that might have allowed the undersized shell to break apart in the most advantageous place to free the growing trilobite inside. Once the carapace was broken along the facial sutures ("zippers" on the trilobite tuxedo), the naked trilobite was able to swim out of the old shell and eventually secrete a new, larger carapace. Most trilobites had between 10 and 30 transverse segments. This means that each trilobite, in the course of its life span, might have produced several dozen preservable skeletons: a shed exoskeleton for each segment of its body, along with the final corpse. This consideration may explain one of the most striking aspects of the trilobite fossils found in Cambrian rocks: they often comprise as much as 90 percent of

the total fossil fauna. This does not mean, of course, that 90 percent of the organisms living in the seas of the Cordilleran Geosyncline were trilobites; it simply means, by virtue of their molting habits, the trilobites produced many more fossils than other kinds of organisms. Nonetheless, even if we make allowances for their ability to produce multiple fossils, trilobites were still the dominant creatures of the Cambrian seas. The Great Basin was crawling (literally!) with trilobites 520 million years ago.

The best Cambrian trilobite fossils in the Great Basin are most often found in fine-grained, fissile or platy shale sequences. The Wheeler Shale of western Utah, the Latham Shale of the Mojave region, the shaly portions of the Carrara Formation around Death Valley, the Dunderburg (or Tybo) Shale of south-central Nevada, the Pioche Shale of eastern Nevada, and the Harkless Formation in the White-Inyo Mountains are only a few of the more notable trilobite-yielding rock formations in the Great Basin. The greenish gray to silvery gray shale in these rock units typically splits easily into thin plates when tapped along the edge with a light hammer. Finding excellent trilobite fossils in the Great Basin is thus not particularly difficult: you merely need to find outcrops of the right rocks, sit down and start splitting the shale layers apart, and carry a large bucket to transport your prehistoric loot at the end of the day. Not surprisingly, specimens of Great Basin trilobites can be found in museums and collections all over the world. Many of the dozens of different species of Cambrian trilobites known from the Great Basin were first identified in this desert region and bear names reflective of their discovery locale: the genera *Nevadella* (Nevada), *Eurekia* (Eureka District, Nevada), *Housia* (House Range, Utah), *Bristolia* (Bristol Mountains, Mojave region), and *Utaspis* (Utah) are just a few of the many examples of Cambrian trilobites that carry epithets inspired by Great Basin localities.

An interesting twist on the story of Great Basin trilobites involves Native American perceptions of these fossils, and, at the same time, illustrates how prominent they are as an element of the regional natural history. In 1976, geologists Michael Taylor and Richard Robison (see Chapter 3 references), who both have spent many years studying the Cambrian rocks of the Great Basin, documented a fascinating episode involving Frank Beckwith, former editor of the *Millard County* (Utah) *Chronicle*, who died in 1951. According to the geologists, Beckwith found a specimen of the trilobite *Elrathia kingii*, a species common in the House Range region, associated with an Indian burial site that was excavated in the early 1900s. The trilobite had a single, clean hole drilled through the semicircular cephalon or "head." The trilobite from the burial site was associated with a leather strap and, Beckwith surmised, was carried around by the deceased individual as some sort of amulet or charm. The reasons for the use of fossil trilobites by the ancient Native Americans, along with the cultural significance of the fossils, remain uncertain. Nonetheless, in the early 1900s, there were still many members of the Pahvant Ute tribe living in the vicinity of Delta, Utah, where Beckwith resided. Beckwith showed the specimen from the excavation to Joseph Pickyavit, a Native American friend, who referred to it in the Ute dialect by the name "*timpe khanitza pachavee.*" Loosely translated by Beckwith, that phrase means "little water bug like stone house in." A water bug in a stone house! Scientists might refer to it as *Elrathia kingii*, but no one can dispute the essential accuracy of the Ute term. It seems that the organic nature of the trilobite fossils, the similarities between them and living arthropods, and their occurrence in rock strata that originated as marine mud were all obvious to the Utes of western Utah long before modern science had verified these perceptions. None of that is particularly surprising, of course, given the traditional intimacy between the Pahvant Utes and the natural features of the their homeland, the eastern Great Basin. Perhaps we should not find it so surprising that science and traditional native wisdom should arrive in the same place. After all, if two different, but equally valid, paths of enlightenment ascend the same mountain, should they not join at the summit of edification?

Cambrian Reefs of the Great Basin

Trilobites may be the icon of the Cambrian Period, but they certainly were not the only bizarre group of marine organisms living of the sea floor of the Great Basin more than 500 million years ago. Beneath the sparkling water of the tropical seas, there were literally hordes of wormlike creatures churning through the mud, clamlike brachiopods resting on the bottom, and a great variety of what were perhaps tiny molluscs with odd shells shaped like simple tubes or conical caps. Some of the hard-shelled organisms that accompanied the trilobites on the Cordilleran miogeocline grew in such profusion as to build large, reeflike masses on the shallow sea. But these creatures were not the corals or other calcareous organisms of the modern tropical seas. Instead, the Cambrian reef-builders on the miogeocline were some of the most bizarre creatures known from the fossil record. So strange, in comparison to the inhabitants of modern coral reefs, were some of the Cambrian reef-builders that scientists are still not sure about their relationships with modern marine animals. But, however strange these reef-forming creatures may seem, they nonetheless played an extremely important role in the ecology of the Cambrian seas of the Great Basin.

By far, the most important reef-building organisms during the Cambrian Period were the group known as the **archaeocyathids** (see Figure 3.11A). The fossils of the these odd creatures are constructed of two perforated calcite ($CaCO_3$) baskets, nested one inside the other, and joined together with small tubes, struts, and walls. Archaeocyathid fossils are generally conical in shape, but may be so gently tapering as to appear tubular in some fragmentary specimens. Complete archaeocyathid fossils are very rare, but enough have been found to demonstrate that the tip of the double-cone commonly possessed some sort of broad base or network of rootlike tubes that must have anchored the animals to the sea floor. Complete archaeocyathid fossils may be several inches long, but they are usually much smaller, about the size of a human thumb. In terms of their general size, shape, and presumed appearance,

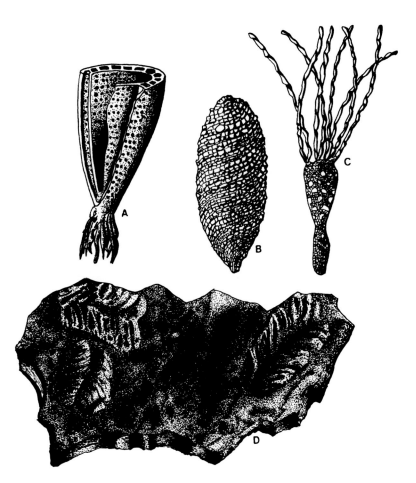

Figure 3.11 Non-Trilobite Cambrian Fossils of the Great Basin.

A, a generalized sketch of an archaeocyathid, revealing the double perforated walls;

B, *Helicoplacus*, a helicoplacoid, slightly more than one inch in length, with the characteristic spiral rows of plates;

C, *Gogia*, an eocrinoid that sometimes exceeds 15 cm (about 6 inches) in length;

D, *Cruziana*, a trace fossil presumably made by trilobites crawling and burrowing through the mud and silt of sea floor. Length of slab = 21 cm (8.5 inches).

cyathids played this role in the marine ecosystem. Other sorts of organic build-ups, such as stromatolites and microbial mounds, existed prior to the Cambrian Period, but they don't qualify as true reefs because they consist primarily of sediment trapped by organic mechanisms, not the actual skeletal tissues of the creatures. Archaeocyathids were the first group of organisms to form large masses on the sea floor through the sheer profusion of their growth. Many archaeocyathid fossils have been found firmly cemented together with calcite that was obviously secreted while the organisms were alive. In the Great Basin, such archaeocyathid reefs are best known from the White-Inyo Mountains and adjacent localities in Esmeralda County, Nevada. In this area, the reefs occur primarily in the limestone of the Poleta and lower Harkless Formations. Elsewhere in the region, archaeocyathids have been found in the Cambrian strata of the Toyiabe and Battle Mountain Ranges (Gold Hill and Scott Canyon Formations, respectively).

As odd as the archaeocyathids seem to be, they are not the only weird element in the Cambrian fauna of the Great Basin. Living in and around the archaeocyathid reefs were some of the most mysterious prehistoric marine creatures ever discovered in the fossil record. For example, some of the same rocks that preserve archaeocyathid reefs have also yielded the fossils of creatures known as **helicoplacoids** (Figure 3.11B). These baffling organisms evidently possessed a elongated shell that tapered to a point on either end, like a tiny football. When found as fossils, helicoplacoid shells are usually crushed flat along a bedding plane in layered rock, so the outline appears oval or elliptical. The shell, usually about an inch or two long (but sometimes larger), was constructed of spiral rows of plates that wrapped around the body. Between the rows of plates there were three grooves that led to an opening, thought to represent the "mouth" of the creature, positioned on the surface of the body. Because helicoplacoids have a shell constructed of plates, many scientists consider them to be related to the living echinoderms (Phylum Echinodermata), such as sea urchins, sand dollars, and

archaeocyathids are strikingly similar to certain modern sponges. However, no living sponge possesses the double-wall calcite cone that is so characteristic of the archaeocyathids. Perhaps these bizarre fossils represent primitive, now extinct, members of the sponge family (Phylum Porifera). While most scientists favor a sponge kinship for the archaeocyathids, some have even suggested that they might actually be related to corals, or even to calcite-secreting algae. The controversy concerning their proper classification reflects one of the few aspects of archaeocyathids about which there is universal agreement: nothing quite like them has ever been found in the modern seas.

Whatever type of creature the archaeocyathids might have been, they were the first animals to construct true reefs. By "true" reefs, we mean a mass of durable skeletal material bound together during the life of the organisms that secreted it. In today's oceans, the colonial corals are the primary reef-building organisms, but in the Cambrian Period the strange archaeo-

sea stars. But, all living echinoderms have a five-fold symmetry, of which there is no hint in the helicoplacoids. The life habits of the helicoplacoids are as great a mystery as is their relationship to currently living creatures. Some paleontologists feel that they were burrowing animals, using the spiral plate rows and grooves to twist their way through the mud of the Cambrian sea floor. Others feel that these enigmatic beasts lived on, not under, the sea floor, anchoring the tip of their "football" bodies to the muddy bottom, while microscopic food was channeled to the mouth via the spiral grooves. We are so befuddled by the helicoplacoids because, among the great multitude of living things in modern seas, no one has ever seen anything remotely similar to these freaks from the Cambrian.

A little less exotic than the helicoplacoids, but still very strange, are the small, stalked Cambrian fossils known as the **eocrinoids** (Figure 3.11C). Though these unique fossils have been found in a dozen or so localities in the Great Basin, they are most abundant and best preserved in the Wheeler Shale and Spence Shale of western Utah and the northern Wasatch Mountains region, respectively. Several different species of the eocrinoids have been identified from fossils found in these rocks, but they are all constructed on a more or less similar body plan. The eocrinoids had a small, bulb-shaped body (the calyx) composed of a mosaic of calcite plates. A stubby shaft extended an inch or two, like a short root, from the lower end of the body. Rising from the upper portion of the calyx was of brace of "arms," composed of tiny rods or plates of calcite. The arms of some eocrinoids appear to have been loosely coiled spirals, and look something like twisted ribbons when flattened and compressed as fossils. In other species, the arms may be somewhat wavy or, on occasion, perfectly straight, appearing more like spines than "arms." The overall anatomy of these odd creatures is more or less similar to the group of echinoderms known as crinoids, or "sea lilies." The crinoids (Phylum Echinodermata, Class Crinoidea) are extremely common marine fossils in later periods of the Paleozoic Era all over the world. There can be little doubt that the eocrinoids, whose name literally

Figure 3.12 A light-colored archaeocyathid reef preserved in Cambrian limestone in Esmeralda County, Nevada.

means "dawn crinoids," are related to their more specialized and advanced kin that would evolve millions of years after the Cambrian Period ended. Still, the eocrinoids are so primitive that they appear almost primeval in comparison to their later Paleozoic descendants.

A complete listing of all the strange fossils known from the Cambrian strata of the Great Basin would constitute a lengthy litany of multisyllabic tongue twisters: joining the trilobites, archaeocyathids, and helicoplacoids would be the inarticulate brachiopods, hyolithids, phyllocarids, stylophorans, and priapulids, among others. Such names are, to the say the least, unfamiliar to most people. This is because many of these groups of organisms have no living members; they are mostly categories created specifically for the creatures which thrived in the Cambrian seas, but which now have been extinct for hundreds of millions of years. All these strange names for the strange fossils preserved in Cambrian rocks bespeak a very odd menagerie of life squirming through the mud and ooze of the Cordilleran miogeocline. Sometimes we find evidence of the movement of some of these unknown creatures in the form of tracks and trails made in the soft mud and preserved as what are called **trace fossils** in rock (Figure 3.11D). Although some trace fossils can be linked to a particular group of organisms that might have left the track or trail, scientists generally cannot identify the trace-makers with precision. Nonetheless, the abundance of trace fossils in the Cambrian rocks of the Great Basin can be quite impressive, suggesting that the sea floor was literally churning with life. Normally, the fossilization of actual body parts is, unfortunately, a highly random and improbable process. Hence, we know from direct fossil evidence only a very small fraction of the total array of weird creatures that populated the Cambrian sea floor of the Great Basin. The fossils that have been collected and studied to date represent just a meager sample of what must have been an opulent assortment of marine life. We can only imagine what bizarre creatures we might have beheld if we could have inspected the Cambrian sea floor firsthand. What additional strangeness might we have discovered if our view was not clouded by the myopia of fossilization? In recent years, this question has become a little less rhetorical. This is because, among the hundreds of Cambrian fossil localities in the Great Basin, there are a couple of places where some very unusual events led to the spectacular preservation of fossils in terms of both quality and quantity. And, as these unique sites have been explored by paleontologists, an astonishing assortment creatures has been revealed.

Fossil *Lagerstätten*: Great Basin Treasures

The term *fossil Lagerstätten* (*Lagerstätte,* singular) was invented by German geologist Adolph Seilacher and co-workers in 1985 to describe occurrences of exceptionally abundant and/or well-preserved fossils. Originally, the term was used by German miners to describe the richest, most productive part of an ore body. Like many Germanic words and phrases, *Lagerstätten* is difficult to translate directly to English, but it means essentially the same thing as the term "mother lode," which has been used traditionally by American miners to describe the principal source, the main concentration, of some valuable metal. A *fossil Lagerstätte* is thus a "mother lode" of fossils, a site where fossils occur in astonishing concentrations or with spectacular preservation. Since the introduction of term, paleontologists have documented dozens of *fossil Lagerstätten* all over the world. In current usage, the word *fossil* is usually dropped from the original phrase; *Lagerstätten* alone now refers to sites that yield fabulous paleontological riches.

Lagerstätten may arise through the conse-quences of several different factors. When contemplating such remarkable fossil occurrences, it is important to remember that the preservation of organic remains as fossils is always a rare event. Wherever and whenever an organism dies, the physical remains are almost always completely decomposed, and nothing remains to be fossilized. Whether the creature possessed bone, or shell, or wood, eventually all traces of it vanish after scavenging critters and decomposing microbes are finished with the carcass. In addition to these extrinsic sources of disintegration, all organisms possess internal sources of degradation, as well. All living things have autolytic ("self destroying") enzymes that are used to dismantle old cells as new ones continuously are generated to replace them. After death, of course, the growth of replacement tissue is impossible, but the autolytic enzymes may still attack the cells that remain in the carcass. This is why, in experimental settings, organic remains always experience some degradation, even when they are placed in a sterile environment. For any portion of an organism to become preserved as a fossil, something has to insulate the carcass from external marauders and limit the activity the internal agents of self-destruction before they have time to complete their task of annihilation. In nature this almost never happens; each fossil that we find represents a fortunate (for us, anyway) exception to nature's rule. For a *Lagerstätte* to form, this inexorable dust-to-dust progression must stop dead in its tracks. But, how?

Much of what it known about *Lagerstätten*, and the conditions under which they can form, is based on the study of the Burgess Shale, a Cambrian rock exposed high in the mountains of British Columbia. The Burgess Shale is world renowned for the amazingly abundant and exquisitely preserved fossils of Cambrian organisms that have been collected from it for more than a century. Many of these organisms were evidently entirely soft-bodied—lacking shells, or bones, or teeth. Yet, their delicate remains are preserved in magnificent detail as impressions and films flattened along the bedding planes in the Burgess Shale. The Burgess Shale is a classic *Lagerstätte*, the type of fossil occurrence for which the term was invented. While it is beyond the scope of our discussion to fully explore the content and nature of the Burgess Shale fauna, there is a very important link between it and the Cambrian fossils of the Great Basin: at the time that the Burgess sediments were accumulating on the sea floor in British Columbia, similar mud was settling onto the miogeocline of western Utah. And, as recent discoveries are beginning to reveal, the Great Basin sediments sometimes preserve similar fossils, in nearly equal abundance and quality. The Burgess Shale is not the only Cambrian *Lagerstätte* in western North America. Behold the wonders of the Wheeler Shale!

The Wheeler Shale: A *Lagerstätte* Revealed

The Wheeler Shale is widely exposed throughout the Cambrian wonderland of west-central Utah. In places like the House Range, the Drum Mountains, and the Fish Springs Range, the shale weathers into gentle slopes littered with silvery gray tablets of thinly layered stone. Here, the Wheeler Shale is commonly about 500 feet thick, sandwiched between thicker limestone units above and below. The drab shale exhibits a monotonously smooth appearance, reflecting a composition of mostly microscopic particles of rock, mixed with clay minerals of various kinds. Some portions of the Wheeler Shale consist of thin layers of silty limestone (composed largely of calcite, $CaCO_3$), rather than true shale, which contains little calcite. In the northern Wasatch Mountains, in the vicinity of Brigham City and Wellsville, Cambrian rocks very similar to the Wheeler Shale are referred to as the Spence Shale, a part of the Langston Formation. Fossils occur almost anywhere in the Wheeler and Spence Shales, but in certain horizons they can be collected, literally and without exaggeration, by the bucketfuls! Since the first trilobite remains were discovered in the late 1800s, paleontologists have collected so many fossils from these two units of Cambrian shale that some have yet to be completely identified or thoroughly studied. Scientists have currently identified more than seventy species of trilobites, and nearly the same number of other types of organisms, from fossils found in the Wheeler

Figure 3.13 Platy weathering of the Wheeler Shale in the House Range of western Utah.

Shale. Furthermore, the quality of fossils from these formations is commonly as impressive as is their abundance. Many complete trilobite specimens, with every fine bump or groove on the exoskeleton intact, have been gathered from Wheeler and Spence localities. Some specimens of *Elrathia kingii*, the most common trilobite from the Wheeler Shale, look as if they crawled out of the sea and died yesterday! Thus, it is no mystery why the Wheeler Shale has become one of the most popular fossil-hunting targets in the world for amateurs, professionals, and even commercial collectors. There are actual "trilobite mines" in some areas of western Utah, where thousands of exquisite fossils have been recovered. Consequently, fossils from the Wheeler (and Spence) Shale have achieved a global distribution: they now reside in museum cabinets and galleries all over the world. Every rock shop seems to have fossil specimens from the Wheeler Shale on its shelves, offering half-billion-year-old remnants of life in the Great Basin for a dollar or two. To refer to the most fossiliferous horizons in the Wheeler Shale and, to a lesser degree, the Spence Shale as true *Lagerstätten* is certainly no overstatement: these formations are, in fact, the "mother lodes" of Cambrian fossils in the Great Basin.

Experiencing the richness of the Wheeler and Spence fossil beds is a stunning experience. But, eventually, after filling their pockets with trilobite specimens, it is natural for people to wonder how it all came about. Scientists are no

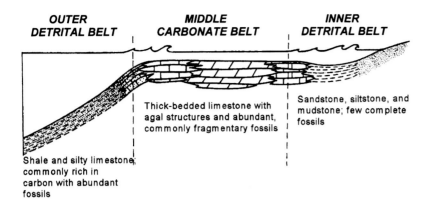

OUTER DETRITAL BELT

MIDDLE CARBONATE BELT

INNER DETRITAL BELT

Thick-bedded limestone with algal structures and abundant, commonly fragmentary fossils

Sandstone, siltstone, and mudstone; few complete fossils

Shale and silty limestone, commonly rich in carbon with abundant fossils

Figure 3.14 Onshore-offshore belts of Cambrian sedimentation in the Great Basin. The Wheeler and Spence Shale represent the outer detrital belt, where fine-grained mud accumulated in calm, stagnant basins beneath relatively deep water. The Cambrian *Lagerstätten* of the Great Basin occur in such deposits.

different from other people; the exquisite paleontological treasures of the middle Cambrian *Lagerstätten* have attracted the attention of some of the world's most illustrious geologists (see chapter references). In addition to the fossil organisms, investigators have studied the nature of the enclosing sediment, the structure of the layered sequences of rock, and the distribution and lateral variations of both the Wheeler and Spence Shale. The results of these investigations can be integrated to formulate a general model of the marine environment in which the Wheeler and Spence sediments were deposited. This, in turn, provides some hints about the mechanisms that resulted in the Wheeler and Spence *Lagerstätten*. Like many other aspects of Great Basin geology, the knowledge gleaned from the middle Cambrian fossil sites can be combined with information from other regions to reveal some fascinating aspects of the Cambrian world. And, compared to the modern sea floor, what an amazing world it seems to have been!

Recipe for a *Lagerstätten*

During the middle Cambrian, when the Wheeler (and Spence) Shale was deposited, sediment accumulated on the shallow miogeocline of the Great Basin in three broad zones, aligned north-to-south, parallel to the shoreline (Figure 3.14). Nearest the shore, sand and silt derived from land exposed to the east accumulated in very shallow water, vigorously agitated by currents and waves. This zone, called the **inner detrital belt,** is where sandy and muddy sediments such as the Tintic Quartzite and the Ophir Shale, both widespread in west-central Utah, collected. For the most part, organic remains in

the inner detrital belt were destroyed by the strong currents and wave action. Consequently, fossils are neither abundant nor well preserved in the rocks of this zone. Offshore from the inner detrital belt, fine-grained, calcareous (rich in calcite) mud and ooze settled on the shallow sea floor. This sediment was almost pure calcite, precipitated either directly from seawater or secreted as shell material by Cambrian organisms living on the miogeocline. In many places in the eastern Great Basin, this calcareous ooze was trapped and bound together by mats and films of algae that carpeted the ocean bottom beneath the swirling surf of the Cambrian sea. Eventually, this fine sediment hardened into thick layers of limestone that commonly preserve algal structures and fragmentary fossils. These limestone sequences comprise the **middle carbonate belt**, and include such formations as the Dome, Swasey, and Orr Formations of western Utah.

Farthest offshore, and of critical importance to the origin of Cambrian *Lagerstätten*, was the **outer detrital belt.** In this zone, very fine-grained sediment settled to the sea floor in relatively deep water, little flustered by waves and currents. In the still depths of the outer detrital zone, muddy ooze collected in thin layers from a variety of sources. Some of the sediment drifted into deeper water from the waves that pounded and pulverized the limestone at the outer edge of the middle carbonate belt. Some of the microscopic silt grains were carried to the outer detrital belt by offshore sea breezes from land exposed to the east. In addition, some of the fine muck may have arrived in the outer detrital belt via oceanic currents or could have been precipitated directly from the deep water. Wherever it came from, the fine sludge collected slowly in thin layers that eventually hardened into gray fissile shale or platy siltstone. Such sediments are fairly widespread in western Utah and eastern Nevada, including the famous Wheeler Shale and Spence Shale. It is in the sediments of the outer detrital belt that the treasure chests of Cambrian fossils are housed. There must have been something very unusual about the sea floor of the outer detrital belt that resulted in the Cambrian *Lagerstätten* of the Great Basin.

Several clues to the nature of the ocean floor in the outer detrital belt are preserved in the Wheeler Shale and other coeval rock units. For example, the closely spaced laminations in the rock, which account for the ease with which it splits into thin plates, are rarely disturbed by any organic structures, such as worm or clam burrows. Normally, the mud that accumulates on the ocean floor is literally churning with bottom-dwelling marine life; burrowing sea urchins, starfish, clams, slugs, "worms," and crabs stir the sediment continuously. Under such conditions, any original layering in the sediment would be disrupted quickly and eventually obliterated beyond recognition. But not in the Wheeler Shale. The thin, even layering in these rocks is preserved nearly as well as are the exquisite fossils. From this, we may assume that there were few organisms thrashing through the mud when the remains comprising the *Lagerstätten* were buried. In addition, on a freshly broken surface, the Wheeler and Spence Shale is often very dark gray in color; to many people it appears almost black. Even though weathering under the desert sun lightens the leaden shade considerably, the dark primary color reflects the great abundance of organic carbon in these sediments: they are nearly black because large amounts of carbon residue accumulated in the sediment while it collected on the sea floor more than 500 million years ago. Organic carbon on the sea floor today usually disappears quickly, either through the destructive attack by oxygen (oxidation) or via its absorption by other living things. The accumulation of carbon in the sediment of the Wheeler Shale seems to signify either the absence of organisms on the sea floor, a limited supply of oxygen in the water, or both. Collectively, the preservation of fine laminations in the shale and the abundance of organic carbon both suggest a deep pocket of stagnant, oxygen-depleted water on the sea floor at the time the Wheeler Shale was deposited. Under such conditions, the populations of scavenging and decomposing organisms must have been minimal. Any potential fossil material that arrived in this basin would not have decayed quickly or completely.

These biologically hostile conditions may

have enhanced the preservation of fossils, but they also would have limited the populations of the organisms could have survived on the sea floor. This, in turn, poses an interesting conundrum: if the *Lagerstätten* reflect the ecological hostility of the sea floor (which promoted fossilization by creating nearly sterile conditions), how did so many organisms become concentrated at that spot? Traditionally, paleontologists have answered this question with the assertion that the trilobites and other organisms in the *Lagerstätten* were not preserved where they lived. Instead, the skeletal remains were thought to have been transported to the deep oceanic crypt after the death of the organisms somewhere else, presumably in a more hospitable environment. The most likely habitat for the Cambrian organisms preserved in the Spence and Wheeler Shale was the outer edge of the middle carbonate belt. There, not far east of the deeper Wheeler basin, hordes of trilobites, eocrinoids, sponges, and braciopods were envisioned to have populated the shallow, warm, sunlit, and well-oxygenated water. It was, it seemed, the perfect solution to the preservation mystery: optimal shallow marine habitat juxtaposed near an aseptic sag. A literal "rain" of skeletal remains—dead trilobites, shed exoskeletons, sponge pieces, concave brachiopod shells—was envisioned to have drifted from the shallows into the deeper basin in the outer detrital belt. Once there, the remains were gently buried under soft ooze, free from the destructive effects of microbes, worms, and scavengers. This explanation of the Great Basin *Lagerstätten* may have some merit, but it is, at best, an incomplete interpretation of these remarkable fossil sites. The "stagnant basin model" does fit well with what is know about the paleoecology of the Cambrian miogeocline, but two problems with it imply that there is probably more to the story of these spectacular fossil beds.

First, some astounding Cambrian fossils have been found, many of them in just the past twenty years or so, that represent very delicate remains that could not have been transported any significant distance into the Wheeler or Spence basins. Fossil sponges have been discovered that are nearly complete, with even the tiny spicules that "floated" in the tissue of the sponge arranged just as they were in life. In addition, the fossils of flimsy fronds of algae, soft "jelly-fish," and the delicate spine-bearing shells of creatures called hyolithids have also been found. It is highly unlikely that such frail remains as these could have survived intact if they had been transported any appreciable distance across the sea floor. These remarkably fragile fossils must represent organisms that actually lived very close to where their remains became buried after death. In other words, the bottom conditions in the outer detrital belt may have been somewhat less sterile than was originally thought: the soft-bodied or scantily shelled creatures managed somehow to adapt to the ecological hostilities of the deep sea floor in the outer detrital belt. Of course, the mere presence of in-place fossils of bottom-dwelling organisms in the Wheeler and Spence Shale does not mean that life was prolific in the dark, stagnant Cambrian basin. It simply suggests that the sterility of the sea floor was probably less acute than some scientists originally envisioned.

Furthermore, it is now known that even if organic tissues are completely isolated and protected from microbial attack after death, they still experience some self-inflicted decomposition due to the lingering presence of autolytic enzymes. Thus, even if the trilobite carcasses drifted into a perfectly sterile basin, they still would have decayed, at least to some extent. A number of scientists, including Nicholas Butterfield of the University of Western Ontario, have recently suggested that the chemistry of the fossil-bearing sediment may play an important role in the development of *Lagerstätten*. Specifically, certain clay minerals such as smectite and montmorillonite are known to have the ability to absorb and/or deactivate many organic compounds, including autolytic enzymes. The autolytic enzymes that remain in the tissues of an organism after death could be rendered ineffective if the remains become buried in an ooze laden with just the right clay minerals. This is precisely the type of sediment that is represented by the interbedded, clay-rich shale and silty limestone in the formations that preserve the famous Cambrian *Lagerstätten* of the Great

Basin. Thus, by neutralizing autolytic enzymes, the clay that comprises the Wheeler, Spence, and Burgess formations helped to prevent the decomposition of whatever organic material was buried while it accumulated in the Cordilleran Geosyncline. The fortunate coincidence of stagnant bottom conditions and deposition of clay-rich sediment did not exist everywhere across the miogeocline during the Cambrian Period. Only in a few places did the magical combination exist; western Canada and the eastern Great Basin are among those special places. Elsewhere, the seawater was less stagnant and fully oxygenated, or the sediment was composed of sandy or carbonate material lacking significant amounts of clay. In these areas, Cambrian life may be recorded as sparse, isolated, and fragmentary fossils in limestone or sandstone. Such fossils pale in comparison to the sensational preservation that characterizes the *Lagerstätten*.

Whether their remains occur as part of a *Lagerstätte*, or as isolated specimens in limestone, the Cambrian organisms comprise a unique marine fauna compared to the array of organisms that later descended from them. So distinctive is the life of the Cambrian Period that paleontologists the world over use the term "Cambrian Fauna" to describe the odd collection of marine creatures that populated the sea floor at that time. This array was unique in that it was strongly dominated by bottom-dwelling arthropods (trilobites and other segmented organisms with jointed legs) that fed primarily on the organic matter in the mud and ooze of the sea floor. Such organisms make up, on average, about 75–80 percent of the fossils found in Cambrian rocks. Other types of organisms belonging to the Cambrian Fauna, such as the clam-like brachiopods, archaeocyathids, helicoplacoids, or "worms" were either much less common, or were short-lived, becoming extinct well before the end of the Cambrian. The Cambrian fossil record also provides little evidence of "suspension-feeding" creatures, those that sift food (microscopic plankton, for the most part) from seawater. Today, and in all prehistoric ages after the Cambrian, suspension-feeding organisms constitute a major segment of the invertebrate communities in both deep and shallow marine habitats. Finally, many of the soft-bodied creatures preserved in the Cambrian *Lagerstätten* are so dissimilar to anything alive today that scientists are still uncertain about their classification and ancestry. In short, the Cambrian Fauna is, in the Great Basin and everywhere else on the earth, a numerically skewed, ecologically unbalanced, and taxonomically curious array of critters.

As we move forward in time (by exploring the younger, post-Cambrian rock sequences in Utah and Nevada), the life that populated the Paleozoic sea floor in the Great Basin becomes noticeably different from its Cambrian state.

And it was not just the marine life that was evolving during the later Paleozoic Era: the land and the sea itself were beginning to respond to geological forces that would result in significant environmental changes. The geological history of the Great Basin in the later Paleozoic Era is a wild roller-coaster ride of changing landscapes and surging seas. These transformations begin slowly following the close of the Cambrian Period, but they accelerate dramatically in the middle Paleozoic Era. By the end of the Paleozoic Era, the Great Basin is virtually seething with geological chaos. Buckle up, because we're headed that way!

Field geology is a rigorous enterprise that continually challenges the body with physical obstacles while it taunts the mind with intractable three-dimensional puzzles. It takes a powerful intellectual resolve and a fair measure of physical strength to overcome these dual enumbrances and produce sound scientific interpretations and new insight. Successful field geologists are also distinguished from their laboratory-bound kin by a certain passion, bordering on spiritual fanaticism, for the difficulties of their task. I've heard it said that unless you've climbed 3,000 feet, dodged a half-dozen rattlesnakes, had your eyebrows singed by a lightning bolt, accurately mapped three square miles, and returned to camp two hours after dark, then you haven't put in a full day. Perhaps that is why field geology has always bred its own unique kind of lore, where science becomes embroi-dered with legends that stretch the truth of the natural world a trifle, and add a blush of color to the rigid exertion of field research. One of my favorite field fables concerns the Ordovician rocks of the Great Basin. For decades, I have used them as a rallying cry to improve the morale of students when bad weather spoiled a field trip to such western Utah localities as Skull Rock Pass, Blind Valley, Fossil Mountain, and the Ibex area. In these places, thousands of feet of Ordovician shale and limestone are exposed in bold escarpments that rise above the sage-covered valleys. These strata represent sediment that was deposited some 500 million years ago on the shallow sea floor of the Cordilleran Geosyncline. At that time, the western portion of North America was positioned close to the equator, and the modern Great Basin was a tropical paradise. Thus, the entire Ordovician succession in western Utah accumulated under the influence of warm tropical seas. In rhythm with the tides and currents, the shimmering seas surged back and forth over and around coral reefs while the warm sun floated high overhead. While I understand that the modern weather in the Great Basin has nothing at all to do with this ancient tropical paradise, the fanciful notion that the Ordovician rocks had some mystical meteorological influence often helped dispel a weather-incited field mutiny before it became a serious obstacle to learning geology. Whenever the mood of students darkened to match the skies, I encouraged them to stick it out with the incantation, "The sun always shines on the Ordovician!"

Furthermore, that capricious adage was not entirely baseless. After more than twenty years of leading field trips to western Utah, I have often noticed that the weather seemed to improve in the proximity to Ordovician exposures, no matter how bad it might have been anywhere else. The consistency of fair weather near the Ordovician outcrops is, of course, the consequence of innumerable and complex atmospheric factors, combined with undeniably good luck on my part. The tropical sediments exposed in the mountains exert no supernatural influence on the weather, but it always made a good campfire story to explain how the calcium ions in the Ordovician seas joined with bicarbonate molecules to carry a load of good-weather karma to the sea floor, preserving it as limestone and dolomite. Today, hundreds of millions of years later, these exposed rock layers seem to have the ability to evaporate snow and rain, to calm fierce winds, and to raise the temperature a few degrees. And, that is why the "sun always shines on the Ordovician": foul weather is dispelled by the cabalistic warmth emanating

CHAPTER FOUR

THE SUN ALWAYS SHINES ON THE ORDO- VICIAN

from the residue of an ancient paradise. Or so the story goes.

On one particular spring trip to the Confusion Mountains, the aphorism met its greatest test. Driving west on Highway 50 toward Skull Rock Pass, the snow was flying horizontally across the road and the sky was a churning mass of knotted gray clouds. The students in the vehicles following my lead truck had been in a somber mood all day, fearing the bitterly cold and uncomfortable night that seemed to await us. We always stopped at Skull Rock Pass, between the House and Confusion Ranges, because the low road cuts offered convenient exposures of the fossiliferous limestone and shale of the Fillmore Formation. When we arrived at the pass on that blustery March afternoon, the students reluctantly piled out of the vehicles parked along the shoulder of the road and walked through the cold wind to the outcrops on the opposite side of the asphalt. They endured my abbreviated lecture with heads bowed against the weather. No one took notes because, I assumed, removing gloves in the blizzard would have been too painful. Suffering through the lecture was bad enough. When I suggested that we split a few shale layers apart to find fossils, the huddled group remained for the most part motionless. I wasn't too thrilled with the idea either, but, to set an example, I grabbed my single-jack and started flailing at the rocks. A few hardy souls stood nearby to watch, while most of the group leaned against the road cut, hoping to gain some relief from the squall. We didn't find any fossils and, when I lost my hat to the wind, I decided to abandon the road cuts and head for camp. One of the students retrieved my hat, handing it to me with numb fingers and a nose reddened by the cold. "Are we done now?" she asked.

Forty minutes later we arrived at our campsite, a sandy juniper grove at the foot of Fossil Mountain. We built a large fire, unloaded the trucks, but waited to set up camp, hoping that the wind and snow would diminish soon. Then, one of the students huddled around the fire nodded to the north and suggested that we should all find better shelter, and soon. He was

right: a wall of angry clouds was moving down the valley, absorbing the desert into an opaque veil of snowflakes as it approached. I suggested we all pile into the trucks and wait for the squall to pass before we set up camp. For the first time that day, the students responded with enthusiasm. I wound up in the rear cargo area of one of the Suburbans, along with four students. From inside the truck, we all watched the storm approach through windows that became foggier by the minute. I kept a close check on the fire outside, and was pleased to see it building into a decent bonfire in spite of the cold bluster. Somebody in our truck told a joke, which started a cascade of laughter, and more jokes, rolling through the vehicle. I was pleased to see the positive mood-swing in our small group. I took it as a sign that the students had given up any thought of flight and were resigned to making the best of a bad weather situation. For several minutes we were distracted by humor, most of which centered on our own plight. We were, after all, a pretty comical outfit: huddled shoulder to shoulder in a metal box, squinting through foggy windows at one of the grandest wilderness landscapes in the world. Someone then suggested that the air in the truck was similar to the atmosphere of Venus, bringing on more hilarity. During a brief lull in the laughter, I wiped the condensation from the side window and tried to check on the storm. It was difficult to see much of anything, but it seemed that the squall line had not moved any closer to us for several minutes. The fire was still raging, but I noticed that the flames were no longer dancing wildly in the wind. When one of the students suggested that the sky had brightened slightly, I decided to leave the truck for a better view. We lowered the rear window in the Suburban, dropped the tailgate, and climbed out to survey the sky above us.

Standing on the tailgate, we could almost taste the freshness of the cold air. Amazingly, the wind had diminished to the point that it was barely noticeable. The sky had, in fact, become lighter; the clouds had lifted enough to reveal the snow-dusted slopes of the Confusion Mountains stretching several miles north of our camp. It was still cold, but in the absence of the wind, conditions were much more pleasant than they had been less than a hour earlier, when we all took shelter in the trucks. One by one, the doors of the other vehicles opened, and the refugees emerged with eyes directed to the sky. Grins widened when the group realized that the storm had dissipated just as it seemed poised to over-run our camp. When a small patch of blue sky appeared in the west, the morale of the group became nothing less than euphoric. Supper was late that night but had a special festive atmos-phere, even though it was too cold for our nor-mal late-night campfire frivolity. Yet again, the magic of the Ordovician had saved us from a threatening tempest. One more story to strengthen the legend; one more yarn to be related to future students.

Whether or not the weather-calming influ-ence is real, the Ordovician rocks of the Great Basin are magical for another reason as well: they extend the earlier histories recorded in older rocks into a new era. By perusing the Or-dovician rocks of the Great Basin, a new dimen-sion in our understanding of the distant past emerges. And what an incredible episode it is! Let's have a look.

THE ORDOVICIAN PERIOD

The Ordovician Period (about 490–443 million years ago), was named in 1879 by Charles Lap-worth, then a Scottish educator with a keen in-terest in fossils and rocks. In the rolling hills of his native southern Scotland, Lapworth had col-lected some very odd fossils called **graptolites** that look like small pieces of a hacksaw blade (see Figure 4.16B) buried along a bedding plane in layered rock. The rocks that produced the curious fossils had been largely overlooked by nineteenth century British geologists, who defined the Cambrian and Silurian "systems" as the first two divisions of the Paleozoic rock record. A system is a package of rock layers, de-fined by a unique array of fossils, deposited dur-ing some specific period of geological time. So, the Cambrian *System* is the physical (rock) record of events that occurred, and life that ex-isted, during the Cambrian *Period* of time. To state the distinction another way, a "system" is a

concrete reality consisting of solid rock, while a "period" is a comparatively vague abstraction, as are all human perceptions of the thing we call time. For the most part, the original European "systems" became the basis for the periods of time on the modern, standardized geological time scale. Hence, most of the periods of geo-logic time have European names, such as Cam-brian (for Cambria, a portion of Wales) and Devonian (for Devonshire). Furthermore, a hi-erarchical subdivision of both the rock record and the geologic time scale developed as geol-ogists later subdivided systems into smaller packages of rock known as *series*. A series, in turn, can be split in several *stages*. A series of rock corresponds to an *epoch* of time (a portion of a period), while a stage is the physical equi-valent of an *age* (a subdivision of an epoch). Be-cause much of the early work in assembling a uniform scale of geological time was conducted in Europe, it is not surprising that the names of stages and series, along with epochs and ages, are also mostly of Old World derivation. But, as we shall see in the case of the Ordovician, the use of some of the traditional European names has occasionally caused confusion on this side of the Atlantic Ocean.

Prior to Lapworth's discovery of the grapto-lites, the strata that rendered them were thought to be of early Silurian age. Lapworth demon-strated that the graptolites, along with several other kinds of fossils he found with them, repre-sented a unique fauna, distinct from either the trilobite-dominated array from the earlier Cam-brian System (and Period), or the brachiopod-rich assemblage from the later Silurian. So, on the basis of the distinctive fossils he had found, Lapworth proposed a new system and period for the Paleozoic Era—the Ordovician—sand-wiched between the Cambrian and the Silurian. The name was inspired by the *Ordovices*, a tribe of ancestral inhabitants of the British Isles. Lap-worth was so thorough in his documentation of the fossils that characterized his new interval that, by the early 1900s, the Ordovician System had been firmly inserted between the Cambrian and Silurian by geologists on both sides of the Atlantic. The Ordovician was eventually sub-divided into smaller components (stages/ ages) known by such obviously Scottish designations as "Arenigian" and "Llanvirian."

While all of this was happening in Europe, geologists in the fledgling United States (principally affiliated with the New York State Geological Survey) were mapping the strata of the northern Appalachian region and adjacent parts of Pennsylvania and Ohio. These geologists developed their own set of terms to describe strata between the Cambrian and Silurian Systems. Such terms as "Cincinnatian" (for the upper Ordovician System) and "Champlainian" (for the lower part of the system) were established and became widely used in the northeastern part of North America. Soon after these series names were established for the North American Ordovician, they were subdivided into stages (e.g, "Canadian," "Chazyan," etc.), and geologists attempted to correlate them with their European equivalents. While Lapworth's concept of the Ordovician Period was well established, the nomenclature for subdividing this system in North America and Europe had evolved into a confusing tangle of local and provincial names. Adding to the uncertainty were the imperfect correlations between local sequences of Ordovician rocks, numerous gaps in the layered successions, and the sparsely fossiliferous nature of portions of the rock sequence. In the late 1800s, a great deal of effort was expended in correlating the British strata, the global standard, with those of northeastern North America. Eventually, North American terms for the Ordovician series and stages were established, paralleling the subdivisions that had been formulated in Europe. The North American terminology was based primarily on rock sequences exposed in Canada, New York, and Ohio. Of course, no one knew, at that time, what rocks lurked in the vast, unexplored wilderness of the Great Basin.

Ordovician rocks in the Great Basin first came under scrutiny by geologists in the late 1800s, during the great era of exploration in the western United States. Such government surveys as those of Captain John Simpson in 1859 and F.V. Hayden in 1872 entered portions of the Great Basin, and included geologists who made fossil collections and documented the geological features of the virtually unexplored region.

These observations were made well before the Ordovician Period had been established as a legitimate interval on the geologic time scale. Consequently, most the Ordovician strata of the Great Basin were originally described as "Silurian," the period (and system) that was then thought to follow the Cambrian. The sketchy information recorded by the government surveys simply demonstrated the existence of "Silurian" rocks in the Great Basin. It was neither sufficiently detailed nor adequately complete to allow a firm correlation between the Great Basin strata and the coeval sequences of rock that were being examined in detail in the well-settled northeastern parts of North America. Eventually, as American and Canadian geologists worked out a standard nomenclature for the Ordovician rocks in North America, none of the strata in the Great Basin was considered. When the template for continentwide subdivision of Ordovician rocks finally emerged in the late 1800s, it included the Canadian and Champlainian "series," along with the later Cincinnatian. In part because geologists needed to eliminate the confusion in Ordovician nomenclature, these terms became deeply ingrained in published studies of Ordovician rocks everywhere in North America.

It wasn't until the discovery of rich ores, and the explosive growth of mining that it fostered, that geologists returned to the Great Basin to further investigate the geology of the region. In the 1880s and 1890s, geologists flooded into such mining camps as Eureka, Nevada, to map the ore bodies and conduct detailed geologic studies. These geologists brought the recently established concept of the Ordovician Period with them and, not surprisingly, they found rocks to which it could be applied. Stimulated by the search for rich ores, knowledge of the rock successions in the Great Basin improved rapidly. In the course of these investigations, Ordovician and Silurian rocks were mapped from the Pioche District of eastern Nevada to the Tintic District of west-central Utah, and in many places in between. The rock succession in the mining districts was analyzed and described in detail, broken into various formations and members, matched up (or *correlated*, as geologists refer to the process) with previously studied rocks in other regions. The rocks of Silurian and Ordovician age were assigned designations reflecting the standard northeastern terminology. Terms such as "Niagaran," "Canadian," and "Champlainian" found their way into the Great Basin, affixed to Silurian and Ordovician formations such as the Lone Mountain Dolomite, the Ely Springs Dolomite, and the Eureka Quartzite.

It wasn't until the mid-1900s that geologists began to realize that the Ordovician and Silurian strata of the Great Basin comprise one of the best-exposed and most complete successions of this age in the world. Today, some of the original terms used to subdivide the Ordovician-Silurian "systems" in North America have been replaced and enhanced by terms reflecting Great Basin localities. In particular, the Ibexian Series (named for the Ibex area of the southern House Range of western Utah) and the Whiterockian Series (from rocks exposed in Whiterock Canyon in the northern Monitor Range of Nevada) have become well established for the lower and middle portion of the Ordovician rock succession (Figure 4.1). Because the Great Basin strata comprising these subdivisions are thick, well exposed, and richly fossiliferous, they have become the North American standard references (or **stratotypes**) for the parts of the Ordovician succession they represent. The Great Basin Ibexian and Whiterockian Series have received almost universal endorsement as a replacement for the older "Canadian Series" and "Chazyan" subdivisions that were established for the strata exposed in northeastern North America. In the Great Basin, this lower and middle portion of the Ordovician rock succession is more than 2,600 feet thick, brazenly exposed along steep mountain slopes, and absolutely packed with fossils. Hundreds of different species of trilobites, brachiopods, conodonts, nautiloids, sponges, and other marine creatures have been identified from the thick Ordovician succession. It is not surprising that the Ibexian and Whiterockian strata in the Great Basin have been designated to serve as the stratotype for this portion of the North America rock record. In fact, subdivision of these "series" has now brought additional Great Basin place names to

European Subdivisions	NE North Amer. Subdivisions		Great Basin Additions	
Ashgilllian Series	Cincinnatian Series			
Caradocian Series				
Llandelian Stage	Mohawkian	C H A M P L A N I A N	WHITEROCKIAN SERIES	
Llanvirnian Stage	Chazyan			
Arenigian Stage	Canadian "Series"		IBEXIAN SERIES	Blackhillsian Stage
				Tulean Stage
Tremadocian Stage				Stairsian Stage
				Skullrockian Stage

(ORDOVICIAN)

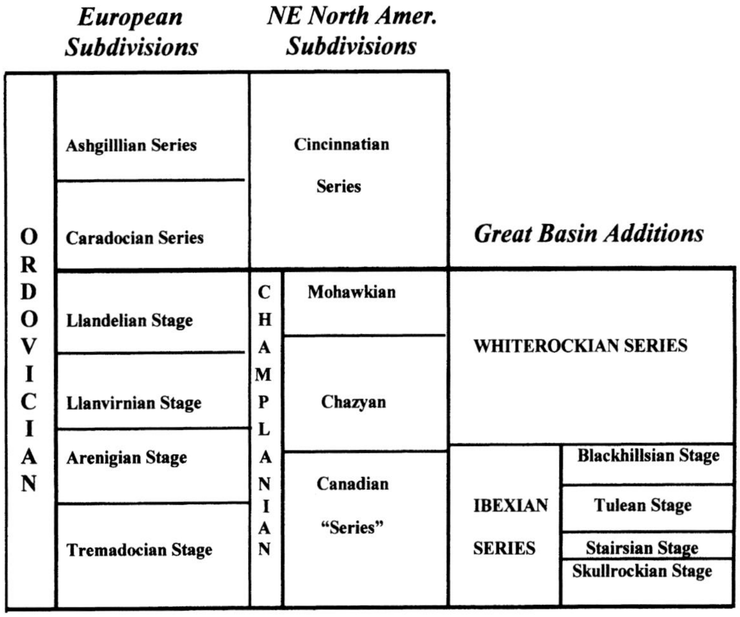

Figure 4.1 Subdivisions of the Ordovician Period.

geological prominence. The Ibexian Series, for example, has now been divided into four "stages": the Skullrockian (from Skull Rock Pass, Utah), the Stairsian (for the "Stairs," ledges of rock west of Skull Rock Pass), the Tulean (based on rocks exposed along the east side of Tule Valley, west of Utah's House Range), and the Blackhillsian (for the low, dark-colored hills at the south end of Utah's House Range). With more than sixty-four formal rock units (formations and members) of Ordovician age identified thus far, the Great Basin is, among all the other things it might be called, a natural museum of the Ordovician Period!

ORDOVICIAN ROCKS OF THE GREAT BASIN

The best exposures of Ordovician strata in the Great Basin are found in the eastern and central portions of the province. The Confusion Range area (western Millard County, Utah), the northern Wasatch and Bear River Ranges (Cache County, Utah), and several mountain ranges in east-central Nevada (such as the Hot Creek, Fish Creek, Grant, and Paranagat Ranges) all have excellent displays of Ordovician rocks. In these areas, the total thickness of the Ordovician strata may exceed 5,000 feet. The Ordovician rocks are generally best seen in rugged canyon walls or across the face of steep mountain slopes, sometimes producing spectacular

Figure 4.2 Ordovician strata (dark layers in the middle slope) exposed in the Pahranagat Range of central Nevada.

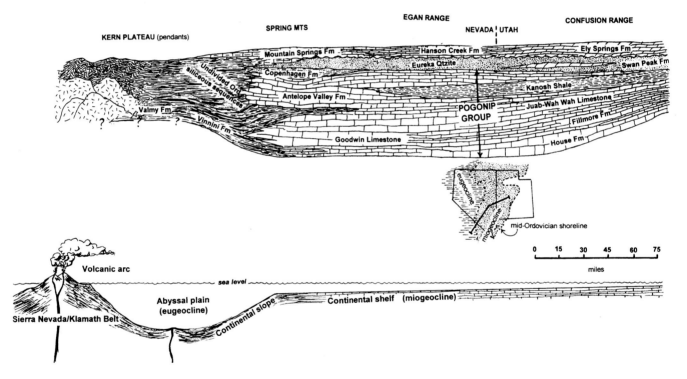

Figure 4.3 Ordovician rock units (top) and ancient environments (bottom) of the Great Basin.

banded cliffs (Figure 4.2). Such thick successions of rock layers record as much as 60 million years of geological history that followed the last days of the Cambrian Period.

From east to west across the Great Basin, Ordovician rocks exhibit a transition from shallow-water, miogeoclinal sediments (limestone and quartzite) to deep-water deposits (chert, shale, and "greenstones"). Thus the east-west dichotomy (miogeocline vs. eugeocline) that first developed in the Cordilleran Geosyncline during the Cambrian Period persists into

the Ordovician with little change (see Figures 4.3 and 4.12). In western Nevada, where the eugeoclinal materials accumulated, rock units such as the Palmetto and Valmy Formations formed. These units consist primarily of dark-colored shale, representing fine ooze that settled to the deep sea floor, interlayered with less-abundant chert beds and, occasionally, with volcanic units. The shale sometimes produces graptolite fossils, which are valuable in establishing the age of the formations and in correlating them with rock successions that formed at the same time on the miogeoclinal tract far to the east. The volcanic units are commonly altered into blocky, massive rock known as greenstone. Crude lumpy masses, known as pillow structures, can sometimes be seen in the greenstone. These, in turn, indicate that the volcanic eruptions occurred on the sea floor, beneath deep, cold water. The eugeoclinal rocks of Valmy and Palmetto Formations are commonly crumpled and shattered in patterns of bewildering complexity. These rocks have been so strongly affected by post-Ordovician geological disturbances that it is usually difficult to directly measure the total thickness of the sequences, or to observe the original pattern of layers. However, it appears that about 2,500 feet of rock accumulated in the eugeocline during the Ordovician. This is only about half the thickness of rock that formed on the shallow miogeocline to the east, but it is consistent with the low rates of sedi-ment accumulation that have been measured on the deep floor of today's oceans. Because thicker sequences formed on the miogeocline, and those strata are much less deformed by later events, most of what is known about the Ordo-vician (and later) history of the Great Basin has been gleaned from miogeoclinal strata. And it is from those rocks that some very interesting twists on the story of the Cordilleran Geosyncline have been revealed.

The Ordovician Miogeocline

In most places in the eastern and central Great Basin, shallow-water marine deposits accumulated throughout the Ordovician Period. The rock sequence that formed under these conditions is generally divisible into three parts: 1) a lower portion composed primarily of lime-stone and shale packed with fossils; 2) a middle section dominated by well-cemented and light-colored quartzite, and 3) an uppermost package of dark gray dolomite layers with relatively sparse fossils (see Figure 4.4 and Plate 7). This tripartite nature of the Ordovician rock record is widespread across the miogeocline, and can usually be recognized wherever Ordovician strata are exposed from south-central Nevada to central Utah. Even

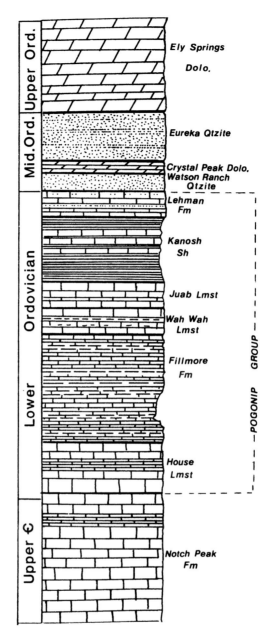

Figure 4.4 Ordovician strata of the Confusion Range Area, Millard County, Utah. As is true across the Cordilleran miogeocline, the Ordovician rocks in western Utah are divided into three packages: the lowermost limestone and shale of the Pogonip Group; the middle quartzites (Eureka and Watson Ranch); and the upper dolomite succession.

though the entire miogeoclinal sequence is of shallow marine origin, the distinctive sediments that comprise each of the three parts reveal some significant differences in the nature of the sea floor at various stages of the Ordovician. Let's explore these differences more closely, be-cause they suggest some interesting details about the conditions on the Ordovician miogeocline.

The Lower Ordovician: Shallow Seas in Turmoil

The lowermost portion of the Ordovician rock record throughout the eastern Great Basin is

Figure 4.5 The lower slopes of Fossil Mountain in the Confusion Range of western Utah expose several different formations of the Pogonip Group.

dominated by a thick and heterogeneous sequence of olive-gray shale, dark gray mottled limestone, massive bluish gray limestone, and limestone-pebble conglomerate. Collectively, this sequence of rock is generally known as the Pogonip Group, which consists of several different formations that vary in name, thickness, and character from central Nevada to western Utah. In south-central Nevada, from Antelope Valley to the Death Valley area, the Pogonip Group is more than 2,500 feet thick and consists principally of the Goodwin Limestone, Ninemile Formation, and Antelope Valley Formation. To the east, in the Confusion Range area of western Utah, the Pogonip Group is more than 3,000 feet thick and is divided into the House Limestone, the Fillmore Formation, the Juab Limestone, the Kanosh Shale, and the Lehman Formation (see Figures 4.4 and Figure 4.5).

Within the Pogonip Group are some very peculiar types of sedimentary rocks that reveal a great deal about the movement of water across the miogeocline in the early Ordovician Period.

Figure 4.6 Intraformational conglomerate of the Fillmore Formation of western Utah.

Intraformational conglomerate, for example, is a type of limestone that consists of flat pebbles of limestone, deposited as a sort of coarse, seafloor gravel, firmly bound together by dull gray limestone. The term "intraformational" indicates that the limestone pebbles are composed of sediment similar to that which surrounds them as a matrix. Because the pebbles commonly have a flat, tabular shape, the intraformational conglomerates have sometimes been referred to as "flat-pebble conglomerates." Whichever name is used, the rock has a very distinctive fragmentary appearance, with varicolored "chips" and plates of limestone amassed into something that looks a little like concrete (Figure 4.6). Intraformational (or flat-pebble) conglomerate occurs within several of the formations that comprise the Pogonip Group, and it is commonly arranged as lenticular packets sandwiched between normal limestone or shale layers.

Such odd rubble in the limestone-dominated rock sequence suggests strong storm or tidal surges across the miogeocline. Sediment very similar to the Ordovician intraformational conglomerates forms today in modern tropical seas when the powerful tidal surges or storm events rip up semiconsolidated mud from the shallow sea floor and wash the individual pieces into a depression, where they pile up in a jumbled arrangement. Later, the fragments of stiff (but not fully hardened) sediment become buried under fine lime mud that continues to accumulate on the sea floor under quiet-water conditions (Figure 4.7). Once the whole mass of sediment is fully lithified, the resulting rock has the same cluttered quality as the intraformational conglomerates of the Great Basin. Recent studies of the flat-pebble conglomerate in the Fillmore Formation of western Utah have indicated, on the basis of pebble orientation and geometric shape of the conglomerate masses, that storm events were the primary mechanism for the deposition of these odd sediments. Evidently, vigorous storms rivaling modern hurricanes developed over the Great Basin miogeocline in the Ordovician, scattering pebbly rubble over hundreds of square miles of the continental shelf. Some of the intraformational con-

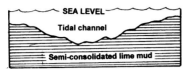
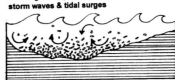

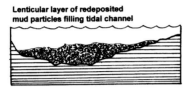
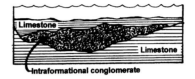

Figure 4.7 Origin of Intraformational Conglomerate. Lime mud accumulates on the shallow floor of the miogeocline in, and adjacent to, a tidal channel (upper left). During storm events, or tidal surges, swift water rips up pieces of the semiconsolidated lime mud (upper right), and redeposits the fragments in low depressions when calm conditions return (lower left). Later, the lime mud fragments are buried beneath additional layers of limestone sediment (lower right). Once the entire mass of limestone is fully lithified, a tapering (lenticular) layer of intraformational conglomerate is sandwiched between normal limestone layers.

Figure 4.8 A large reef-like mud mound in the Ordovician Antelope Valley Formation at Bare Mountain, Nevada. The mound is preserved as the tapering mass of light-colored limestone near the skyline.

glomerate also might have been deposited in tidal channels, as the daily forward and backward surges of water washed pebbles into shallow tidal basins. Whatever caused the water agitation—storms, tides, or both—it is clear that the tropical seas of the Great Basin were subject to occasional episodes of powerful torrents swirling over the shallow floor of the miogeocline.

Another indication of vigorous water movement in the Pogonip sediments are the many **mud mounds** that are found in some of the limestone units that comprise the group. Best known from the Fillmore Formation, Lehman Formation, and Antelope Valley Limestone, these elongated hemispherical structures can be thought of as humps of sediment that were trapped in and around reef-like masses of coral, bryozoa, sponges, and algae. The "thickets"

composed of the skeletons and colonies of these organisms evidently served to capture small grains of sediment that were being driven along the sea floor by strong bottom currents. The grains eventually buried the reef in a mound of mud several feet high. The mounds are typically elongated in the direction of the prevailing bottom currents. Some of the organisms in the reef, particularly the algae, secreted calcium carbonate that helped to bind the soft mud into a solid mass that eventually hardened into dark gray, highly fossiliferous limestone. The mud mounds are usually about 3–5 feet high, 6 feet wide, and about 100 feet long, tapering in the down-current direction. At Meikeljohn Peak in the Bare Mountain area of southern Nevada, however, a reef-like mound in the Antelope Valley Lime-stone is 270 feet thick (Figure 4.8)! The lime-stone near the top of this mound has

Figure 4.9 Fenestral textured Ordovician limestone from Meikeljohn Peak, Nevada.

numerous small pockets filled with white crystalline calcite. This suggests that the top of the mound was very near the surface of the sea, where some of the limestone was dissolved by seawater that had been diluted to a low salinity by rainfall. The irregular pockets and cavities dissolved in this manner were later filled by white calcite that crystallized from normal marine fluids that migrated through the open spaces. The result is that the limestone at the top of the Meikeljohn Peak mound has a distinctive mottled appearance, with blotches of white calcite scattered through the gray limestone (Figure 4.9). Geologists use the term **fenestral** to describe the fabric of such dappled limestone, and it is usually interpreted as the result of exposure to fresh, or very low salinity, water at or above the surface of the sea. Fenestral fabrics are common in many of the lower Ordovician limestone bodies in the Great Basin. Though there are several mechanisms by which fenestral limestone forms, exposure to fresh (or freshened) water while the sediment was still soft seems most likely for such rocks in the Pogonip Group. If so, then the thickness of the Meikeljohn Peak mud mound, which has very well developed fenestral limestone at the top, represents the maximum depth of the early Ordovician seas at that point in southern Nevada. Imagine how gentle the slope of the miogeocline must have been: some 275 miles offshore of the shoreline in central Utah, the sea was only 270 feet deep!

Intraformational conglomerate, elongated mud mounds, and fenestral limestone all signify the presence of powerful currents surging across the very shallow miogeocline during the time that the Pogonip sediments were deposited. Some of these currents were generated by storm

events, while others may have been related the general pattern of circulation in the early Ordovician seas. In addition, tidal fluctuations would have resulted in significant additional agitation. It is almost certain that the tides were more pronounced in the Ordovician than they are now because the moon was somewhat closer and exerted a more powerful gravitational pull on the seas. Thus, in addition to currents and storm-generated waves, great torrents of water must have flushed back and forth across the nearly flat miogeocline twice a day. Pockets of calm water may have existed here and there on the miogeocline, but they were probably very rare. The primitive marine organisms that inhabited these swirling tropical seas had to be well adapted to the swift currents, tumbling cascades, and periods of temporary exposure that would have prevailed across the lower Ordovician miogeocline. We will return to the life of the Ordovician later in this chapter, but there are still more details to decipher from the rocks that bear the fossils.

Mid-Ordovician Quartzite: A Flood of Sand
Wherever the top of the Pogonip Group is exposed in the Great Basin, the limestone and shale it contains are overlain by one of the most distinctive formations in the entire region: the Eureka Quartzite. Named in 1883 for a locality near Eureka, Nevada, this rock formation has a striking rusty white to light orange-tan color, while the rocks above (late Ordovician dolomite) and below (the Pogonip Group and equivalents) have drab gray to black hues. The Eureka Quartzite sometimes appears from a distance as an electrifying stripe of light-colored rock, 300–500 feet thick, slashing between thick and monolithic stacks of somber gray strata above and below. Unlike the dull-colored strata that envelop it, the Eureka sediments are very hard and resistant to weathering. For this reason, the white band of the Eureka Quartzite usually forms a long cliff or bench that juts out prominently from an otherwise smooth mountain slope. The visual and topographic uniqueness of the Eureka Quartzite make it hard to miss in the eastern Great Basin: there is no other rock unit quite like it in the entire Paleozoic succession.

No wonder it was one of the first rock sequences to be named by the pioneering geologists of the late 1800s!

The term **quartzite** can mean two different things in geology: 1) a quartz-rich metamorphic rock that forms from the alteration of sandstone by the effects of heat, pressure, and chemically active fluids and vapors; or 2) a very fine-grained, well-washed sandstone (a sedimentary rock) that consists of tiny quartz (SiO_2) grains that have been thoroughly cemented into a hard, durable mass. Technically, a metamorphic quartzite should be referred to as a **meta-quartzite**, while the sedimentary version would be more accurately described as an **orthoquartzite.** The Eureka is the latter of these two types of quartzite; it is a body of sandstone that is very fine-grained, consisting of well-rounded quartz grains about 0.1 mm in size, and cemented with silica (also SiO_2) so that the rock is composed almost entirely of SiO_2. The purity of the Eureka Quartzite is one of the reasons it is so easily identified throughout the eastern Great Basin.

The Eureka Quartzite is widely distributed across the miogeocline. This formation not only can be identified in Ordovician successions from central Utah to southwest Nevada, but also more or less similar materials are found along the fringes of the ancient oceanic platform. In northeast Utah, for example, rocks of about the same age as the Eureka Quartzite comprise parts of the Swan Peak Formation, a sequence of mudstone and impure sandstone. To the north, in Idaho, sandy mid-Ordovician deposits are known as the Kinnikinic Formation. Eastward from the Great Basin, in Colorado and Wyoming, geologists have identified additional mid-Ordovician sand deposits as the Harding Sandstone. West of the outer edge of the miogeoclinal shelf, in western Nevada and the Mojave region, no sandy sediments appear in the Ordovician record, but there is evidence of a gap, or unconformity, about the time that the Eureka Quartzite would have been accumulating. Considering the broad distribution of the Eureka Quartzite and correlative units, coupled with its average (and remarkably uniform) thickness of about 350 feet, it becomes obvious

Figure 4.10 Eureka Quartzite outcrop (light-colored upper cliff) in the Confusion Range, western Utah. The thinner light band below the Eureka is the Watson Ranch Quartzite (see Figure 4.4).

that this formation records an influx of an enormous amount of clean quartz sand across the mid-Ordovician miogeocline. And it all appears to have happened relatively suddenly, starting about 455 million years ago. What events could have caused such a dramatic change in the type of sediment that accumulated on the miogeocline? What might have triggered the great mid-Ordovician "sand flood"?

Like all rocks, the Eureka Quartzite tells its own story in the subtle details of its texture, composition, and structure. We know, for instance, that most of the sand that appeared in the Great Basin in the mid-Ordovician came from a great distance. This is because the sand grains comprising the Eureka Quartzite and related rock units are very small and well-rounded by extensive abrasion. If the sand were derived from local sources, we would expect larger grains, and less rounding of them, than are observed in the Eureka deposits. Also, except where the sand was mixed with muddy sediment (as in parts of the Swan Peak Formation), the mid-Ordovician quartzites of the Great Basin are very pure, consisting almost entirely of a single mineral, quartz. This is also a reflection of extensive transport, which generally leads to the destruction of other minerals less durable and more chemically reactive than quartz. Geologists often refer to deposits of fine-grained, well-sorted sand like those we find in the Eureka Quartzite as "mature": they represent a lengthy period of transportation prior to deposition. So,

Figure 4.11 Cross-bedding in the Eureka Quartzite.

in pondering the nature of the Eureka Quartzite, geologists must look beyond the Great Basin for the source of so much clean, fine-grained sand. But, in what direction should we look?

Fortunately, the structures preserved in the Eureka Quartzite provide some hints about the direction from which the flood of sand grains came. Many exposures of the Eureka Quartzite reveal **cross-bedding**, a style of stratification in which small-scale strata are inclined at an angle to the larger-scale primary bedding (Figure 4.11). Cross-bedding develops whenever sediment grains accumulate under the influence of a current. The geometry, orientation, and scale of the inclined secondary layers can be used to determine the direction and, to some degree, the strength of the currents that transported and deposited the sediment. The cross-bedding in the Eureka Quartzite suggests that the flood of sand came primarily from the northeast, and spread out across the miogeocline in a southwesterly direction. This conclusion has been strengthened recently by the discovery of tiny grains of the mineral zircon within the quartz grains that comprise the Eureka Quartzite. These grains have been traced, on the basis of age and chemistry, to zircon-bearing rocks in the Peace River Arch region of Canada. In general, the mid-Ordovician shoreline was oriented northeast-southwest (Figure 4.12), so the sand that smothered the miogeocline appears to have been washed in from the north and northeast by currents moving parallel to the shore. Such currents are called **longshore** currents, and they must have been persistent and powerful along the inner miogeocline during the mid-Ordovician.

The longshore currents that deposited the

Eureka Quartzite were no doubt related to the geographic positioning of North America during the Ordovician Period. At the time that the Eureka sand flood was in progress, the equator ran approximately from west Texas to the Great Lakes region (Figure 4.12). The Great Basin region was situated within a few degrees of the Ordovician equator and would have been influenced by the ancient trade winds. Today, these winds blow toward the equator in both the northern and southern hemisphere, but they veer sharply to the west as they move to lower latitudes. In terms of the modern geography of the Great Basin, this means that the prevailing mid-Ordovician winds would have swept over the shallow water of the miogeocline to the southwest, in the same general direction that the Eureka sand was dispersed. The longshore current that brought the sand flood to the Great Basin was part of the Ordovician equatorial current, a strong global current that, to this day, is still generated by the trade winds.

But there are still mysteries to address concerning the Eureka Quartzite. If the sand came from the northeast, what was its source? And why did it appear so suddenly in the Great Basin in the mid-Ordovician and, as we shall soon see, vanish in the late Ordovician? These uncertainties can be partially resolved by the observation that the Eureka Quartzite is generally associated with an unconformity, or gap, in the rock record across the Great Basin. In most places in eastern Nevada and western Utah, the Eureka is separated from the underlying Pogonip Group by an irregular, wavy surface that was produced when the uppermost Pogonip sediments were exposed to erosion prior to the accumulation of the sand grains. In some places, the Eureka Quartzite appears to have filled deep cavities developed in the underlying limestone during the time of exposure. Less commonly, the unconformity appears to be situated within the Eureka Quartzite, rather than at its base. These indications of erosion associated with the deposition of the mid-Ordovician quartzite suggest that the "sand flood" occurred at a time of falling sea level and shallowing water over the miogeocline. Prior to, or even during, the infusion of sand grains, some of the sediment on the miogeocline was washed away from the exposed por-tions. Else-where on the miogeocline, the bottom currents were probably intensified to the point that sediment could have been scoured away, even in places that remained submerged. The unconformity produced by the drop in sea level is sometimes so subtle a feature of the Ordovician rock record in the Great Basin that it is easily missed. Nonetheless, the gap has been substantiated across the Great Basin by the detailed dating, based on fossils, of the rock units above and below it. As much as 10 million years of Ordovician history is missing along the unconformity. Geologists are still not certain why the sea level fell in the mid-Ordovician. It may have been related to climatic changes or to minor geological movements near the miogeocline. In any case, recall that the seas of the Great Basin region were extremely shallow in the Ordovician Period; it wouldn't have required a major drop in sea level to expose parts of such a shallow sea floor.

As the sea level dropped in mid-Ordovician time, the shoreline would have migrated seaward along western margin of North America. This westerly retreat of the shore would have exposed vast tracts of previously deposited sediment along the continental edge. North and northeast of the Great Basin, in the direction from which the sand flood came, Cambrian sandstones would have been left high and dry by the retreating seas. These sandstones would have weathered to produced enormous quantities of sand that could have been transported thousands of miles into the Great Basin region by the powerful longshore currents. Thus the sand that arrived in the Great Basin during the mid-Ordovician had experienced two cycles of weathering and transportation: one in the Cambrian in the interior of North America, and one in the Ordovician along the western margin of the continent. No wonder that when the Eureka sand arrived in the Great Basin it was so fine-grained, well rounded, and pure!

The Late Ordovician:
Beginning of the "Dolomite Interval"

Almost everywhere in the Great Basin where the Eureka Quartzite is present, it is overlain by a series of coal-black layers of rock known as the Fish Haven Dolomite (in the northeastern Great Basin) or the Ely Springs Dolomite (in the central and southeastern parts of the region). By virtue of their characteristic raven color, which contrasts strongly with the lighter shades of formations above and below, these formations are easy to identify as a bold black stripe slashing across the face of desert mountains (see Plate 7). The late Ordovician dolomite formations mark the beginning of a long period of time when similar strata continued to form on the miogeocline. In fact, for more than 60 million years, from the late Ordovician Period, throughout the following Silurian, and into the early part of the ensuing Devonian, dolomite was the primary rock formed across the inner zone of the miogeocline in the Great Basin. We will call this interval of time (late Ordovician through early Devonian) the "Great Basin Dolomite Interval," for the primary type of rock present in shallow marine formations of this age. This is a very distinctive phase of Great Basin history; nothing like the Great Basin Dolomite Interval ever happened before or since the mid-Paleozoic Era, anywhere in the region. What is dolomite, and what does it tells us about conditions in the Great Basin during the time when it was so prevalent? Let's begin by considering the nature

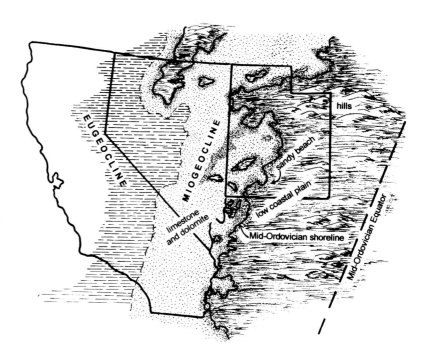

Figure 4.12 Mid-Ordovician Paleogeography of the Great Basin Region. The shallow miogeocline extended from the shoreline in central Utah to the edge of the continental shelf in central Nevada. Because the Great Basin was situated a few degrees north of the mid-Ordovician equator, the shallow seas covering the miogeocline were warm and tropical in character.

of dolomite, in general, and the late Ordovician dolomite formations, in particular.

Dolomite is a carbonate mineral of the composition $CaMg(CO_3)_2$. Technically, a rock composed mostly of this mineral should be called a **dolostone**, and some geologists today refer to rock formations such as the Ely Springs and Fish Haven by this term. However, when these rock units were first defined (in 1932 and 1913, respectively), the term dolomite, not the currently more popular dolostone, was affixed to the formation names. This is because it was common practice in the early 1900s to use a single name for the both the mineral and a rock composed dominantly of it. Since the dual use of these names is well ingrained in the geological literature of the Great Basin, we will continue that time-honored tradition here, even though in doing so we're bending a linguistic rule slightly. Besides, the first term applied to a formation always has priority in geology, unless it has been formally changed through established protocols. This practice avoids the confusion that might result from using two different names for the same rock sequence.

While it has an obvious chemical similarity to calcite ($CaCO_3$), the chief component of limestone, dolomite differs from limestone in some important ways. Dolomite is generally less soluble than calcite, so it tends to be more resistant to weathering than limestone. As a result of this, the exposures of the Ely Springs and Fish Haven Dolomite in the Great Basin are often marked by nearly vertical cliffs, or by an abrupt steepening of the slope of the land. Dolomite is also less reactive to acids than is calcite. Thus, the effervescence of carbon dioxide that occurs when acids are applied to most carbonate minerals is very weak in dolomite. Only when dolomite is powdered (which increases the surface area for chemical reactions to take place) does the effervescence in acid become noticeable. This explains why the standard field test for dolomite, which might be called the "scratch-and-squirt" test, works so well. Simply pulverize a bit of an outcrop with a rock hammer, and squirt a few drops of acid on the powder—if the powder bubbles, and the rock didn't, it's dolomite. Limestone, on the other hand, will produce a vigorous bubbling effect without being powdered.

Dolomite also differs from limestone texturally. Limestone is extremely variable in appearance; it may appear smooth and fine textured, or it might be silty, oolitic (containing tiny egg-shaped masses of calcite), fragmental, or even microcrystalline. Dolomite, on the other hand, is often coarsely crystalline and commonly looks "sugary" to the unaided eye. This crystalline nature of dolomite appears to be a secondary feature, one that developed in the rock after the original sediment was deposited. In terms of color, dolomite may be dark or light gray, brownish, light pink, or even golden-hued. In addition, the process of recrystallization that commonly affects dolomite often obliterates or masks any organic remains that might be present. Consequently, though many of the Great Basin dolomite successions have produced identifiable fossils, they are usually less fossiliferous than are limestone sequences, and the fossils they produce are often poorly preserved. For the same reason, internal structures such as fine laminations or delicate cross-bedding are generally more difficult to observe in dolomite than in limestone layers. Because of these characteristics, typical outcrops of Great Basin dolomite are often rather uninspiring: brown to black blocky cliffs composed of more or less nondescript and monotonously crystalline rock, with few fossils or distinctive structures.

In spite of their drab appearance, the dolomite units of the Great Basin are very intriguing rocks. This is because the origin of the dolomite has been, and still is, one of the greatest riddles of Great Basin geology. As we have already mentioned, the recrystallized texture of most dolomite, coupled with the presence of transitional rock (partially "dolomitized" limestone), suggests that dolomite is formed through the transformation of original limestone sediment. This deduction is supported by the well-documented relationship between the amount of dolomite and geologic age: modern dolomite is virtually nonexistent, but it becomes more common as we look farther into the geologic past. At first blush, the formation of dolomite would seem to be a fairly simple matter of adding a little magnesium to calcite ($CaCO_3$) to form a new compound, $CaMg(CO_3)_2$. However, the actual process is not nearly this simple. The magnesium in dolomite is not scattered through the atomic framework in a haphazard fashion. Each magnesium atom in dolomite occupies a specific site among other atoms, and this pattern is repeated regularly throughout the atomic maze. When magnesium is added to calcium carbonate in the laboratory, some of it may substitute for calcium atoms, but the orderly pattern of dolomite does not develop. We get magnesium-rich calcite, not true, atomically ordered dolomite from such experiments. Many attempts have been made in the laboratory to simulate the dolomitization process under different temperatures and chemical conditions, but all have failed to produce significant quantities of dolomite.

Even more intriguing is the observation that, in the modern world, very little dolomite has ever been identified. Thin dolomitic crusts have been observed forming in tidal salt flats (known as "sabkhas") in the Persian Gulf region, in near-shore lagoons along the southeast coast of Australia, and on dried surfaces of mud washed by storms into inland basins in the Bahamas. The common factor in all these sites is a warm climate with high evaporation rates. When seawater becomes isolated from the open ocean in a tidal basin or restricted lagoon under such climatic conditions it can become extremely saline (**hypersaline**) as water is evaporated and the dissolved minerals become more concentrated in the residual fluid. As more and more water evaporates, mineral deposits begin to form in the basins that contain the withering lakes and ponds. The first minerals to crystallize from the hypersaline fluids are calcium-rich minerals such as calcite and gypsum (a hydrated calcium sulfate). Eventually, salt (or **halite**, the mineral name for sodium chloride) will form a crusty layer as the bodies of water continue to shrink. As the various minerals are deposited from the dwindling water, the concentration of magnesium relative to other elements in the residual fluid increases rapidly. This is because calcium, sodium, and other elements are removed from the water to form the crystals of early formed

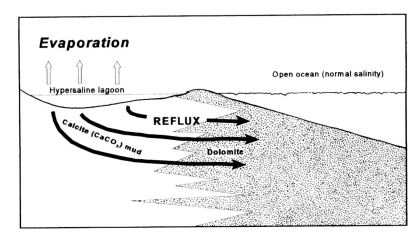

minerals, but magnesium is not. Magnesium tends to remain in the water until the Mg:Ca ratio in the brine rises from a normal value of about 5:1 to more than 8:1. Because of the extremely high salinity, this residual liquid is a very dense fluid and tends to sink downward through the mud at the bottom of the lagoon. Dolomite can form when these slowly moving, magnesium-laced fluids react with calcite in the previously deposited mud. This hypothetical method of forming dolomite has been called the evaporative reflux model (Figure 4.13), and it has often been suggested as the process by which the Great Basin dolomite formed. We know that the evaporative reflux model is theoretically valid, and, moreover, small amounts of dolomite have been identified in places where similar conditions exist in the modern world. Furthermore, throughout the entire Great Basin Dolomite Interval, the dolomite deposits formed only on the inner, near-shore tracts of the miogeocline, just as one would expect from the evaporative reflux model. In Figure 4.13, note that far offshore of the hypersaline lagoons, the lime mud would be unaffected by the dolomitizing fluids. We would thus expect that the Great Basin dolomite deposits would grade offshore into normal limestone sequences. This is exactly what is found when the dolomite formations are traced to the west, into the deeper water portion of the Cordilleran Geosyncline (Figure 4.14). For example, the Ely Springs Dolomite of western Utah is roughly equivalent to the Hanson Creek Formation of east-central Nevada (mostly dolomite, with some limestone) and the Caesar Canyon Formation of the

Figure 4.13 The Evaporative Reflux Model for the Origin of Great Basin Dolomite. Dense and hypersaline fluids seep downward and offshore from a restricted lagoon through calcite mud, forming dolomite in the process. Without this process, the calcite mud would have hardened into normal limestone. This hypothetical near-shore process is compatible with the restriction of Great Basin dolomite to the eastern portions of the Paleozoic miogeocline.

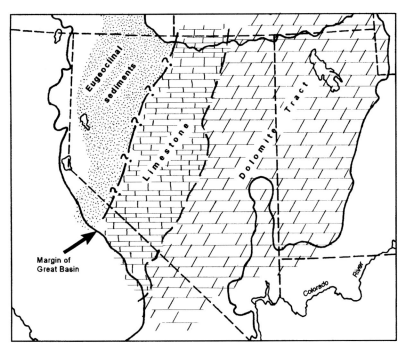

Figure 4.14 Distribution of dolomite and limestone during the Great Basin Dolomite Interval. Dolomite formed on the inner, near-shore portion of the miogeocline, while normal limestone sediments accumulated farther offshore.

Toquima Range of west-central Nevada, which consists mostly of limestone. This same trend becomes even more evident in the Silurian and early Devonian strata, which also formed during the Great Basin Dolomite Interval. It is clear that whatever process produced the Great Basin dolomite sequences, it was most active nearest the shoreline of the mid-Paleozoic ocean. In addition, given the location of the Great Basin several degrees north of the equator during the Dolomite Interval, it is almost certain that the climate was warm and that evaporation rates were high at the time the dolomite was forming.

Despite the appeal of the evaporative reflux model for the origin of the Great Basin dolomite formations, it is not a perfect explanation. One difficulty with the concept is that in places where similar conditions exist today the volume of dolomite that forms is minuscule, compared to the massive quantities that exist in the Great Basin. Even if we allow for a span of 60 million years (the duration of the Great Basin Dolomite Interval) to elapse, the rate at which dolomite forms from the reflux process today is too low to account for all the rock that formed on the miogeocline. In addition, we would expect to find, somewhere landward of the dolomite tract, deposits of gypsum and salt that formed as the water was evaporated from the restricted lagoons and basins. No such rocks have been

identified in central Utah, or anywhere else along the area of the mid-Paleozoic shoreline. Although subsequent erosion could have removed the gypsum and salt deposits, it seems unlikely they all traces of them would vanish in this manner. Because of these difficulties, several other mechanisms for the origin of Great Basin dolomite have been proposed. Some of them involve complex chemical reactions that may have taken place between magnesium-rich seawater seeping through mud and interacting with low-salinity groundwater percolating seaward from land. Perhaps, as other geologists have suggested, these chemical reactions occurred when seawater was squeezed out of mud on the sea floor as it became compacted under the weight of additional layers. It is well beyond the scope of our survey to review all the theories for the formation of dolomite in the Great Basin. However, none of them fully, and to the satisfaction of all scientists, explains the dominance of this type of rock from late Ordovician through early Devonian time. Of all the theoretical mechanisms that have been advanced as explanations for the dolomite conundrum, the evaporative reflux model seems to have the most appeal. But, there is still much to learn about the nature of this very baffling episode in the history of the Great Basin. Currently, the dolomite dilemma remains one of the most implacable riddles of Great Basin geology, but it is also one of its greatest charms. After all, everybody loves a mystery. Even geologists!

ORDOVICIAN LIFE: A REVOLUTION ON THE SEA FLOOR

In certain portions of the Ordovician rock succession of the Great Basin, fossils of marine creatures that inhabited the miogeocline are spectacularly abundant. The most opulent fossil beds occur in several different formations of the Pogonip Group. In particular, fabulous fossil collections have been amassed from such units as the Kanosh Shale, Wah Wah Formation, Lehman Formation, Fillmore Formation, and the Antelope Valley Limestone. Some of the horizons in these formations in western Utah, where the most prolific fossil-producing localities are situated, are so densely packed with fossils that they can only be described as **coquinas**, or "shell beds." In these strata, more than 90 percent of the limestone is composed of nothing but fossil shell material (Figure 4.15). The middle and upper parts of the Ordovician sequence are usually far less fossiliferous. The Watson Ranch and Eureka Quartzite units, for example, sometimes yield trace fossils (burrowing structures or trackways) of marine invertebrates, but otherwise these strata are essentially barren of fossils. Evidently, the strong bottom currents that swept the sand across the miogeocline in mid-Ordovician time also destroyed most of the potential fossil material, or somehow prevented large populations of organisms from inhabiting the sea floor, or both. The dolomite formations, because of the pervasive recrystallization that has affected them, produce few and generally poorly preserved fossils compared to the limestone and shale of the Pogonip Group. The late Ordovician marine fauna might have been as rich and varied as that of the older Pogonip sediments, but the dolomitization process appears to have obliterated much of the fossil evidence.

If we compare the overall character the Ordovician fossil assemblages to those of the preceding Cambrian Period, some rather striking differences emerge. As we have already discovered, the Cambrian fauna in the Great Basin is numerically dominated by trilobites and related arthropods. About 500 million years ago, near the beginning of Ordovician time, this unique array of creatures began to decline in both abundance and diversity. At virtually the same instant of geological time, a new assemblage began to proliferate across the shallow marine habitats on a global scale. The new assemblage was a much more varied assortment of creatures than was the Cambrian fauna, and was not dominated by any single taxonomic group. Instead, it was comprised primarily of brachiopods (Phylum Brachiopoda), crinoids (Phylum Echinodermata), corals (Phylum Cnidaria), ostracods (a group of specialized arthropods), and cephalopods (Phylum Mollusca), in addition to other, less abundant, groups such as the gastropods (snails), conodonts, and the

Figure 4.15 A remnant of an Ordovician shell bed. This rock, part of the Kanosh Shale of western Utah, consists almost entirely of shell fragments from several different groups of marine organisms.

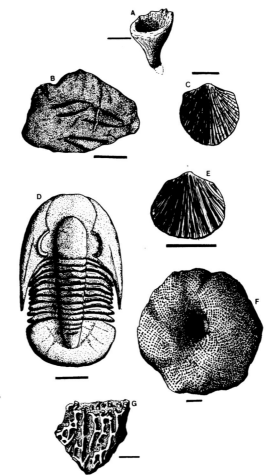

Figure 4.16 Common Ordovician and Silurian Fossils from the Great Basin.
A, *Streptelasma*, from the Laketown Dolomite;
B, *Didymograptus*, a widespread graptolite;
C, *Orthambonites*, a brachiopod from the Kanosh Shale;
D, *Bathyurellus*, a trilobite common in the Pogonip Group;
E, *Platystrophia*, from the Fish Haven Dolomite;
F, *Receptaculites*, a sponge-like organism from the Pogonip Group;
G, *Halysites*, a Silurian "tube-coral" from the Laketown Dolomite.
Scale bar = 1 cm in all figures.

aforementioned graptolites (Figure 4.16). Paleontologists have estimated that, during Cambrian time, the sea floor was populated by a maximum of about 150 different families of marine creatures. By the end of the Ordovician, that number had exploded to more than 400. In addition to being more varied than the Cambrian fauna, the Ordovician congregation of fossils is also much more abundant. Even though fossils may be abundant in the Cambrian strata of the Great Basin, few exposures of that age can match the famous Ordovician localities for the sheer profusion (and diversity) of organic artifacts. Even in the Cambrian *lagerstätten*, where fossils occur in great concentrations, the assemblages tend to be rather monotonous mixtures compared to the eclectic Ordovician assortments.

Making matters even more intriguing is indisputable fossil evidence that this great transition in marine life was not restricted to the sea floor of the Great Basin. It was clearly a global phenomena. So conspicuous and widespread are differences between the Cambrian fauna and the new menagerie that emerges in the Ordovician that, in the early 1980s, paleontologist Jack Sepkoski formally described the latter as the **Paleozoic fauna**. The Paleozoic fauna was so named because after it became established on the sea floor in the Ordovician Period, the overall character of marine life changed very little during the remainder of the Paleozoic Era, a span of more than 200 million years! Obviously, something very interesting happened on the shallow sea floor in Ordovician time. It caused the organisms that lived there to begin a world-wide evolutionary riot that had lasting effects on the marine ecosystem. What happened?

The Cambrian-Ordovician Transition and Beyond

The emergence of the Paleozoic fauna, and the contemporaneous decline of the Cambrian fauna, seems to reflect the interplay of several different factors. First, near the end of Cambrian time, the trilobites in the Great Basin (and elsewhere) experienced an interesting series of extinction events. At least three extinctions affected the trilobites in western North America during the last 15 million years of the Cambrian Period. Each extinction event appears to have been rather sudden, perhaps occurring in as little as a few thousand years. Dozens of species of trilobites abruptly vanished during each of the extinctions, but there were always a few survivors that repopulated the sea floor. In each case, the survivors eventually gave way to new species that replaced the forms extinguished during the extinction. A few million years after the recovery, a new wave of

extinction would begin, and the whole cycle would be repeated. The survivors of the periodic late Cambrian trilobite extinctions appear, in most cases, to have been trilobites that were adapted to relatively cold and deep water beyond the edge of the miogeocline. For this reason, many scientists have suggested that the periodic trilobite extinctions might have been related to episodic coolings of the climate near the end of the Cambrian Period. However, other factors may also have been involved in these cyclic extinctions. Curiously, only those trilobites adapted to the warm, shallow water of the miogeocline seem to have been strongly affected by the extinctions. The non-trilobite organisms in the Cambrian fauna were, for some reason, not so severely affected by any of these late Cambrian extinction events.

Whatever the cause of the trilobite extinctions, the long-term effect was very important. As the trilobites struggled with the cycles of decline and recovery, they gradually lost their firm grip on the ecological niches they had dominated throughout the Cambrian. Each wave of extinction created new adaptive opportunities for non-trilobite creatures, and several different lineages took advantage of the trilobites' misfortune. The echinoderms, including the earliest starfish, along with the cephalopods, brachiopods, and snails, seem to have been particularly successful in exploiting the niches left vacant by expired trilobites. Of course, during the recovery that followed each extinction, the trilobites reclaimed some of the lost ground, but not all of it. As a group, the trilobites survived until the very end of the Paleozoic Era; but after the transition from Cambrian to Ordovician time, their dominance on the sea floor steadily declined. By the time that the last few species of trilobites vanished 250 million years ago, they comprised only an insignificant element in the extravagant Paleozoic fauna.

The trilobite extinctions of late Cambrian may have set the stage for the advent of the Paleozoic fauna, but something else seems to have accelerated the transition. The organisms that comprise the Paleozoic fauna feed primarily by straining plankton, larvae, and other microscopic food items from seawater. Such marine organisms are known as suspension, or "filter," feeders, and include brachiopods, some clams, crinoids, and bryozoans. In addition, the corals, a very prominent component of the Paleozoic fauna, should be added to this list. Corals are, technically speaking, predators because they use stinging cells to immobilize food organisms that come with reach of their tentacled weaponry. However, corals are sessile creatures that are dependent on currents or the self-propulsion of their prey to bring food to them. This is true of both the colonial corals of the modern world and of the solitary corals in the Paleozoic fauna. Thus, in a way, corals are also "filter" feeders because they are anchored to the sea floor and do not actively pursue prey like other predators. Compared to the preceding Cambrian fauna, the Paleozoic fauna is notably rich in suspension-feeding organisms. In fact, as Paleozoic time progresses, the various creatures in the Paleozoic fauna become increasingly specialized for this mode of feeding. The brachiopods (along with bryozoans, their distant kin), crin-oids, bivalves, and sponges all developed different, but equally elaborate, straining devices. In addition, by the end of Ordovician time, the marine communities of the sea floor had developed a layered, or "tiered," structure, as various groups of suspension-feeders adapted to filtering seawater at different heights above the bottom. The brachiopods, encrusting bryozoans, and the sponges typically occupied the lowest zone, while the branching types of bryozoans, crinoids, and various corals could rise upward into the water column by as much as several meters above the muddy sea floor. In the Cambrian fauna, suspension-feeding organisms are present, but the most dominant organisms are those that feed by ingesting sediment (detritus feeders), grazing algae, or scavenging the remains of dead creatures. So conspicuous is the suspension feeding dominance of post-Cambrian fossil assemblages in the Great Basin, that it is virtually impossible to mistake Cambrian strata for Ordovician rocks, once a few fossils have been found. Starting in the Ordovician Period, something clearly happened in the Cordilleran Geosyncline (and elsewhere) to bestow an advantage on suspension-feeding organisms.

Moreover, once the suspension feeders rose to dominance in the Ordovician, they maintained that dominance until the close of the Paleozoic Era, a span of more than 200 million years.

The advent of the Paleozoic fauna, with its profuse array of suspension feeders, is at least in part a probable reflection of the continuing fragmentation of land masses into smaller continents. As we have already learned, this rifting began with the break-up of Rodinia near the beginning of the Cambrian Period. By the early Ordovician, numerous small continents were scattered across the globe, though they appear to have been concentrated near the equator. There were only a few large land masses in the early Ordovician, and much of the earth was similar to the modern southwest Pacific Ocean, where scores of small islands (e.g., Java, Sumatra, and the Philippines) stand above the waves of the tropical sea. As an illustration of the small size of Ordovician land masses, consider the fact that the western portion of North America, including the Great Basin, was submerged as far east as the modern Rocky Mountains. At the same time, eastern North America was also submerged, all the way from the modern Atlantic seaboard to the upper Midwest. Thus, there is approximately three times more land exposed now in North America than there was in the Ordovician Period. Most of the other land masses on the earth were equally small and widely dispersed in Ordovician time.

The fragmented nature of the Ordovician world, along with the concentration of land masses in the tropics, must have had a significant effect on the global climate. Under these conditions, the differences between marine and continental air masses would have been minimal, and fierce winter storms would have been infrequent, at least in the tropics, where most of the land and shallow sea floor was located. Also, because of the steady warmth and uniform climatic conditions, the global climate was probably much less seasonal than it is today. The differences between summer and winter during the Ordovician were no doubt very subtle, perhaps even negligible. These uniform conditions would have resulted in rather monotonous weather for most of the Ordovician land areas huddled along the equator. While there was probably no terrestrial life (or at least very little) during the Ordovician, the shallow seas that surrounded each small land mass would have experienced an equally uniform climate. There was probably little variation in the environmental factors that affected the marine ecosystem. The strength and duration of sunlight, the water temperature, the amount of run-off from land, and the level of nutrients in the water probably varied little from season to season, year to year. Consequently, the levels of biological activity in the sea also remained nearly constant for long periods of time. The populations of microscopic plankton would have been stable and uniform. As these tiny plants and animals died, their remains drifted toward the sea floor as a "rain" of food particles, sinking toward the suspension feeders below. With constant climatic conditions, the supply of food to the shallow ocean floor was a steady drizzle, subject neither to cloudbursts of biological productivity nor to winter droughts. The advantage of suspension feeding under these conditions is obvious: it allows organisms to intercept food before it gets to the sea floor, where competition for it must have been intense. The suspension-feeding creatures of the Paleozoic fauna prospered for so long because they evolved the adaptations necessary to exploit the most favorable ecological niches under the uniform and steady conditions of the Ordovician. The deposit feeders, the algae grazers, and the scavengers survived during this time, but only as relatively small populations of creatures living on the leftover food particles that escaped the filtering devices of the brachiopods, crinoids, bryozoans, and corals. So, the establishment of the Paleozoic fauna was probably driven primarily by a geological event, the extended fragmentation of Rodinia, and the climatic changes that it fostered. But, don't despair over the mud eaters, they were destined for a later time (as we will see); and there was a bump in the road approaching even for the suspension-feeders. Once again, the fossils from the Great Basin clearly document the next phase in the environmental history of the Cordilleran Geosyncline.

Late Ordovician Extinctions: A Long Winter Arrives

The Ordovician rocks and fossils of the Great Basin evince a rather idyllic scene on the miogeocline 480 or so million years ago. The warm sun floated in a blue sky high above what must have been a tropical, though treeless, paradise. Below, warm aquamarine water swirled through tidal channels and pools filled with an amazing variety of primitive sea life. Reefs, built primarily by coral, bryozoans, and encrusting algae, lay just beneath the shimmering water. Schools of cephalopods, looking a bit like squid with long conical shells, cruised the tranquil shoals looking for shell beds on which to feed. Graptolite colonies drifted along like tiny clumps of seaweed, bobbing up and down in the roiling surf. Even the winter days were pleasant, and storm clouds rarely gathered in this Paleozoic Shangri-La. Life, at least for those suspension-feeding organisms that dominated the Paleozoic fauna, was very good indeed.

Near the end of the Ordovician Period, approximately 445 million years ago, something happened to disrupt the ecological bliss that began with the advent of the Paleozoic fauna. A great wave of extinction swept through the shallow seas of the earth, resulting in the first of five great mass extinctions that have occurred over the past half-billion years (Figure 4.17). On a global scale, the late Ordovician mass extinction claimed approximately 60 percent of the genera of marine organisms that populated the sea floor. In North America, more than 100 different families of creatures that were present in the Ordovician didn't survive into the following Silurian Period. In the Great Basin, the effects of this mass extinction are most notable when the diverse and rich Ordovician fauna is compared to the fossil assemblages from younger Paleozoic strata. Silurian strata in the region, for example, yield a comparatively scant fossil fauna, composed almost entirely of just two phyla: the brachiopods and the corals. Organisms that exhibit a strong reduction, or a profound transition in character, in the post-Ordovician fossil assemblages of the Great Basin include the crinoids, bryozoans, sea stars, bivalves, sponges, ostracods, and cephalopods. Not until the late Devonian Period, about 375 million years ago and more than 50 million years after the mass extinction, did life on the miogeocline fully recover from the Ordovician decimation. Moreover, it was not just the suspension feeders anchored to the shallow sea floor that suffered. Fossils collected recently from the deeper-water deposits of the Vinini and Hanson Creek Formations, deposited at the outer edge of miogeocline, indicate a stepped-pattern, but still abrupt, decline in the numbers of graptolites, **conodonts** (tiny toothlike fossils from a primitive chordate animal), and archaic planktonic animals known as **chitinozoans**. All of these organisms floated, or swam, in the water column well above the floor of the shallow seas, suggesting that the ecological distress affected the entire ocean, not just the habitats of the bottom dwellers. At the close of Ordovician time, something disturbed the global marine ecosystem so significantly that the effects have been documented almost everywhere that fossil-bearing Ordovician strata exist.

The causes of mass extinctions can be very difficult to identify because the global ecosystem is a complex entity, consisting of numerous interdependent and interacting elements. In the case of the Ordovician extinction, we can glimpse the possible causes by considering the pattern and timing of the event, along with some trends observed in the sediment that was deposited at the time. The fossil evidence seems to suggest that the extinctions occurred abruptly over a period of between 500,000 and 1,000,000 years. However, the extinctions were clearly not instantaneous, for, even within the Great Basin, some groups of organisms declined before others. The diachronous nature of the extinctions seems to rule out any short-term catastrophe or other sudden cause, but the relative abruptness of the event still seems to signify some rapid deterioration in the marine ecosystem. It is also clear from the fossil record of the extinction that the groups of organisms most affected were those belonging to tropical reef communities. Sea creatures that lived in temperate water, though they were less abundant and diverse than tropical forms, appear to have been much

less affected by the late Ordovician extinction event. Near the Ordovician extinction horizon, layers of silty or sandy sediment commonly appear in many of the Great Basin rock successions. This sudden influx of particles and grains from land suggests a rapid decease in the depth of water that covered the miogeocline. If the sea level fell abruptly, then the ancient shoreline would migrate westward toward the deeper parts of the miogeocline. Rivers would then drain across the exposed sea floor to reach the receding shore, eventually discharging silt-laden muddy water into the miogeocline to bury reefs in fine sand and silt. Moreover, there is commonly a prominent gap, or unconformity, that separates late Ordovician strata from early Silurian layers in most areas of the inner miogeocline. The Ordovician-Silurian unconformity probably was produced when the sea level fell enough to temporarily expose the sea floor, resulting in the erosion of some of the sediment that had accumulated there.

We know that such sea level fluctuations can be related to cooling trends in the global climate. This is because more water is transferred to land as snow and ice, and less is returned immediately to sea, when temperatures fall. In addition, water, like most other substances, contracts as it cools, albeit by a small amount. Thus,

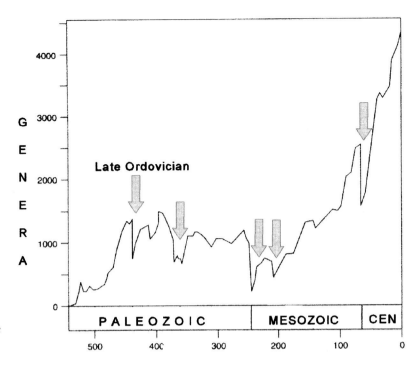

Figure 4.17 The Five Major Extinctions. The late Ordovician extinction is the first of five major crises that affected life during the past 500 million years. This graph, modified from the work of the late paleontologist Jack Sepkoski, illustrates the decline in life (number of genera) at each of five extinction events (arrows). The rapid increase in diversity that precedes the Ordovician event reflects the development of the Paleozoic fauna.

the water that remained in the ocean basins as the global temperature fell occupied less space, causing the sea level to drop even farther. The reduction in water depth in the Great Basin is accompanied by an equally conspicuous trend in the pattern of global sediment accumulation: for several million years after the mid-Ordovician, less limestone and more shale and siltstone accumulated in the world's oceans. Since the precipitation of calcite, the major component of limestone, is favored by high water temperature, the reduction in this type of sediment is further evidence of global cooling in late Ordovician time. Remember also that it was the tropical organisms, living in relatively warm water, that suffered the most during the late Ordovician extinctions. This is, no doubt, because the organisms that dominated the extravagant tropical reefs were all well adapted to warm-water conditions and a constant supply of abundant food. When the climate began to deteriorate, the supply of planktonic food for these creatures became less plentiful and reliable. Many suspension feeders living in and around the reefs vanished as the world slipped into the first major ice age of the Paleozoic Era.

The indications of global cooling in the late Ordovician rock record are so prevalent that geologists have formally identified this event as the **Hirnantian Glaciation**, named for the final slice of Ordovician time (the Hirnantian Age) that was originally discerned in strata exposed around Hirnant, North Wales. The pattern of extinction, the types of organisms affected, and the supporting evidence for climatic degradation, all leave little doubt that rapid and profound cooling was the cause of the ecological chaos in the late Ordovician. What, in turn, might have caused the precipitous climate change that led to the Hirnantian Glaciation? The answer seems to involve the movement of the largest Ordovician continent, **Gondwana**, over the South Pole near the end of the Ordovician Period. Though the continental masses were generally smaller in the Ordovician than they are today, Gondwana was the largest, consisting of the central cores (but not the entirety) of modern Africa, Australia, Antarctica, and South America. This relatively large land mass,

an oddity in the fragmented geography of the Ordovician world, slipped steadily southward from the tropics for millions of years near the end of Ordovician time.

As Gondwana moved poleward into regions of colder climate, glaciers began to grow from excess snow and ice that collected on its surface. Sea level began to drop as water was climatically transferred from the oceans to the glaciers. By the end of the Ordovician, a large portion of Gondwana was covered by an extensive ice cap, and the actual South Pole was located in what is now northwest Africa. Such a vast tract of blinding white snow and ice reflected so much of the heat energy arriving from the sun that global temperatures began to fall rapidly, and a worldwide ice age was underway. Responding to these changes in climate, the populations of planktonic organisms in the sea began to decline, causing the "rain" of food particles to the sea floor to slacken to a light sprinkle. Deposit feeders fared better than suspension feeders during this crisis because the organic matter buried in the sea-floor sediment provided them with an alternative source of food. Even among the suspension feeders, some temperate forms were more tolerant of cool-water conditions and fared much better than their tropical kin. One by one, genus by genus, the inhabitants of the tropical reefs disappeared. In less than a million years or so, the lush abundance of the Ordovician sea floor was gone. The rich carpet of life on the miogeocline of the Great Basin thinned dramatically as the world slipped into its first major mass extinction.

Paradise Regained:
The Silurian and Early Devonian Periods

In spite of its severity, the late Ordovician extinction was short-lived. The climate change that initiated the crisis was reversed within a few million years at most, as Gondwana continued to move over the South Pole and back toward tropical latitudes. Of course, any movement of anything at the pole will carry it toward the equator. Thus, it was inevitable that the Ordovician ice age should end: the very motion of Gondwana that caused it also limited its duration. With this large land mass moving back toward the tropics, the glacial ice began to melt and less of the sun's energy was reflected from the surface. The earth warmed, the glaciers receded, and sea level began to rise as meltwater drained back into the ocean. In the Great Basin, carbonate mud began to accumulate across the miogeocline as the rising sea submerged areas along the shore that were formerly exposed. As we have already seen, much of this mud initially formed limestone but was soon converted to dolomite. The Great Basin Dolomite Interval was still in progress after the great extinction was over. Silurian (443 to 417 million years ago) rocks in the Great Basin are of rather uniform dolomitic character and have been subdivided by geologists into only a few widespread rock units. Among these, the Lone Mountain Dolomite of central Nevada and the Laketown Dolomite of western Utah are the best-known formations. In addition, the lower portion of the Roberts Mountains Formation of central Nevada consists of carbonate sediments deposited at the outer edge of the miogeocline during the Silurian Period. Most of the Silurian dolomite in the Great Basin is light tan to gray, massive to blocky rock with a finely crystalline appearance that is often described as "sugary." These dolomite sequences are generally rather monotonous and produce few fossils. Sometimes, however, well-preserved brachiopod shells and coral skeletons can be found in certain layers where conditions for preservation were optimal. These fossils demonstrate the return of tropical conditions in the miogeocline, and the recovery of suspension feeders from the Ordovician extinction event. The relatively low abundance of Silurian fossils in the Great Basin is more attributable to the recrystallized nature of the dolomite than it is to low populations of organisms on the shallow sea floor. Nonetheless, even allowing for the poor potential for preservation, the diversity of Silurian fossil assemblages is much lower than is that of the rich Ordovician fauna. Moreover, the types of brachiopods present in the Silurian are much different from those found in Ordovician strata. In particular, the large and elongated pentamerid brachiopods (Figure 4.18) found in

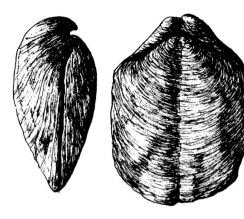

Figure 4.18
Penatameroides, a Silurian brachiopod from the Great Basin. This pentamerid brachiopod, seen in side (left) and "top," or dorsal, (right) views, may be more than 1.5 inches long. The large pentamerid brachiopods are especially characteristic of the Silurian strata in the Great Basin.

Silurian rocks of the Great Basin represent a type unknown from the older Ordovician strata. So, while the climatic and biotic recovery from the Ordovician extinction event was well underway in Silurian time, it involved new types of organisms and took million of years to complete the full restoration of the shallow marine ecosystem in the Cordilleran Geosyncline.

The Devonian Period began about 410 million years ago and, when it arrived in the Great Basin, nothing seems to have changed very much. Dolomite continued to accumulate in the inner portions of the miogeocline, occasionally burying brachiopods, corals, and crinoids that populated the shallow sea floor. Early Devonian dolomite sequences on the inner miogeocline are so similar to the underlying Silurian deposits that the two are sometimes difficult to distinguish in isolated outcrops. Units such as the Sevy Dolomite and Simonson Dolomite of western Utah and eastern Nevada, along with the lower part of the Sultan Formation of the Mojave Desert region, are comprised mostly of finely crystalline, drab-colored dolomite almost identical to Silurian rocks. Along the early Devonian shoreline, which passed north-south through central Utah, the dolomite sediments accumulated along with mud and silt derived from nearby land. This "muddy" dolomite is represented by the Water Canyon Formation of northeast Utah. The Water Canyon sediments are especially interesting to paleontologists because they produce fragmentary remains of primitive fishlike creatures known as **ostracoderms**. The ostracoderms represent several different groups (orders) of the jawless vertebrates, distant relatives of the modern hagfish and

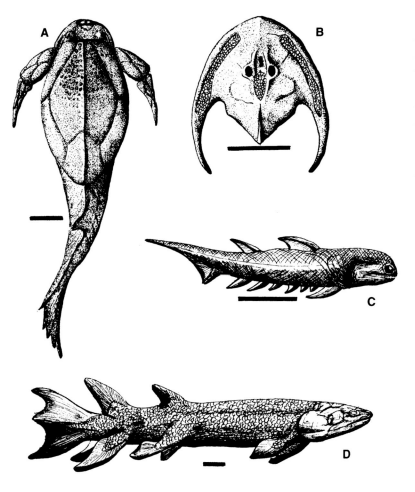

Figure 4.19 Devonian Fish of the Great Basin. A, reconstruction of *Asterolepis*, a placoderm from central Nevada; B, the head shield of the ostracoderm *Cephalaspis utahensis*, from the Water Canyon Formation; C, *Climatius*, a Devonian acanthodian fish similar to those that may have left the spines in the Denay Limestone of Nevada; D, *Eusthenopteron*, a large (more than a meter long) lobe-fin fish known from the Devonian strata of the central Great Basin. Scale bar in all sketches = 5 cm (2 inches)

lampreys. Most ostracoderms were small animals (generally only a few inches long), with spikelike "fins," a rudimentary tail, and a circular mouth that lacked a moveable jaw. In addition, the ostracoderms were armored with relatively thick bony plates, particularly in the head region. Despite their fishlike appearance, ostracoderms were poor swimmers that probably ingested mud through their jawless mouths, straining it for small food particles. They were so primitive that calling them fish is a bit of an exaggeration, but they certainly would have been seen wiggling through the mud in the shallows of the early Devonian miogeocline.

It is primarily the bone plates from the head region, or fragments of them, that are found as fossils in the Water Canyon Formation. Several different species of ostracoderms have been identified from these remains. *Cephalaspis utahensis* (Figure 4.19B), *Protaspis*, and *Cardipeltis* are among the most characteristic forms of ostracoderms known from the Water Canyon For-

mation. These "fish" are probably the descendants of even more primitive creatures that probably originated in Cambrian time, even though their oldest remains found in the Great Basin are early Devonian in age. In addition to the ostracoderms, early and middle Devonian strata in the Great Basin have also produced the remains of several other types of primitive fish. **Placoderms** are also prehistoric armored fish, but they differ from the ostracoderms in possessing a hinged jaw and an improved fin system that was better designed for active swimming. In addition, many placoderms grew to several feet in length, making them much larger than the ostracoderms. Most placoderms were clearly predators and had sharp-edged bony plates along the jaws that functioned much like the teeth of modern carnivorous fish, though they are structurally different. Several different types of placoderms, including *Asterolepis* (Figure 4.19A), have been found in the Denay Limestone of central Nevada. *Asterolepis* was only about two feet long but was heavily armored with bony plates, especially in the head region. The pectoral "fins" of *Asterolepis* were really just spiky appendages that seem better suited for digging through the sea floor mud than for swimming. The tail fin was formed from a flap of stiffened tissue beneath the up-turned tip of the tail. *Asterolepis* was probably not a strong swimmer, and appears to have wiggled along the bottom of the Devonian seas, searching for edible items such as brachiopods, sponges, or "worms" buried in the mud.

In addition to the armored ostracoderms and placoderms, the tiny spines and scales of small spiny fish known as **acanthodians** (Figure 4.19C) have been found in the Devonian and Silurian rocks in north-central Nevada. Acanthodian fossils occur in the Devonian Windmill Limestone of the Simpson Park Range, as well as in the Late Silurian Roberts Mountains Formation of the Roberts Mountains region. Acanthodians evolved in the Silurian and were the earliest of the jawed fishes. They were usually only a few inches long but had a well-developed system of paired fins. Each fin was supported along the anterior (forward) edge by a robust spine that is easily preserved as a small fossil, even when the remainder of the fish carcass was completely decomposed. The spines of acanthodians may have been used to control the orientation of the fin surface, allowing these small fish with maneuver with agility. Such nimbleness in the water is observed in many fish that inhabit coral reefs of the modern world, and the acanthodians may have occupied a similar ecological niche in the Devonian. In addition to their swimming functions, the spines of the acanthodians may also have deterred large predators from attacking these relatively small fish. The Devonian tropical seas were also home to several types of "lobe-fin" fish, formally known as **sarcopterygians**. The base of the fins in sarcopterygian fish was supported by a fleshy lobe of muscle in which were embedded several bony elements that connected the fin to the body of the fish. The outer edges of the fins were formed from fine rays of bone that emanated from the muscular lobe at the base of the fin. The design of the fins in the sarcopterygians resulted in a quick and powerful stroke that allowed these carnivorous fish to strike with a startling suddenness. *Eusthenopteron* (Figure 4.19D) is one of the best-known Devonian lobe-fins in the world and has been identified in the Great Basin from remains preserved in the late Devonian strata at Red Hill in the Simpson Park Range of north-central Nevada. Eusthenopteron was a large predatory fish with a powerful three-forked tail for propulsion and well-developed system of lateral lobe-fins for maneuverability. This fish belongs to a specialized group of sarcopterygians (known as the **rhipidistians**) that are thought to be the ancestors of later land-living vertebrates. However, it seems that even as *Eusthenopteron* was terrorizing the fish of the Devonian seas in the Great Basin, another type of lobe-fin may have been flirting with land on the edge of the Cordilleran miogeocline.

Lungfish remains recently have been identified from Devonian beds in central Nevada. Technically known as **dipnoans**, lungfish are also members of the sarcopterygian clan of fish, but today are found in rivers, lakes, and lagoons very near land. Lungfish living in the streams of modern Australia and Africa are famous for their ability to withstand drought by excavating burrows in the mud of drying rivers, sealing themselves inside water-tight cavities until rains restore the flow of their watery habitat. The Devonian lungfish of the Great Basin are known primarily from very scrappy, but distinctive, fossil material such as tooth plates and jaw fragments. Thus, we can't be exactly sure about the appearance of the entire animal or how, in comparison to living lungfish, they lived. The sparse lungfish remains do occur with fossils of normal marine invertebrates such as brachiopods, corals, and crinoids, indicating that the Devonian lungfish could clearly tolerate the salinity of the open ocean. Perhaps, they also were able to swim upstream in the rivers that drained into the Great Basin seas from the west. In this regard, it is important to note that the remains of Devonian lungfish in the Great Basin tend to occur in silty and muddy sediments that appear to have been deposited in quiet, shallow water very close to land. It is also possible that the lungfish lived exclusively *in* the rivers, just as modern lungfish do, but that their fossils were preserved when the fragmentary remains were washed out to the adjacent sea floor, where they were buried under near-shore marine sediments.

The numerous fish fossils found in the Silurian and Devonian rocks of the Great Basin suggest that, unlike the marine faunas of earlier times, the tropical seas 360 million years ago were teeming with vertebrates. The coral reefs of the miogeocline must have been home to an amazing variety of small, brightly colored, tropical fish (the acanthodians) that hovered in schools above the sparkling sea floor. The ostracoderms and placoderms probably "swam" along the bottom, plowing through the mud and coral detritus, searching for a tender morsel such as a buried clam or brachiopod. Darting from crevices and pockets in the reef, *Eusthenopteron* and perhaps other sarcopterygian fish flashed though the schools of smaller fish, scattering them in panic. The lungfish were probably less-active swimmers and might have been concentrated in the calm waters of lagoons and pools near the shoreline, or in the mouths of rivers entering the sea. Based on fossil

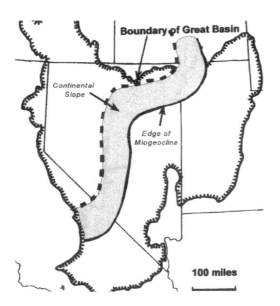

Figure 4.20 Position of the Continental Slope in late Silurian and early Devonian time. Note the prominent eastward bend in the boundary between the miogeocline and the continental slope south of the modern Idaho border.

occurrences elsewhere, it is likely that other types of fish—sharks and cod-like bony fish— also inhabited the tropical seas of the Great Basin, though their remains have yet to be identified. Because of the great abundance and variety of fish fossils found in the Devonian strata, this period of geological time is known worldwide as the great "Age of Fishes." It was certainly that way in the Great Basin! Moreover, all of the post-Devonian marine faunas of the Great Basin include vertebrate animals, and we will encounter the distant descendants of the Silurian-Devonian fish in later chapters.

DEFORMATION ON THE MIOGEOCLINE: GEOLOGICAL STORM CLOUDS

While the dolomitic mud was forming among the reef debris on the inner portion of the serene miogeocline during the Silurian and early Devonian Periods, a somewhat different situation existed farther offshore. Along the outer edge of the continental shelf, where the miogeocline began its westward plunge into deep water, the dolomitization process was less effective. Here, thin layers of muddy limestone, shale, and dolomite mark the break between the continental shelf and the continental slope (see Figure 4.20). These outer shelf and slope deposits are restricted to a belt passing through central Nevada and are typified by such units as the Roberts Mountains Formation. The impure limestones of the outer miogeocline and shelf

commonly exhibit distorted layers that formed when masses of partially hardened lime mud slid down the steep slope and piled up in deeper water. In addition, some of these rocks in the Roberts Mountains Formation clearly show a decrease in grain size from the bottom of a layer toward its top. This type of layering is called **graded bedding**, and it is a good indication of great submarine landslides that carried sediment down the steep continental slope. Rocks that have well-developed graded bedding are commonly called "turbidites," because the primary mechanism for their formation involves **turbidity currents**, the technical name for an under-sea mudslide. In the modern world, turbidity currents can be generated by fierce storms, volcanic eruptions, rapid deposition of sediment, or earthquakes. The presence of turbidites in the Roberts Mountains, along with the crumpled masses of slumped sediment, provides evidence of some geological instability along the outer edge of the miogeocline near the end of Silurian time.

Since sediment on the ocean floor may be destabilized by several nongeological factors, the turbidites and deformed layers in the Roberts Mountains Formation do not necessarily prove that the outer shelf was affected by powerful earthquakes or volcanic eruptions as Devonian time began. However, other simultaneous events can be recognized in the rock record that clearly indicate some significant geological ruckus had commenced in the Cordilleran Geosyncline. For example, using the distinctive character of sediments deposited at the continental shelf/slope break, geologists can easily map the edge of the miogeocline by tracing the distribution of the Roberts Mountains Formation and similar rocks. The outer edge the miogeocline identified in this way passes north-south through central Nevada, makes an abrupt right-angle bend just south of the Nevada-Idaho line, and jogs again along the Nevada-Utah border to extend northward into central Idaho (Figure 4.20). This sharp deflection of the edge of the ancient miogeocline seems to become most prominent in Silurian-Devonian time, and might be the result of faulting along a zone of instability. The faulting that produced the offset of the outer edge of the miogeocline also might have generated powerful earthquakes that shook the sea floor with enough energy to dislodge masses of mud along the shelf/slope break, creating the turbidity currents that surged down the ramp into deeper water.

The evidence for Ordovician–early Devonian geological unrest in the Cordilleran Geosyncline is relatively subtle. This was primarily a time of serenity in the warm, tropical seas that covered the miogeocline. The presence of volcanic rocks in the eugeoclinal deposits father west, along with the inception of faulting along the margin of the miogeocline, suggest that the lithospheric plates were beginning to interact in ways that would generate stresses and volcanic activity. As the end of the Devonian Period approached, these geological rumblings intensified until, across the entire miogeocline, tranquility turned to turmoil. The transition from Devonian to Mississippian time in the Great Basin brought with it geological upheavals so profound that they deserve a chapter of their own. An astounding transformation awaits us as we continue our odyssey through time in the Great Basin. And, as always, the stories are engraved in the rocks.

Scientists call them *Puma concolor*. Most people call them cougars. Others prefer the term puma, mountain lion, or even catamount. However, most of my Great Basin friends—an assortment of cowboys, geologists, artists, teachers, and renegades—simply refer to them as "lions," these large cats that secretively prowl the mountainous regions of the West. I saw my first Great Basin lion on a hot, midsummer day more than thirty years ago and, simultaneously, came face to face with the geological commotion that affected the region at the end of the Devonian Period. Such a juxtaposition, between modern wildlife and ancient upheavals, may seem odd at first, but reflects yet another confluence of nature that can befall a geologist in places like the rocky ridges of the Roberts Mountains in Eureka County, Nevada.

At the time of my dual geological and zoological revelation, I was a young student mapping the rocks in a portion of the Roberts Mountains (Plate 8) as part of a summer field course that was required for geology majors. My two mapping partners and I had just finished lunch under the mottled shade cast by a large and shaggy juniper tree growing along the upper slope of a mountain canyon. It was a great spot for lunch, with a good view across and down the canyon. In the sloping walls of the canyon, tilted rock layers could be traced for considerable distances, allowing us to engage in what we called "power mapping": mapping a large area from a distance without actually walking the rock outcrops. It was (and still is) a frowned-upon practice in field geology, but also one that is always employed, at least to some degree, by every field geologist in the world. As was our habit following lunch, all three of us were in the middle of our siesta, a half-hour

doze that facilitated digestion (our excuse for the practice) and helped us avoid some of the worst midafternoon heat and sun (the real motivation). For some reason, I awoke first on that particular day. As I sat up and reached for my map board, I tried to stay quiet while my partners still dozed with sweat-stained hats positioned over their faces. I silently opened my map case, and began reconnoitering the terrain in a reluctant prelude to the end of our lunch break.

I was gazing across the canyon, not focusing on anything in particular, when I spotted them. Two dusty gray-brown objects were perched motionless on a ledge of ugly red-brown rock exposed on the opposite side of the canyon. At first, I thought they were small tree stumps, until the slumber left my eyes to reveal four ear tips rising in two pairs, one pair from each head. Lions! Two of them, still as statues, and staring right at me, as I was at them. The two were almost identical in size, and a little smaller than I had imagined them to be, based on what I had previously read and seen in zoos and nature films. Wondering if they might be young siblings, I strained to sharpen my eyes on them, trying all the while to remain motionless. Then, remembering how difficult it is to judge size across distances in the Great Basin, I began to wonder if the lions were really as small as they seemed. If they were, in fact, large lions, were they studying us as potential lunch? I could feel just a touch of primal fear well up within me as I sat immobile on the ground returning the feline stares. I wanted to wake my partners to share the moment with them, but I was hesitant to make any motion or sound for fear of scaring the lions off. Perhaps, it was fear of something else that kept me frozen on the canyon slope.

CHAPTER FIVE

Mississippian Mayhem

The Antler Orogeny and other Upheavals

Then, as suddenly as I had noticed them, the two lions wheeled in unison, scampered up the slope to the top of the canyon wall, and disappeared over the ridge with tails flailing as they ran. They were as silent as a feather in their motion, and I still marvel at the way they seemed to float over the rocky soil without touching it. Even before they vanished, I tried to wake my comrades, but neither was able to rise soon enough to see the cats. Of course, they questioned my report of the two beasts, when I explained what I had seen. Was I sure? Could the animals have been deer? Or coyotes? Where, exactly, had I seen them? It was while answering that last question that I noticed the geological significance of my lion sighting. Pointing toward the position of the cats on the opposite side of the canyon, I described the ledge of ugly brown rock about two-thirds of the way from the bottom of the canyon to the top of the ridge. Trying to locate the ledge I was describing, one of my partners looked across the canyon and said, "Oh, you mean right there where the dip changes?" Exactly!

We shouldered our packs, climbed down the slope, across the canyon bottom, and back up the other side to investigate the ledge of rock for evidence of the cougars. The change in orientation of rock layers (or "dip," as geologists de--scribe it), from strongly tilted below the knobby brown mass to nearly horizontal above it, became less apparent as we approached the outcrop than it was from across the canyon. One of my partners reached the ledge first and reported no sign of wildlife, cats or otherwise. He was right; there were no prints left in the soil, no droppings on the ledge, no tufts of cat fur anywhere. As we scanned the ground for lion sign, I, of course, became the brunt of several jokes concerning my expertise in zoology ("Are you sure they weren't pack rats?"), my fondness for beer ("Couldn't wait for happy hour, eh?"), and my suitability for field work in the desert ("How long have you been hallucinating?"). I defended my observation as best I could, but, in the absence of any corroborating evidence, the lion incident became a running joke among the three of us. Eventually, the ribbing subsided to the point that our interest shifted from the alleged

lions to the rock on which I (really!) saw them. Up close, the red-brown ledge was a gnarled mass of shattered chert, laced with tiny fractures filled with calcite, and full of pockets and cavities. The reddish rock was hard but very brittle, and it splintered like glass when it was smacked with a rock hammer. Though the ledge was tilted slightly, like a shelf embedded in the canyon slope, there was no discernable layering within it. We measured the orientation of the well-stratified rock above and below the red ledge and found that there was a change, but it was less than I would have imagined from the opposite side of the canyon. The rocks below the chert ledge dipped to the northeast at an angle of about 25°, while the rocks above the ledge were tilted southeast by about 10°. We knew that such a change in orientation is often a result of low-angle faulting, where one slab of rock layers is forced along a low-angle fracture over a tilted sequence below. Such fractures are commonly called **thrust faults** (Figure 5.1), or just "thrusts" for short, a fitting name that reflects the compressive forces that cause fracturing and rock motion of this type.

Along the actual plane of a thrust fault, a zone of pulverized rock is often produced by the immense pressures that accompany the deformation. Such shattered masses are squeezed between the slabs of moving rock, and later can be cemented by minerals carried in water that migrates along the fault plane. The result is a hard mass of crushed rock, thoroughly cemented, and often replaced with secondary minerals, such as red-brown oxides, calcite, and silica. Rock masses like the one that served as a lion perch in the Roberts Mountains are generally known as **jasperoids**, and they are one of the most obvious manifestations of thrusting. In fact, geologists often use the jasperoid zones to map the extent of thrust faults in central Nevada. Some of the best-known (though, as we will see later, somewhat controversial) thrust faults in the Great Basin were produced during the late Devonian and early Mississippian, when colossal geological forces began to rupture the stillness of the peaceful miogeocline. The Roberts Mountains Thrust is marked by numerous jasperoid zones and has long been considered to be one of the most prominent geological features of the central Great Basin. It, along with other thrusts, represents a major period of geological unrest that geologists have traditionally described as the Antler Orogeny. The Antler Orogeny, as we shall see, was not the only geological squall to occur in the Great Basin near the Devonian-Mississippian transition, but it was, as geologist Bill Fiero has referred to it, a "major rock-busting event." Since the term was introduced more than a half-century ago by Ralph J. Roberts of the U.S. Geological Survey, the Antler Orogeny has become a venerable element of the geological terminology of the Great Basin. For generations, it has been regarded as the cornerstone of the late Paleozoic history of the region and *the* event that set the stage for the remainder of the Paleozoic Era. The rock contortions it produced, first noted by Roberts on the slopes of Antler Peak, south of Battle Mountain, Nevada, have traditionally been regarded as evidence of the earliest major mountain-building episode in the Cordilleran Geosyncline. In recent years, however, both new information concerning rock ages and reinterpretations of the complex relationships between rock masses in central Nevada have led some geologists to question the timing, mechanics, and even the significance of the Antler Orogeny. Moreover, as we will see in this chapter, the Antler Orogeny was certainly not the only geological disturbance to affect the Great Basin region during the last 100 million years of the Paleozoic Era. Nonetheless, the events that occurred about 350 million years ago, near the beginning of the Mississippian Period, heralded a new geological domain utterly different from anything that preceded it. Among these events, the Antler Orogeny is the most notable, as well as the most contentious. It is a perfect beginning for our exploration of the chaos that occurred during the mid-Paleozoic Era.

THE ANTLER OROGENY

Among geologists, there is almost universal agreement that the Antler Orogeny was a pivotal mountain-building event in the geologic history of the Great Basin. Beyond that point, though,

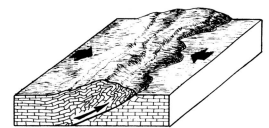

Figure 5.1 A thrust fault forms when compression in the earth's crust forces one slab of rock over another along a low-angle fault plane. Note the deformation in the allochthon, the mass of rocks above the fault, and the linear mountain system that results from thrusting.

there still exists significant disagreement about the precise style of rock deformation that raised the ancient mountains, the fundamental plate tectonic causes of the unrest, and the exact timing of the event. In the mid-1980s, I attended a conference of geologists that gathered to discuss various models for the nature and pattern of mountain building during the Antler Orogeny. As a nonspecialist in structural geology, I was anxious to learn from the delegates which of three or four then-prevalent models for the event was the most plausible. After several days of discussion of the latest research, I left the conference with *eight* different interpretations, each supported by at least some evidence! More than twenty years of subsequent research has not greatly sharpened our ideas on the basic nature of the Antler Orogeny. The Antler Orogeny still generates considerable debate among geologists working to resolve the mid-Paleozoic history of the Great Basin. For our purposes, it is probably best to first review the traditional description of the Antler Orogeny before we examine the uncertainties about what type of deformation occurred during this event, when it happened, and where. Bear in mind, of course, that all three of these aspects of the Antler Orogeny are still being questioned by at least some geologists. Universal scientific agreement (sometimes, even a general consensus) still evades us on several important attributes of the Antler Orogeny. We know from the geological evidence that *something* important happened in the Great Basin in mid-Paleozoic time, we're just not sure exactly what it was.

The overall pattern of deformation that occurred during the Antler Orogeny, at least as it has traditionally been envisioned, is actually surprisingly easy to describe, if we view the event from a "before-and-after" paleogeographic perspective. Prior to the end of Devonian time, about 350 million years ago, the subsiding

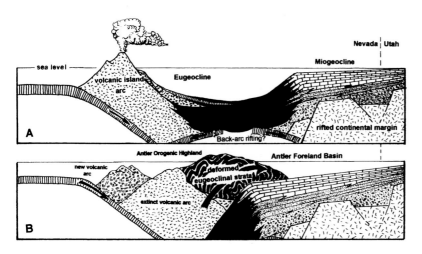

Figure 5.2 The Traditional View of the Antler Orogeny. A, The geologic setting of the Cordilleran Geosyncline prior to the late Devonian–early Mississippian Antler Orogeny. An offshore volcanic island arc existed west of the eugeocline in what is now the northern Sierra Nevada and Klamath Mountains. B, The Antler Orogeny resulted from the thrusting of eugeoclinal rocks over the edge of the miogeocline along thrust faults. The compression that drove this mountain building may have been related to the collision between the volcanic arc and the edge of the continental shelf.

Cordilleran Geosyncline separated the stable portion of North America (called the craton) from an offshore volcanic island arc (see Figure 5.2A). As we have already seen, the intervening oceanic trough was divided into a shallow part (the miogeocline) and a deeper portion (the eugeocline), separated from each other by the steep continental slope. The presence of a volcanic mass to the west of the eugeocline is evinced by such rock units as the Shoo Fly Formation and the Sierra Buttes Formation of the northern Sierra Nevada. These two formations either contain, or consist mostly of, late Devonian volcanic materials, mildly to strongly metamorphosed by later events. The volcanic rocks in these formations appear to have been erupted in an island-arc setting. Presumably, the volcanic activity was the result of **subduction**, a process in which one plate of rock descended to the east (Figure 5.2A), beneath a slab of overriding crust moving in the opposite direction. This ancient subduction zone was probably very similar to the plate tectonic interactions occurring today under the modern Philippine Islands in the southwest Pacific Ocean. Similar, but older, volcanic materials are known from the Sierra Nevada and Klamath Mountains, indicating that the volcanic activity, and the island arc itself, probably originated well before the Devonian Period. In the eugeocline behind the volcanic islands, Devonian deep-water sediments, such as the lower Slaven Chert and correlative units of central Nevada, are very similar to the shale-chert-volcanic "western assemblage" rocks that accumulated in the same general area during

earlier periods of the Paleozoic Era. The volcanic components of these eugeoclinal rocks are especially interesting and have been the focus of much geological study in recent years. The volcanic materials are commonly interbedded with sediments, usually chert and/or shale, and generally metamorphosed to blocky rocks known as **greenstones**. This term reflects the characteristic pale color of these metamorphosed volcanic rocks, produced by greenish-colored minerals, such as chlorite and epidote, that formed during the alteration of normal submarine lava flows. The greenstones are chemically similar, and sometimes texturally comparable, to rocks erupted from modern submarine volcanoes associated with the extension of oceanic crust. If this interpretation of the pre-Antler volcanic rocks of the eugeocline is correct, it suggests that the deep marine basin west of the continental shelf sank as the crust beneath it was stretched and extended. Thus the foundering of the eugeocline may have been the consequence of the oceanic rifting: the spreading of oceanic plates was accompanied by a rapid sinking of the sea floor. Geologists are still uncertain how long such rifting controlled the subsidence of the eugeocline. Great Basin greenstones are associated with eugeoclinal strata as old as the Ordovician. Perhaps the rifting of the sea floor in the eugeocline began as soon as continental rifting was completed in the Cambrian, or it might have become activated sometime during the later Ordovician or Silurian periods. In any event, the Devonian greenstones prove that rift-related volcanic eruptions were evidently fairly common on the deep floor of the eugeocline during this period. During lulls in the submarine volcanic activity, normal deep-sea mud buried the hardened lava to create the chert-shale-greenstone association typical of the lower Slaven Formation. The deep sea floor of the eugeocline rose to the east, up the continental slope and to the outermost edge of the miogeocline, located in central Nevada. Eastward from the shelf edge, the shallow sea floor extended well into Colorado and the adjacent Rocky Mountain region during most of the Devonian Period. This pattern was not drastically different from the general setting that characterized the

Cordilleran Geosyncline during the earlier periods of the Paleozoic Era.

The Antler Orogeny resulted in a radical change in the geological setting of the Cordilleran Geosyncline and, consequently, modified the ancient geographic patterns dramatically, and forever. Beginning near the end of Devonian time, the deep marine rocks of the eugeoclinal suite appear, for some reason, to have been pushed along thrust faults up and over the continental slope, onto the outer edge of the miogeocline (Figure 5.2B). The mixed assemblages of black shale, chert, dirty limestone, and greenstone were shattered and crumpled as they were driven to the east above thrust faults that extended, according to the classical view, from the Mojave region, through central Nevada, into Idaho as far north as the Canadian border (Figure 5.3). For decades, the Roberts Mountains Thrust has been regarded to be the most prominent of these thrust faults, but several others of similar age and origin also developed during the Antler Orogeny. Eugeoclinal rocks ranging in age from Cambrian through early Devonian were transported as much as 100 miles to the east, eventually piling up in a contorted and splintered mass above the relatively undisturbed limestone and dolomite of the outer miogeocline. As the eugeoclinal strata were driven east, they not only were bent and broken but also were "shuffled," as older strata sometimes slid above younger layers, while younger units were sometimes jammed beneath older beds. In addition, "slivers" of miogeoclinal limestone were sometimes ripped up and incorporated into the deformed rock heap. This mass of jumbled and shuffled eugeoclinal rock transported east above thrust faults during the Antler Orogeny has been described by geologists as the Antler **allochthon**, a heterogeneous rock body of bewilderingly complex internal structure. Several different rock formations, such as the Harmony Formation (Cambrian), Vinnini and Valmy Formations (Ordovician), and the Elder Sandstone (Silurian) have been recognized in the allochthon, but the internal complexity of this tangled mass makes it difficult to use such names consistently, or with clarity. Today, after millions of years of erosion, the

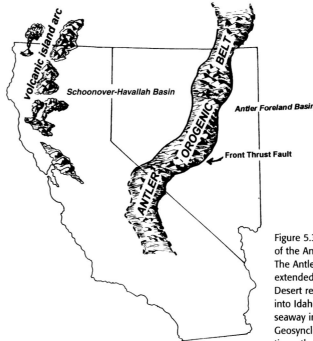

Figure 5.3 Paleogeography of the Antler Orogenic Belt. The Antler highland extended from the Mojave Desert region northward into Idaho, splitting the seaway in the Cordilleran Geosyncline into two portions: the Antler Foreland Basin toward the continent and the Schoonover-Havallah Basin to the west.

remnant allochthon is in places more than 15,000 feet thick. It must have been even thicker in the Paleozoic Era, prior to its partial erosion. As the Antler allochthon developed over the edge of the miogeocline, it steadily built an ever-larger pile of displaced eugeoclinal rock on the shallow sea floor. Eventually this mass of rock rose above sea level and land appeared at the edge of the miogeocline. This land, known as the Antler Orogenic Belt, was a narrow strip of emergent sea floor that slashed across the Great Basin region from the Mojave Desert to Idaho, splitting the Cordilleran Geosyncline into two marine basins, one on either side. The inner, or easternmost, marine basin, between the Antler Orogenic Belt and the shoreline, is known as the **Antler foreland basin.** A comparable basin developed to the west, between the Antler highlands and a new volcanic arc farther offshore. Due to the lack of good exposures and the complications arising from later geologic events in the area, geologists know less about the western basin than they do about the Antler foreland basin. Nonetheless, the western basin has been referred as the Schoonover, Havallah, or **Schoonover-Havallah basin**, for the two best-known rock formations that accumulated in it after the Antler Orogeny.

Despite its impressive north-south extent, the

Antler Orogenic Belt was not a particularly long-lived feature. As the shattered rocks of the Antler allochthon emerged above sea level, the land surface was immediately attacked by the forces of erosion. Rock worn away from the summits of the rising mountains was shed as sediment to the east, into the foreland basin, as well as to the west, into the Schoonover-Havallah basin. After about 25 million years of exposure and erosion, the narrow Antler highlands were worn back down to sea level and submerged again, just prior to the end of the Mississippian Period, about 320 million years ago. Beginning in late Mississippian time, following the complete erosion of the Antler highlands, marine sediment accumulated on the eroded surface of the allochthon, where the Antler Orogenic Belt had formerly stood. The long linear mountain chain, with summits that probably stood thousands of feet above sea level, had vanished almost as suddenly as it had appeared. The boundary between the older, partly eroded, and deformed rocks of the allochthon and the overlying strata that were unaffected by the Antler Orogeny is today recognized as a prominent unconformity, or gap, in the rock succession throughout central Nevada. East and west of the Antler belt, the unconformity is less noticeable, or even nonexistent, because sediment was deposited continuously in areas that remained submerged during the Antler disturbance. The rocks above the unconformity, those that buried the eroded remnant of the Antler allochthon, are commonly referred to as the **Antler overlap sequence**. Mapping the extent of the unconformity beneath the Antler overlap sequence is one way that geologists can estimate the size, location, and trend of the Antler highlands.

What caused the Antler Orogeny? This question still evokes many controversies among geologists. Accordingly, we will reexamine the question later in this chapter, but one of the most popular traditional explanations for the Antler Orogeny involves the collision between North America and the volcanic island arc that existed to the west (Figure 5.2B). The generation of the thrust faults, and the severe folding of the rock layers in the Antler allochthon, both require extreme compressive, or "squeezing," forces. Rock

deformation of the type that occurred during the Antler Orogeny cannot occur under tensional, or "stretching," stress. A convergent plate tectonic event, such as an island arc–continent collision, seems to be a likely mechanism for most of the geological convulsions that occurred during the Antler event. In this traditional scenario, the island arc acted like a gigantic geological vise as it collided with the outer edge of the continental shelf. Eugeoclinal rocks, caught between the arc and the continental shelf, were squeezed to the east as the volcanic mass was driven into the continental slope. With its volcanic fires extinguished, the arc thermally contracted and disappeared into the deep levels of the crust, which explains why no volcanic rocks, other than the aforementioned greenstones, have been observed in the Antler allochthon or the rocks it overrode. However, it appears that a new offshore volcanic arc evolved rapidly to replace the collided one; Mississippian arc-related volcanic rocks and sediments are present again in the northern Sierra Nevada (in the Mississippian Peale and Taylor Formations) and the Klamath Mountains (as part of the Bragdon For-mation). As we will soon see, this traditional explanation suffers from a number of deficiencies, but it has remained prominent because of its simplicity and, until very recently, the lack of any other plausible solution to the riddle of the Antler Orogeny.

The recognition of the Antler Orogeny as the first major mountain-building episode in the geologic history of the Great Basin is based on indisputable evidence that has been compiled by numerous geologists during the past fifty years. What we have described above might be called the traditional model for the events that occurred during this period of upheaval. In fact, not all of the geological evidence supports the scenario of a period of thrusting related to the collision between the edge of the continent and an offshore volcanic island arc. While this concept does explain many of the geological attributes of the Devonian-Mississippian rock record in the Great Basin, detailed studies have revealed some inconsistent, or even contradictory, evidence. There are still several lingering uncertainties that cause many geologists to seek alternative explanations of the Antler Orogeny that are more compatible with all of the geological evidence. This search has produced some novel explanations that are radically different from the traditional view. The most controversial aspects of the Antler Orogeny, as we have thus far discussed it, concern the precise timing of the event, the exact style and causes of the deformation in the allochthon, and the ultimate plate tectonic mechanisms. Admittedly, most of this debate about geological details would only concern, or even interest, professional geologists. But, in our quest to comprehend the deep history of the Great Basin, it is worth our time to briefly review the current controversies simply because they may, in time, lead to a much more accurate perception than is afforded by the orthodox interpretation of the Antler Orogeny. As we will see, some of the uncertainties may also point to a reinterpretation that is utterly, and spectacularly, different from the traditional view. In the process, we will also learn a little about the way geology works, the inherent uncertainties in geological data, and why reconstructing geohistory can sometimes be a laborious effort that requires decades of research.

Puzzles of the Antler Event

Traditionally, the Antler Orogeny has been thought to have occurred in the time of the late Devonian and the early Mississippian Periods. This conclusion was based on the fact that the youngest rocks in the allochthon, considered to be part of the Slaven Chert, were dated by fossil evidence as late Devonian in age. Thus, the eastward thrusting could not have begun any earlier than latest Devonian (about 360 million years ago), because that is the age of the youngest strata carried eastward along thrust faults. However, some of the rocks that accumulated in the foreland basin (specifically, the late Devonian Pilot Formation) east of the Antler belt, appear to consist of reworked eugeoclinal sediment several million years older than the youngest rocks in the allochthon. This poses a thorny dilemma: how could eugeoclinal debris accumulate in the foreland basin *before* the eastward transport of the allochthon began? Either the thrusting

began earlier or the Pilot Shale has been misinterpreted or erroneously dated. In addition, some geologists have suggested that the upper part of the Slaven Chert is as young as earliest Mississippian, which of course would suggest that the Antler Orogeny began sometime *after* the beginning of that period, about 350 million years ago. Many geologists today consider the Antler Orogeny to be strictly a Mississippian phenomenon. The issue of the timing is important because, once it is resolved, it will help establish the duration of the orogeny, a key factor in identifying the exact cause of the Antler event. Some tectonic mechanisms for mountain building require more time than do others.

The minimum age, or the "end," of the Antler Orogeny can be estimated by dating the earliest layers in the overlap sequence, which were deposited after the mountain building was over and the Antler highlands were eroded and submerged. As simple as this sounds, it involves a number of difficulties that make timing the end of the Antler Orogeny even more controversial than determining when it began. For one thing, it is not always clear which rock strata belong to the overlap succession and which do not. Sometimes the unconformity beneath the overlap strata is so subtle as to be all but invisible. Where this is true, there appears to be continuous accumulation of sediment above the allochthon without any major break, or obvious gap. In addition, along the extensive trend of the Antler highlands, some parts stood higher than others. The low-lying segments of the Antler belt might have been barely above sea level. Such areas would experience resubmersion sooner than higher terrain, and "overlap" strata would be deposited relatively soon after the uplift. In this manner, the overlap strata might form in lower areas millions of years before the loftier regions of the Antler belt were eroded enough to be submerged. Furthermore, because the overlap rocks are almost entirely sedimentary (more on this later), their ages can only be established by the use of fossils, rather than by the more precise techniques involving the decay of radioactive elements, which are applicable only to igneous rocks. Fossil dating can be reliable and consistent, but it is also less precise than radiometric techniques and, of course, you have to have the fossils, a requirement that is not universally present in all overlap strata. Finally, many of the rocks comprising the overlap sequence were deformed and faulted during post-Antler disturbances, especially in west-central Nevada. This means that several periods of geological turmoil may have affected parts of the overlap sequence. The effects of such disturbances are cumulative—rock layers faulted and folded during the Antler Orogeny become even more so during later orogenic events. This makes it very difficult to untangle the precise effects of the Antler Orogeny in such multi-mangled rock successions. Small wonder that uncertainty about the "end" of the Antler Orogeny exists!

Geologists have long assumed that the oldest strata in the overlap assemblage of central Nev-ada were represented by the Diamond Peak Formation or the underlying Chainman and White Pine Shale. These strata are of mid-Mississippian age, about 335 million years old. If these rocks really do postdate the ruckus of the Antler Orogeny, then we might estimate that the entire episode encompassed only about 10–15 million years. This is a relatively brief time for the slow interaction of lithospheric plates to construct such an extensive mountain system. However, there is considerable evidence that the geological unrest did not stop after the Dia-mond Peak Formation buried the Antler al-loch-thon: within the overlap sequence, there are numerous folds, faults, and unconformities that clearly postdate the Antler event and were probably developed in the Pennsylvanian, Permian, or even later periods. Perhaps the most famous evidence of post-Antler unrest is the unconformity near the Carlin tunnel on Interstate-80 (Plate 9). There, steeply tilted layers of the Mississippian Diamond Peak Formation are over-lain by much more gently inclined strata of the late Pennsylvanian Strathern Formation. The boundary between the two formations is a classic angular unconformity and has been ogled and photographed by virtually every geologist who has ever traveled on I-80 between Elko and Carlin, Nevada. It clearly indicates that the rocks of the Diamond Peak For-

mation were tilted upward, beveled by erosion, and buried sometime after they were deposited atop the worn-down Antler allochthon. So, what was it that disturbed the Diamond Peak strata? Perhaps it was continued, or renewed, deformation of the Antler Orogeny. Or, it could have been a later, post-Pennsylvanian period of geological unrest unrelated to the Antler event. In either case, can we terminate the Antler Orogeny after mid-Mississippian time when the oldest overlap sediments accumulated on the allochthon? What about those areas, such as the Independence Range of northern Nevada, where geologists have additional and indisputable evidence for deformation and uplift in the middle Pennsyl-vanian Period? Was this deformation part of the Antler Orogeny, or was it a separate episode of unrest million of years later? We are now beginning to glimpse just a hint of the complex controversies regarding the timing of the Antler event.

In addition to uncertainties about the timing and duration of the Antler Orogeny, there is also a considerable difference of opinion among geologists about the precise style of deformation and its ultimate plate tectonic causes. When geologists of the 1950s and 1960s were formulating the earliest notions about the Antler Orogeny, the fact that eugeoclinal rocks—primarily successions of black shale, chert and greenstone—appeared to rest on shallow-water limestone of the miogeocline seemed to be certain indications of eastward thrust faulting. How else could rocks deposited in deep water far to the west become perched atop strata that formed in shallow water 100 miles to the east? In fact, the thrust faults themselves are not at all easy to observe anywhere in the central Great Basin. Their presence was more *inferred,* on the basis of such things as jasperoid zones and changes in orientation of strata, than it was *observed.* (You will note that there are no photographs in this book that clearly show the Roberts Mountains Thrust; in fact, it simply can't be easily seen in any single outcrop, anywhere in the Great Basin!) Furthermore, later deformation in the central Great Basin has tilted and jostled the entire rock succession, intrusion of igneous rocks in the Mesozoic and Cenozoic

Eras has metamorphosed much of the deformed sequence, and erosion has obliterated large portions of the allochthon. Compounding these problems is the simple fact that in many places where the rocks involved in the Antler Orogeny are present, the exposures are just not very good. This is true even in the Antler Peak area, where the concept of the Antler Orogeny first originated. The higher slopes of this portion of Battle Mountain are mostly covered with stunted sage, rock rubble, and soil. The Dewitt Thrust, near Antler Peak, was mapped in the early 1950s and is the first fault ever linked to the Antler Orogeny. In reality, this fault is either concealed or is so poorly exposed for most of its extent that no geologist can actually trace it on the ground for more than a few hundred meters. Most of the other faults attributed to the Antler Orogeny are equally cryptic. If not thrust faults, what else could the boundary between the miogeoclinal strata and the overlying eugeoclinal (or "western assemblage") materials be?

A new idea, one that is attracting a frenzy of discussion among contemporary geologists, is that in many locations the rocks of the allochthon rest on the miogeoclinal succession above a *sedimentary* contact. That is, the deep-water, eugeoclinal strata might have been deposited directly on the shallow-water miogeoclinal succession, rather than having been thrust up and onto the edge of the continental shelf. New and better age determinations have shown that many of the rocks of the Antler allochthon, formerly presumed to be very old, are actually younger than the miogeoclinal strata on which they rest, just as would be expected if the sediment accumulated continuously through time without any thrusting. In many places, the deep-water suite rests on an erosion surface, or disconformity, developed at the top of the shelf limestone. The disconformity that separates the two rock sequences suggests a period of uplift and emergence of the miogeocline prior to the deposition of the eugeoclinal strata. It is this disconformity, the boundary separating overlying eugeoclinal strata from the underlying limestone of the miogeocline, that was originally interpreted as a thrust fault. Many geologists today feel that it makes equally good sense, at

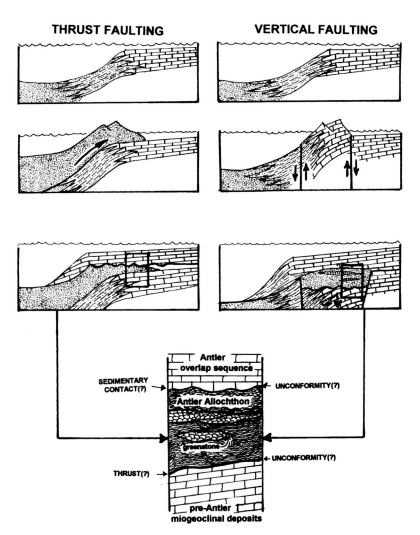

THRUST FAULTING **VERTICAL FAULTING**

Figure 5.4 Tectonics of the Antler Orogeny. The mountain building attributed to the Antler Orogeny may have resulted from thrust faulting (left), the traditional view, or it may reflect vertical faulting (right). The rock succession that records these events (bottom) would appear similar, but the surfaces separating the various components would be interpreted differently. In many Great Basin localities, there is insufficient geological evidence to select one model over the other.

least in some areas of the central Great Basin, to envision the deep-water succession accumulating directly on a foundation of miogeoclinal limestone. In other words, perhaps there was not as much thrusting and contraction of the Cordilleran Geosyncline as geologists initially thought.

If it was not thrust faulting, driven by compressive forces, that initiated the Antler Orogeny, what was the nature of the upheaval during the Mississippian Period? One alternative explanation involves the rapid uplift and subsidence of large blocks of rock along nearly vertical faults (see Figure 5.4). In this scenario, the outer edge of the miogeocline was fractured into numerous large blocks bounded by nearly vertical faults. The fault-bounded blocks were pushed up to create the Antler Orogenic Belt in a series of blocks, each raising layers of miogeoclinal limestone well above sea level. Other

Figure 5.5 Limestone conglomerate in the late Devonian Stansbury Formation of northwest Utah. The large, dark-colored blocks are pre-Devonian limestone and dolomite eroded from the Stansbury Uplift in late Devonian time.

blocks, adjacent to those that rose, appear to have dropped to form localized and restricted basins that bordered the elevated tracts. After several million years of erosion, it appears that the geological forces, at least in some places, were reversed and the uplifted fault blocks dropped rapidly along the same faults that elevated them. The subsiding fault blocks descended in deep water, where the "eugeoclinal" deposits accumulated on their eroded upper surfaces. This alternative view of the Antler Orogeny obviates the need for thrust faults, and it seems to offer better explanations for some of the new dates and structural relationships observed in the Antler allochthon. In addition, there were other portions of the miogeocline that were evidently experiencing similar vertical movements at about the same time that the Antler Orogeny was beginning. The best known of these lesser uplifts of late Devonian age is the Stansbury Uplift west of modern Salt Lake City, Utah. In this area, a small part of the miogeocline was lifted above sea level to shed a very localized mass of conglomerate as part of the Stansbury Formation (Figure 5.5). Unlike the rest of the eastern Great Basin, no Ordovician or Silurian strata are present in this area, and the gap in the rock record signifies the late Devonian erosion that followed the uplift of the Stansbury block above sea level. Currently, there is no consensus among geologists on which of the interpretations of the Antler Orogeny—the traditional thrust belt scenario or the new fault-block model—is most plausible. Smaller and contemporaneous uplifts, such as the Stansbury Uplift, seem to clearly involve vertical movement of the crustal blocks, fostering additional questions about the basic cause of the Antler Orogeny.

Perhaps both mechanisms—deep vertical faulting and lateral thrusting—were at work in different places at different times near the Devonian-Mississippian boundary.

The controversy about the Antler Orogeny may appear to be an overblown argument about geological minutia; but, it has some very important implications for other aspects of Devonian and Mississippian history in the Great Basin. Thrust faults generally develop when compressive forces associated with convergent motion of tectonic plates intensify to the point of fracturing the rocks involved. The deep vertical faults that lead to uplift and subsidence, evoked in the alternative view of the Antler Orogeny, would require a much different plate tectonic setting. Was the disturbance caused by the collision of an island arc, as has long been suspected? If the thrust faults are not thrusts, or at least not as pervasive as once thought, then the concept of an island arc collision becomes much less tenable. So, depending on which model of the Antler disturbance is correct, our interpretations of its fundamental causes will vary dramatically. It is simply impossible to generate an extensive thrust belt with the same geologic forces that are required to induce block faulting and vertical displacement. So, the debate over the Antler Orogeny is an important one because its resolution will strongly affect our perceptions of the Paleozoic history of the Great Basin. But, there's even more reason to ponder the Devonian-Mississippian events in the region: against the background of causal uncertainty about the Antler Orogeny comes a new possibility, one that suggests a greater calamity than any strictly geological scenario could produce.

THE ALAMO MEGABRECCIA: WINDOW TO A CATACLYSM?

The Guilmette Formation is a one of the thickest and most widespread late Devonian rock units in the eastern Great Basin. In most areas of eastern Nevada and western Utah it is composed of 2,000 feet or more of limestone, muddy siltstone, and carbonate rubble. Nearly all of the sediments within the Guilmette Formation accumulated in shallow water, but detailed studies of the strata have revealed a cyclicity that suggests numerous (150, or so) fluctuations of sea level over the 8–10 million years of time recorded by these deposits. In south-central Nevada, a very distinctive rock appears within this thick succession of miogeoclinal deposits. Near the small town of Alamo in the Pahranagat Valley, the Guilmette Formation contains a single bed, more than 300 feet thick, of jumbled blocks and fragments of limestone, heaped in disarray on each other and solidly cemented into a mass of petrified chaos. Named the Alamo Megabreccia, this part of the Guilmette Formation is one of the most distinctive geological horizons in the entire rock succession of the Great Basin. The Alamo Megabreccia is not present in the northern part of the Great Basin, but it can be traced over more than 1,500 square miles in the south-central portion of the province. The considerable lateral extent of the megabreccia, coupled with its average thickness, suggests that it contains approximately 60 cubic miles of rock, all of it in a single layer, evidently deposited during one event! Geologists have traditionally viewed this rubble in the middle of the Guilmette Formation as the result of submarine landslides triggered by earthquakes, or, perhaps, the debris generated by fierce hurricane-strength storms that ravaged the miogeocline. In 1995, however, geologists John Warme and Charles Sandberg discovered some very unusual, and very revealing, characteristics of this unique deposit.

The Alamo Megabreccia seems, on the basis of thickness trends, to have originated at a point source located near Hancock Summit, in western Nye County, Nevada (Figure 5.7). At that location, the Alamo Megabreccia attains its greatest thickness of nearly 400 feet but appears to taper from this point in all directions, at least wherever late Devonian strata can be studied. In addition, the term "megabreccia" is more than fitting. Normal **breccia**, a sedimentary rock consisting of material deposited by landslides or powerful floods, consists mostly of angular rock particles from about 1/10 of an inch (about 2 millimeters) to about 2 inches (50 millimeters) in size. In contrast, some of the blocks in the Alamo Megabreccia are gigantic: more than

Figure 5.6 Outcrops of the Guilmette Formation in the West Pahranagat Range, Nevada.

Figure 5.7 Limestone breccia in the Guilmette Formation near Hancock Summit in south-central Nevada. The dark inclusions are small fragments of reef-deposited limestone. In places, such fragments can be gigantic blocks, hundreds of feet long.

1,000 feet long, in the most extreme cases. Even where such enormous blocks are absent, the Alamo Megabreccia still contains millions of rock fragments the size of boulders or cobbles. It is the prevalence of limestone fragments as large as cars and houses that prompts the addition of the prefix "mega-" to the name of the strange rubble in the Guilmette Formation. In the larger blocks of limestone, the original layering is still discernable and is commonly tilted

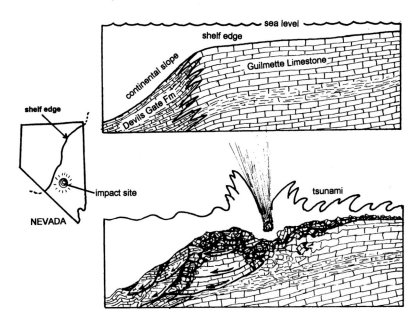

Figure 5.8 The probable origin of the Alamo Megabreccia. The megabreccia may represent rubble created during a violent impact between the outer continental shelf (top) and a meteorite or asteroid. The impact seems to have occurred in south-central Nevada, where it destabilized a large portion of the miogeocline and generated a monstrous tsunami (bottom).

and rotated with respect to the overall orientation of the bed of megabreccia. Such large and disheveled fragments suggest a forceful event, one that was much more powerful than a normal hurricane or sea-floor earthquake. And, that event appears to have been confined to a small area: the point of maximum thickness in the Hancock Summit area.

Warme and Sandberg also discovered even more intriguing aspects of the Alamo Megabreccia. Some of the small grains of sand between the large limestone blocks consist of quartz that shows several zones of crystal damage on the microscopic scale. These shadowy zones in the translucent quartz are known as "shock lamellae," and are very similar to features produced in quartz grains at sites of nuclear explosions or meteorite impacts. The presence of such grains between the larger blocks of limestone in the Alamo Megabreccia suggests an origin involving an explosive event that generated powerful shock waves. Finally, samples from near the top of the Alamo Megabreccia were found to contain two or three times the normal background amount of the rare element iridium, which is virtually nonexistent on the surface of the earth but is much more concentrated in asteroids and meteorites from space. Though the amount of iridium in the Alamo unit is still very low (no greater than 139 parts per trillion), the values do represent a significant increase in the usual concentration of this exotic element.

Was the Alamo Megabreccia produced by the impact of an extraterrestrial body into the shallow sea covering the miogeocline? The evidence outlined above is very suggestive of such an event, but by no means can it be regarded as certain proof of it. Nonetheless, the impact of an asteroid or large meteorite could, in fact, have produced the bizarre kind of rubble observed in the Alamo sequence. It is very difficult to imagine how any other mechanism, such as a hurricane or an earthquake, could have broken and transported such gigantic blocks of limestone and produced the shocked quartz and iridium enrichment, all at the same time. Under the scenario of an asteroid impact (Figure 5.8), the extraterrestrial object initially slammed into the shallow sea near the edge of the miogeocline. The resulting explosion would have shattered the edge of the shelf into numerous blocks of limestone that were propelled, or slid, in all directions. The explosion, and the resulting displacement of mass on the sea floor, would have generated huge waves, or **tsunamis**, that helped to transport the refuse away from the point of impact. The crater blown into the miogeocline was apparently resubmerged after the impact and became filled with debris that settled into it. This, coupled with the effects of later uplift and erosion in the area, may explain why no trace of an actual late Devonian crater exists today at the impact site. It is the very unusual nature of the Alamo Megabreccia that records the impact, and at least some geologists now view it as the residue from an incredibly violent and catastrophic event.

More research on late Devonian strata in the Great Basin is still needed to confirm the impact origin of the Alamo Megabreccia to the satisfaction of all scientists. But, let's assume, for the moment, that the impact did occur, and ponder the ramifications of this event. The estimated time of the impact, based on fossils from rocks above and below the Alamo deposits, is about 367 million years ago. This is close enough to the end of the Devonian Period to possibly be related to the Antler Orogeny and the Stansbury Uplift. If, as many modern geologists have suggested, these disturbances involved mostly vertical movement of blocks of rock along high-angle faults, then could the impact event that blasted the Alamo Megabreccia have initiated the movement? Perhaps the stresses associated with the asteroid detonation reactivated older fractures that originated during the time of continental rifting. If so, blocks at the outer edge of the miogeocline might have lurched up to create the Antler and Stansbury highlands before subsiding again beneath the waves of the Cordilleran seas. Such vertical jostling of rock is compatible with one of the newest theoretical concepts for the Antler disturbance, the "collapsed borderland" model. This scenario, still being evaluated by geologists, involves the piecemeal fracturing, uplift, and collapse of large blocks of rock along the continental margin. The new model is based on the borderland of California, which has experienced a complex pattern of unrest similar to the Antler disturbances during the past 20 or 30 million years. The recent events in California have given rise to the varied terrain along the modern Pacific coast, and a similar sequence could have produced the complex pattern that seems to characterize the much older Antler disturbance in the Great Basin. If the collapsed borderland model is correct, it may well have been a violent event from space that triggered the unrest. It is, of course, still possible that the original interpretation of the surface beneath the Antler allochthon as a thrust fault is accurate, at least in some areas. In that case, something, such as the collision of an island arc or "microcontinent," must have occurred to generate the immense compression needed to activate the thrusts. Conceivably, both events may have occurred near the end of the Devonian Period: two collisions, one from space, one from the sea, could have crushed and shattered the miogeocline to lift the sea floor upward. In this case, we should redefine the Antler Orogeny as a period of widespread uplift that occurred in different places at different times for different reasons. The resulting geologic patterns of such a series of events, after millions of years of erosion and overprinted with features produced during later upheavals, would be extremely complicated and difficult to decipher. That complexity seems to be one of the few attributes of the Antler Orogeny about which there is universal agreement among geologists. Despite the clear indications of mid-Paleozoic mountain building in the rock record of the Great Basin, there are still many unanswered questions about its fundamental nature.

THE ANTLER FORELAND BASIN

Regardless of how the Antler Orogenic Belt was elevated, its emergence drastically changed the pattern of sediment deposition in the Cordilleran Geosyncline. As the rocks exposed in the highlands were eroded, sediment was shed to the east and west of the north-south-trending strip of land. However, we know very little about the patterns of sediment accumulation in the sea west of the Antler belt in western Nevada and eastern California. This is because the sediment that was deposited in that area was severely deformed, metamorphosed, and/or buried by later geologic events. To the east, however, a thick succession of strata was deposited in a broad oceanic basin referred to earlier as the Antler foreland basin. These strata are widely distributed in eastern Nevada and western Utah and have, for the most part, escaped the geologic convolutions that affected contemporaneous layers to the west. So, to continue the story of the Cordilleran Geosyncline, we shift our attention to places such as the Stansbury Range and Lakeside Mountains of Utah, the Diamond and Monitor Ranges of Nevada, and many other localities in between where Mississippian rocks are thick and well exposed.

According to the conventional view, the Antler foreland basin developed soon after eastward thrusting placed eugeoclinal rocks atop the outer edge of the miogeocline. The mass of the Antler allochthon caused the underlying crust to sink into the deeper layers of the subsurface earth, and subsidence of the sea floor accelerated as the foundation beneath it foundered. This behavior, where vertical motion of the crust is induced by a change in the mass that it supports, is an excellent example of what geologists call **isostasy**, a term which describes the balance between gravity, elevation, and mass. The principle of isostasy is very simple to

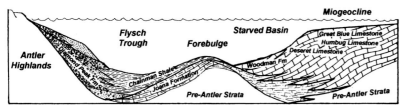

Figure 5.9 The Antler Foreland Basin. During Mississippian time, a great volume of sediment was eroded off the Antler highland and transported east, where it accumulated in the flysch trough and starved basin. Limestone, commonly containing the fossils of reef-building organisms, continued to form in the shallow marine environment of the miogeocline far to the east.

understand by imagining a canoe floating in a lake. When mass is added to the canoe, as when someone steps into it, the canoe responds by sinking a little deeper into the water that supports it. Though the rock of the upper mantle that supports the overlying crust is not liquid, it does support the crust because of its relatively high density. So, when weight is added, to either the canoe or the earth's crust, both respond by sinking. This is probably what caused the Antler foreland basin to develop in the Mississippian Period. When the Roberts Mountains (and other) thrusts transported the rocks of the Antler allochthon to the east, an enormous mass piled up in the "crustal canoe," causing subsidence into the underlying mantle. Of course, if the controversy regarding the reality and significance of thrusting during the Antler Orogeny is resolved in the favor of deep, vertical faults, then some other explanation for the foreland basin will have to be formulated. If most of the Antler "thrusts" are, in reality, just sedimentary contacts, then isostatic adjustments could not have resulted in the foreland basin. In that case, perhaps it was a resumption of extension in the region that caused the subsidence. There is, in any event, no doubt that a deep marine trough developed in the eastern Great Basin following the Antler Orogeny. Proof lies in the stack of rock layers, more than 10,000 feet thick, that accumulated in it!

A heterogenous assemblage of Mississippian

Figure 5.10 Outcrop of pebbly sandstone in the Mississippian Diamond Peak Formation at the south end of the Diamond Range, central Nevada.

and younger deposits accumulated in the Antler foreland basin. The basin was noticeably asymmetric, with the deepest part located closest to the Antler highland, where (presumably) tectonic loading and isostatic subsidence was greatest (Figure 5.9). This part of the basin is sometimes referred to as the Antler **foredeep** or the Antler **flysh** trough. Into this deep basin, pebbles, cobbles, and sand were flushed from the nearby highlands by vigorous streams. Where the rivers met the sea, muddy deltas were constructed from finer grains, such as silt and sand. Farther offshore, in the murky depths of the foredeep, microscopic silt and clay grains settled to the sea floor to accumulate as muddy ooze. These sediments are now hardened into the conglomerate, sandstone, and shale of such well-studied rock units as the Pilot Shale, the Diamond Peak Formation (Figure 5.10), and the Chainman Shale, among others. The term flysh, sometimes used to describe this part of the foreland basin, is actually an old geological term for the finer-grained sediments that are shed from an area experiencing active uplift. While the term is appropriate for parts of the Chainman and Pilot Shale, most geologists today prefer the term foredeep for the western part of the foreland basin, because coarse rubble (now conglomerate and sandstone) also accumulated there. Erosion of the Antler highland must have been vigorous during the Mississippian, because the aggregate thickness of foredeep strata indicates that thousands of feet of sediment was washed into the basin during the relatively brief period of a few million years.

From the foothills of the Antler Orogenic Belt in central Nevada, the foredeep extended eastward into what is now western Utah. There evidently was a small submarine rise, known as the **forebulge** (see Figure 5.9), that separated the foredeep from a less-pronounced sag on the sea floor. This pocket is known as the **starved basin**, because it received less sediment than any other portion of the Antler foreland trough during the Mississippian Period. Scientists have estimated that sediment accumulated in the starved basin at a rate of only about 30 feet per million years, in contrast to the rapid dumping of material into the foredeep. The middle basin was "starved" even in comparison to the shallow platform to the east, where limestone was deposited at rates varying from 100 to more than 300 feet per million years. The starved portion of the Antler foreland basin was deprived of sediment for two reasons. First, the forebulge formed an effective submarine barrier to mud and silt originating from the Antler highlands. In addition, large reef-like masses of limestone to the east, no less than 350 miles wide, allowed little sediment to reach the starved basin from land. Consequently, in northwestern Utah, where the starved basin was located, the middle portions of the Mississippian rock sequence are relatively thin, constituting what geologist call a "condensed" section—a thin succession of strata that encompasses a relatively long span of time. The best examples of such condensed sections occur in the Deseret Limestone and Woodman Formations. In both of these formations, starved-basin deposits are represented by black banded chert, phosphatic shale, and calcareous mudstone and siltstone (Figure 5.11)

The easternmost component of the Antler foreland basin was a broad platform, representing the portion of the Cordilleran miogeocline that was most distant from, and least affected by, the Antler Orogeny. On this surface, shallow-water conditions prevailed throughout the Mississippian Period, though the water might have been a little deeper than in the Devonian Period. Located astride the ancient equator, the seas that covered the miogeocline were clear, warm, and tropical in character. As the miogeocline slowly subsided with the rest of the foreland basin, great quantities of limestone sediment accumulated on the platform, commonly as large reef complexes constructed by primitive corals, sponges, and other organisms. Thick-bedded and highly fossiliferous limestone sequences, such as those that comprise most of the Great Blue Limestone and parts of the Humbug Formation of northern Utah, contrast strongly with the deeper-water sediments that formed elsewhere in the foreland basin. The buildup of sediment at the outer edge of the platform resulted in the westward growth of the miogeocline over the starved basin as Mississippian time progressed. Eventu-

Figure 5.11 Tilted layers of Mississippian starved-basin deposits in the Deseret Limestone at Flux, Utah, in the northern Stansbury Mountains.

ally, limestone and other types of shallow marine sediment were deposited over most of the basinal sediments in the entire foreland region. To the east, the shallow miogeocline rose gently, like an enormous tilted ramp, toward the shore in central Colorado, more than 600 miles from the Antler highlands.

Toward the end of the Mississippian Period, the Antler foreland basin became nearly filled with sediment. The distinctions between the foredeep, forebulge, starved basin, and miogeocline gradually disappeared as the deep areas were filled up with sediment. A few million years before the beginning of the Pennsylvanian Period, shallow-water conditions existed across the entire foreland basin, and the Antler highland was nothing more than low rolling hills, partly buried by the "overlap" deposits discussed earlier in this chapter. The final phase of filling the Antler foreland basin involved the deposition of fine-grained clay, silt, and limey mud across the entire trough from the remnant highland in the west to the shoreline to the east. This "cap" of dark shale and silty limestone is represented by the Manning Canyon Shale of Utah and the lower part of the Ely Limestone of eastern Nevada. These sediments accumulated in extremely shallow lagoons and tidal basins that mark the end the Antler foreland basin as a deep, sediment-collecting trough. Both the Manning Canyon Shale and the Ely Limestone span the transition from Mississippian time to the Pennsylvanian Period, when new geologic

Figure 5.12 Plant fossils from the Mississippian Diamond Peak Formation. Except for the horsetail stem at the bottom, these fossils represent sections of the roots, branches, and bark of lycopods, or "scale trees."

Figure 5.13 A coastal forest of the Mississippian. The low land bordering the seas in the Antler foreland basin was covered with lush tropical forests dominated by primitive plants such as lycopods, horsetails, and ferns. Several types of amphibians and fish populated the shallow inlets and lagoons near land, some of the former wiggling ashore as the first vertebrates to ever live on land in the Great Basin region

events in the Great Basin would transform the older patterns brought about by the Antler Orogeny. But, before we review those new patterns, there is still one more fascinating aspect of the Antler foreland basin to explore: its life.

LIFE IN THE ANTLER FORELAND BASIN

The Mississippian deposits of western Utah and eastern Nevada are among the most fossiliferous rocks of the entire Great Basin province. Moreover, because the various kinds of sediments accumulated under widely varying ecological conditions, the Mississippian fossil assemblages are anything but uniform. Within the Antler foreland basin, the foredeep, starved basin, shallow seas, and muddy lagoons were each populated by their own unique array of organisms that were adapted to those specific habitats. For these

reasons, collecting fossils from the Mississippian strata of the Great Basin is an endless adventure that always has the potential to produce something new or unexpected.

Recall that the Antler Orogeny was the first true mountain-building episode in the Cordilleran Geosyncline, and resulted in the creation of a large swath of land in the middle of a shallow sea. No surprisingly, it is in the Mississippian rocks deposited on the periphery of that land that geologists have found the oldest fossils of terrestrial, or land-living, organisms. Perhaps the most intriguing of these terrestrial fossils are the plant remains that occur in the Diamond Peak Formation and the younger Manning Canyon Shale. Collectively, fossils from these formations (Figure 5.12, Figure 5.13) document a rich tropical flora dominated by primitive plants such as ferns, mosses, horsetails, and **lycopods** (also known as lycopsids, or "scale trees"). The lycopods were the largest and most distinctive trees in the Mississippian forests that cloaked the coastal margins of the foreland basin. These trees were up to 100 feet tall and had scaly leaves arranged in spiral or vertical rows around their large trunks. The fossils of lycopod trees are easily identified by the characteristic leaf scars (Photo 5.12) that wind around the trunk. Many different species of lycopod trees have been identified from the plant fossils of the Manning Canyon Shale and the Diamond Peak Formation, indicating a fairly rich flora of tree-sized plants. Such trees were evidently widespread and persistent in the Great Basin region; fossils representing the scale trees have been found in Mississippian and Pennsylvanian rocks from Death Valley to northern Utah. The smaller vegetation appears to have been equally varied. Primitive relatives of the modern horsetails (*Equisetum*) are one of the most abundant groups of smaller plant fossils found in the Mississippian rocks of the Great Basin. Generally known as the genus *Calamites*, these plant fossils are distinguished by their segmented stems and leaf bundles that radiate from the branches at regular intervals. Some of these ancient horsetails grew much larger than their modern counterpart, attaining maximum heights in excess of 15 feet. Other members of the horsetail group (the Sphenopsids) were of small shrub size. The ferns of the Mississippian were very much like modern ferns, except that they were far more abundant then than are current ferns in the modern flora. The floor of the lycopod-*Calamites* forests in the Great Basin region appears, at least in places, to have been carpeted by a dense cover of low-growing ferns. Fern fossils, sometimes with spectacular preservation of the delicate fronds, are so abundant in certain horizons of the Manning Canyon Shale that several localities in western Utah have become world famous for the beautiful fossils they produce. Some of the Mississippian ferns appear to have grown to a very large size, at least in comparison to their living descendants.

The forested shores of the Antler foreland basin also supported a variety of terrestrial animals. Though the fossils of such creatures are not nearly as abundant as are those of plants in the Mississippian strata of the Great Basin, scientists have found enough of them to demonstrate that there was considerable animal life swarming in and around the woodsy lagoons. Fossils of insects similar to modern dragonflies have been found in the Manning Canyon Shale, suggesting that such bugs buzzed through the foliage in the swamps where the muddy sediments accumulated. In addition, the remains of a small salamanderlike amphibian known as *Utaherpeton franklini* were discovered in 1991 in the upper part of the Manning Canyon Shale. This creature was only about 2.5 inches long, with tiny peglike teeth lining its small jaws. With such miniature teeth and small body size, it does not appear that *Utaherpeton* could have been a very effective herbivore, especially in consideration of the size and nature of the foliage of lycopods and horsetails. So, what was this diminutive amphibian eating? Perhaps it preyed upon other small amphibians, or on small ground-dwelling insects, or the eggs and carcasses of either. The fossils of insects and amphibians in the Manning Canyon Shale provide just a glimpse of the terrestrial life of the Mississippian-Pennsylvanian interval in the Great Basin. We still have much to learn about the animals that populated the lowland forests on land adjacent to the Antler foreland basin.

Many more fossils of land-living creatures from the Manning Canyon Shale remain to be discovered, and those fossils will someday shed additional light on the earliest terrestrial ecosystem in the Great Basin.

Offshore from land, normal marine conditions prevailed in the central and eastern parts of the Antler foreland basin. Here, the deposition of mainly limestone sediments continued throughout the Mississippian and Pennsylvanian Periods. The conditions on the shallow sea floor must have been optimal for a variety of marine invertebrate organisms, because rocks of Mississippian age are among the most fossiliferous of any in the entire Great Basin region. Fossils of marine organisms in these rocks are so abundant that they can be found almost anywhere that Mississippian strata are exposed. Most of the fossils belong to reef-building invertebrate creatures, such as corals and bryozoans, or represent the remains of other animals that were well adapted to the reef environment (Figure 5.14). In fact, coral reefs in the eastern part of the Antler foreland basin grew outward so rapidly that, by the end of Mississippian time, thick sequences of limestone "reef rock" buried much of the shale and siltstone that originally had accumulated in the starved basin to the west. In addition to the invertebrate creatures, the coral reefs of the foreland basin were also inhabited by many different types of fish. Most

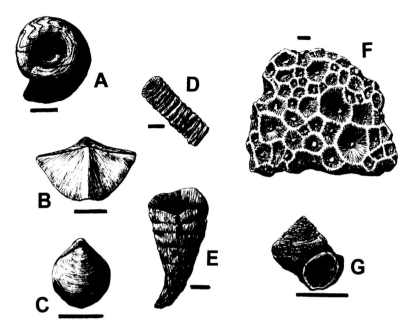

Figure 5.14 Marine Invertebrate Fossils from the Mississippian Strata of the Great Basin.
A, *Cravenoceras*, a cephalopod related to the modern nautilus;
B, a spiriferid brachiopod;
C, *Composita*, another brachiopod;
D, a section of the "stem" of a crinoid, consisting of circular plates stacked in a series;
E, *Caninia*, a solitary coral;
F, a compound coral;
G, *Glabrocingulum*, a gastropod, or "snail."
Scale bar equals 1 cm (0.4 inches) in all sketches.

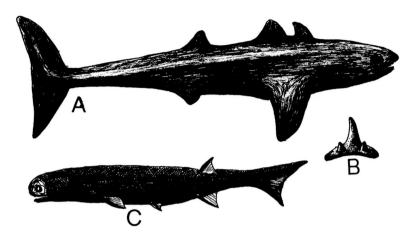

of the sluggish, heavily armored fish of the Devonian had become extinct by late Mississippian time and were replaced by such proficient swimmers as the early members of the shark family. Shark teeth have been found in several formations that were deposited in the Antler foreland basin during the Mississippian and Pennsylvanian Periods, including the Great Blue Limestone and the Manning Canyon Formation of western Utah. From the fossil evidence, it appears that several different kinds of sharks prowled the reefs in the foreland basin, most of them possessing a three-cusped tooth (Figure 5.15B) that is primitive compared to modern shark teeth but still well designed for capturing swimming prey animals. Aside from the teeth, no other fossils of Mississippian and Pennsylvanian sharks have been found in the Great Basin. The largest teeth, a little less than an inch high, suggest a rather small shark, perhaps 2 or 3 feet long. Based on specimens of Mississippian sharks found elsewhere in North America, the Great Basin forms of this age were probably very similar to modern sharks in general appearance, except for their relatively large eyes and primitive tail and fin system (Figure 5.15A). In addition to the shark teeth, well-preserved fossils of a small, primitive fish known as *Utahacanthus guntheri* also have been found in the Manning Canyon Shale of western Utah (Figure 5.15C). These small fish were only about 3 inches long, had tiny diamond-shaped scales covering most of their body, and stout spines on the leading edge of their many fins. *Utahacanthus* belongs to a primitive group of fish known

as the acanthodians that arose in the Silurian and became extinct in the Permian. In late Mississippian, when *Utahacanthus* darted through the shallow water of the Antler foreland sea, the acanthodians were already very rare and well on the path leading to their ultimate extinction. *Utahacanthus* would have been considered a "living fossil" in the Mississippian, a time when the more advanced fishes, such as sharks and the bony fish, were evolving rapidly.

THE PENNSYLVANIAN PERIOD: A TIME OF TRANSITION

The Mississippian Period ended, and the Pennsylvanian Period began, about 320 million years ago. Nothing in the Great Basin seems to have changed dramatically after the beginning of Pennsylvanian time, but the influence of the Antler Orogeny on the geologic setting of western North America was definitely fading. The sediment comprising the rocks of such units as the Manning Canyon Shale and the Ely Limestone continued to accumulate in the foreland basin well into the Pennsylvanian Period. As the basin between the Antler highlands and the coastal reefs filled with sediment, the distinctions between the flysch trough and starved basin disappeared. Eventually, by mid-Pennsylvanian time, shallow marine deposits were forming across the entire eastern Great Basin. The Antler highlands itself, in central Nevada, was deeply eroded at this time, and began to disappear beneath the overlap sequences in many places. From a prominent linear mountain system during the Mississippian, the Antler belt evolved into series of lowlands and islands separated by narrow straits, coves, and bays in the early Pennsylvanian (Figure 5.16). Far to the west, somewhere beyond the California-Nevada border, an active volcanic island arc existed, presumably linked to a subduction zone that descended either to the west or east. By the end of the Paleozoic Era, at the close of the Permian Period around 245 million years ago, this geological setting was completely transformed, as we will learn in Chapter 6. The Pennsylvanian and early Permian Periods were thus times of transition from the patterns produced by the

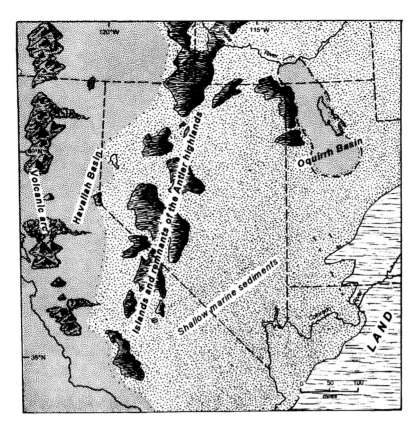

Figure 5.16 Pennsylvanian Paleogeography of the Great Basin Region. Remnants of the Antler highlands, along with small areas uplifted by new forces, created numerous islands separating the shallow shelf to the east from the deep-water Havallah basin to the west. Another deep basin, the Oquirrh basin, developed in western Utah where the sea floor sank faster than mud accumulated on it.

Figure 5.17 (below) Mississippian through Permian strata of the central Great Basin. The Pennsylvanian unconformity, a prominent erosion surface, was produced during a period of uplift that followed the Antler Orogeny and preceded other geological disturbances that occurred near the end of the Paleozoic Era.

Antler event to an utterly different arrangement of geologic features. And what an intriguing transition it was!

No sooner had the rumblings of the Antler Orogeny diminished than new forces began to affect the central Great Basin region. In many places in northern Nevada, a widespread unconformity, developed during a period of Pennsylvanian uplift and erosion, separates strata of this age from rocks above that are millions of years younger (Figure 5.17). This unconformity was produced when portions of the early Pennsylvanian sea floor were lifted above sea level and exposed to the elements of erosion. After a few million years of exposure, a prominent erosion surface was developed as many of the exposed rock layers were stripped away. Eventually, the uplifted rock sequence was worn down and buried under younger sediments. The uneven Pennsylvanian unconformity marks the erosion surface formed during the time of exposure. Sandstone and conglomerate layers in such Great Basin rock units as the Carbon Ridge and Strathern Formations, along with the somewhat younger Arcturus Formation, provide additional evidence of Pennsylvanian uplift and erosion. As

we have already learned, there is considerable debate among geologists over when, precisely, the Antler Orogeny ended. The renewed deformation in the Pennsylvanian may be interpreted as a continuing phase of the Antler Orogeny,

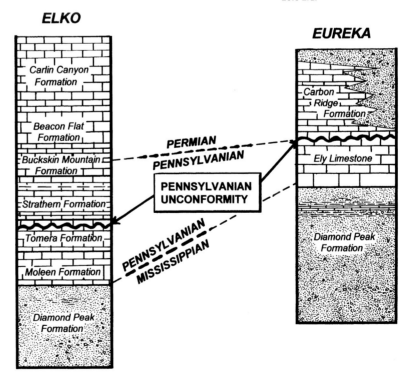

ELKO

Carlin Canyon Formation

Beacon Flat Formation

Buckskin Mountain Formation

Strathern Formation

Tomera Formation

Moleen Formation

Diamond Peak Formation

EUREKA

Carbon Ridge Formation

Ely Limestone

Diamond Peak Formation

PERMIAN
PENNSYLVANIAN

PENNSYLVANIAN UNCONFORMITY

PENNSYLVANIAN
MISSISSIPPIAN

Figure 5.18 Cliff-forming limestone of the Permian Pequop Formation, northeast Nevada.

following a brief respite of tranquility in early Pennsylvanian time. Alternately, the Pennsylvanian activity may signify a new mountain-building episode, following closely on the heels of the Antler thrusting and folding. Supporting the latter view is the appreciable geological evidence that the Pennsylvanian unrest in central Nevada involved mostly vertical motion along high-angle faults, rather than the low-angle thrusting and the more horizontal forces that character-ized at least some parts of the Antler Orogeny. Such distinctive differences in the manner of uplift suggests a unique cause for the Pennsylvanian mountain building that affected areas in the central Great Basin. Such considerations have led some geologists to refer to the Pennsylvanian unrest as the "Humboldt Orogeny," to distinguish it from the Antler disturbance; but the former term is still not widely used in the Great Basin.

Pennsylvanian mountain building appears to be confined to specific, relatively small regions in the central Great Basin. Combined with remnants of the Antler highlands, the uplift of the Pennsylvanian produced a complicated terrain of small islands and shoals rising from the sea along a generally north-south belt in central Nevada (see Figure 5.16). In other places, there is excellent evidence for an opposite sort of geologic unrest: the sinking of deep basins. The Oquirrh basin, in northwest Utah, is by far the most dramatic example of a Pennsylvanian basin in the Great Basin. This region, which ex-

tended to the northwest well into modern Idaho, sank so rapidly during the Pennsylvanian Period that more than 16,000 feet of muddy sediment accumulated in it during that time. Though the Oquirrh basin began to subside in the Mississippian Period, and continued to sink well into the Permian Period, its main phase of activity was during the intervening Pennsylvanian. The rock succession that accumulated in this basin, known as the Oquirrh Formation, is more than three miles thick and consists primarily of muddy limestone and impure sandstone. Elsewhere in the eastern Great Basin, where the sea floor was more stable, Pennsylvanian strata are usually less than 3,000 feet thick in the aggregate. The extreme thickness of the Oquirrh Formation reflects the relatively rapid sinking of the basin and its continuous filling with sea-floor mud. Elsewhere on the shallow platform of the eastern Great Basin, the sea floor sank very slowly and limestone sediments accumulated beneath the clear, warm, and shallow seas throughout the Pennsylvanian and into the succeeding Permian Period. Such shallow marine sediments are typified by the Pequop Formation of northeast Nevada that commonly forms thick walls of massive limestone in the mountainous terrain of the northern Great Basin (Figure 5.18). Similar shallow-water deposits accumulated on the stable platform as far south as the Mojave-Death Valley region, where they are represented by such rock units as the Bird Spring Formation and Keeler Canyon Formation.

In the western Great Basin, beyond the elevated tract of land in central Nevada, another basin was foundering at the same time that the Oquirrh basin was developing. This basin is known as the Havallah basin, and it was situated between the emergent land in central Nevada and an offshore chain of oceanic volcanoes (a volcanic island arc) that was located an uncertain distance to the west. The precise configuration and size of the Havallah basin are unknown because the rocks that formed in it were subsequently crumpled, torn apart, and thrust to the east during post-Pennsylvanian geological disturbances. Nonetheless, the nature of sediments that accumulated in the basin during

Pennsylvanian time clearly indicates deep oceanic conditions. Such materials as gritty, impure limestone, layered chert, coarse-grained turbidites and conglomerates, undersea lava flows, and very fine grained shale and argillite occur in the Pumpernickel and Havallah Formations of western Nevada. To varying degrees, these materials have been metamorphosed, transported along thrust faults, and deformed since they accumulated in the Havallah basin. Consequently, they are very difficult to interpret, and some uncertainty still exists concerning their nature and significance. However, the rocks of these two formations clearly signify the existence of a deep oceanic trough with, or close to, active submarine volcanoes somewhere in the western Nevada area during the Pennsylvanian Period. The exact size, depth, configuration, and extent of the Havallah basin remain obscure, but it undoubtedly existed. Like the contemporary Oquirrh basin, the Havallah basin also appears to have originated in the Mississippian Period and persisted until Permian time.

BUILDING PANGAEA: THE VIEW FROM THE GREAT BASIN

What could have caused the continuation of mountain building after the Antler Orogeny, produced a shift in the styles of rock deformation, and, at the same time, resulted in the sinking of the Havallah and Oquirrh basins? Such multifarious geological events seem to signify an extremely complex tangle of interacting causes. We must account for both the compressional and tensional forces that affected various areas in the Great Basin at different times during the Pennsylvanian-Permian transition. It is obvious that in some places squeezing forces activated thrust faults that lifted jumbled masses of rocks into mountains. But, at other times, in other places, rocks were stretched beyond the breaking point, separating into blocks that subsided rapidly to become the foundations of the deep basins. Certainly, any relatively simple, singular, and straightforward explanation for such geologic complexity would be highly improbable. Or would it? With a few problematic exceptions,

most of the post-Antler Paleozoic events in the Great Basin can be linked to a single fundamental cause, an ongoing plate tectonic episode that was reaching its climax in the late Paleozoic Era: the assembly of a supercontinent.

For almost a century, geologists have used the word **Pangaea** ("all earth") to describe a supercontinent that existed at the end of the Paleozoic Era. Evidence for such a unified land mass was described in the early 1900s, verified in subsequent years, and now is considered beyond dispute. In the past twenty years, techniques have been developed to reconstruct the approximate sizes, configurations, and locations of the modern continents through time. In this manner, we are now able to map the assembly of Pangaea from the time the first small fragments collided in the Cambrian Period to the time the last continent was sutured to the mass in the latest part of the Permian Period. Though the growth of Pangaea was gradual and progressive throughout most of the Paleozoic Era, there was a major event in the Pennsylvanian that is probably linked to the geological unrest that occurred at that time in the Great Basin. It was in Pennsylvanian time that a large southern supercontinent, known as **Gondwana**, collided with the southern edge of North America. Gondwanaland consisted of the main portions of modern South America, Africa, Australia, and Antarctica, all of which had been joined together by earlier Paleozoic collisions. North America was, at the time of its collision with Gondwanaland, joined to an eastern land mass that today makes up the main part of northern Europe. Collectively, the North American–northern European land mass was called **Laurasia**. The collision between Gondwanaland and Laurasia occurred along a broad belt that extended from the southern Appalachian region westward across what is now the Gulf Coast, and into west Texas (Figure 5.19). As these two great masses converged upon each other, rocks along the leading edge of both continents were crushed, faulted, and thrust upward to form a long chain of mountains. So impressive were these mountains that even today, some 300 million years after they were initially lifted, remnants of them still stand in the Ouachita

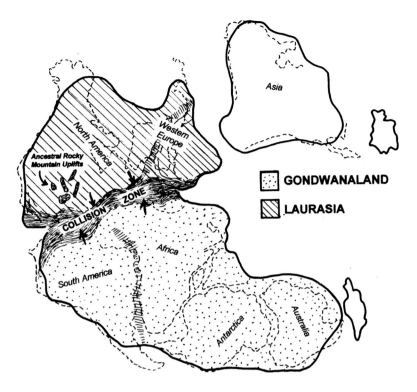

GONDWANALAND

LAURASIA

Figure 5.19 The Assembly of Pangaea. Throughout the late Paleozoic Era, the supercontinent grew continuously through the addition of smaller land masses. During the Pennsylvanian Period, Gondwanaland and Laurasia were joined along a collision zone that extended from the southern Appalachian region to the southwest margin of North America. This collision triggered the uplift of the Ancestral Rocky Mountains and may have played a role in the contemporaneous disturbances in the Great Basin. For reference, the approximate outlines of the modern continents are indicated by the dashed lines.

Mountains of Oklahoma and Arkansas and the Marathon region of west Texas. In Laurasia, this great collision also produced compressional forces that were transmitted as far inland as the Rocky Mountain region, where several mountainous regions developed. These mountains are known as the Ancestral Rocky Mountains, because they rose in the same general area where the modern Rockies now stand. However, the modern Rocky Mountains were elevated much later in time (during the early Cenozoic Era, 40 to 60 million years ago) and resulted from plate tectonic interactions that were utterly different from those that produced their Pennsylvanian ancestors. Nonetheless, there is a connection between the ranges in the two mountain systems. Some of the uplifts of the modern Rockies, such as the Front Range of Colorado and the Uncompahgre Plateau of eastern Utah, rose along Pennsylvanian fractures that were reactivated some 200 million years after they originally developed. Though the Ancestral Rocky Mountain uplifts developed far east of the Great Basin region—primarily in Colorado, Wyoming, Utah, and New Mexico—there does appear to be a connection between the late Paleozoic events in the two regions.

The Pennsylvanian disturbances in the Great

Basin appear to have commenced at the same time the Ancestral Rocky Mountains began to ascend. Though it is difficult to link the collision of Laurasia and Gondwanaland to geological disturbances so far inland, it seems to be more than a coincidence that the Great Basin unrest and the Ancestral Rockies upheavals were so nearly contemporaneous. Perhaps the compressional forces that resulted from the great collision were transmitted into the Great Basin, where older zones of weakness deep in the crust, possibly related to the late Proterozoic rifting, were reactivated. There are some hints that several of the Ancestral Rocky Mountain uplifts developed above such older fracture zones that were mobilized during the Laurasia-Gondwanaland collision. Geologists are less certain about the link between old fracture systems and the Pennsylvanian disturbances in the Great Basin, but such a connection is at least plausible, if not likely. Another similarity between the Pennsylvanian disturbances in the Great Basin and the Ancestral Rockies features is that deep basins seem to develop very close to many of the elevated blocks in both areas. The Paradox Basin in southeast Utah, for example, sank steadily through Pennsylvanian time along the flanks of the rising Uncompahgre Uplift. In a similar manner, the Oquirrh and Havallah basins in the Great Basin appear to have subsided at the same time that uplift was in progress in the nearby Antler belt. The subsiding basins in both areas probably resulted from stretching, or tensional, forces that might have arisen from some sort of "rebound" effect, or perhaps a sliding motion, that occurred as the continents collided. These similarities fortify the concept of a connection between the uplift of the Ancestral Rocky Mountains, the collision of Laurasia with Gondwanaland, and the geological turmoil in the Great Basin during Pennsylvanian time. But the rocks involved in these events in the Great Basin have been further deformed and transported by later orogenic events, making it very difficult to sort out the possible linkages between these coeval phenomena. The precise nature of the nexus between the Ancestral Rocky Mountains and the Pennsylvanian havoc in the Great Basin, and the explanation for it, remains unclear.

Whatever may have caused it, the Pennsylvanian deformation in the Great Basin occurred against the backdrop of significant geographic and environmental changes that were developing on a global scale. When the Permian Period began, about 290 million years ago, Pangaea was completely assembled, oceanic circulation was modified by the enormous supercontinent, and the worldwide climate began to shift toward warmer, somewhat drier, and more seasonal conditions. Shallow seas withered, while great dune fields, some even larger than those of the modern Sahara Desert, advanced in many areas. Throngs of plants and animals were unable to keep pace with these rapid changes, especially in the shallow, tropical oceans. Many groups of marine invertebrate organisms began a rapid decline in diversity and abundance during the Permian Period. Some of them vanished completely, as the greatest extinction event in the history of the earth began to unfold. It was during these times of dramatic global changes, as the Paleozoic Era gave way to the Mesozoic, that a major change occurred in the geologic architecture of the Great Basin. For more than 60 million years, the Antler Orogeny and related disturbances had dominated the patterns of land and life in the Great Basin. But, at the end of the Paleozoic Era, the old blueprint was obliterated during a major episode of geological remodeling known as the Sonoma Orogeny, an event so notable that it deserves a chapter of its own.

As intriguing as the geological stories are in the Great Basin, the human history of the region is equally alluring. Countless legends of colorful characters and their heroic acts, tragic events, monumental accomplishments, and colossal failures exist in every corner of the Great Basin. Many people, myself included, love to wander through this desert in a haze of historical lore, vicariously reliving the stories each time we pass near the site of some notable deed. That the exploits of John Frémont, Kit Carson, Jedediah Smith, Joseph Walker, Bulldog Kate, Hot-eyed Mary, Death Valley Scotty, Claude Dallas, Howard Hughes, and others are sometimes hyperbolically exaggerated makes no difference. Often, the legends have little to do with the true qualities of the real people, or for that matter, the actual substance of their deeds. The most fascinating aspects of the fables are, in many cases, the way they have become magnified through repetition to the point that no one can be sure exactly what happened in the first place. For some reason, the gray desolation of the Great Basin has an amazing ability to transform the mundane into the monumental, leaving plenty of room for imaginative reconstructions of historical reality. No other region, anywhere on the planet, can boast of more exaggerated legends per square mile!

Among the legions of characters immortalized in desert lore, one of my favorites, for no other reason than the adventurous life he led, much of it in the wilds of the Great Basin, is Jack Watson. Unlike many other legendary desert folk, Watson did not achieve any great feats of survival or travel in the Great Basin. Neither did he, as far as I can determine, hang any outlaws, rob any banks, massacre any immigrants, swindle any stockholders in bogus min-

ing ventures, ascend any high peaks, or write any poetry about which contemporary scholars could ruminate. In fact, it was not until 1997, when Michael Kelsey compiled Jack's story (see chapter references) from his diaries and notes, that Watson became known to more than just the few people who wandered into the desert wilderness of western Utah. In other published accounts of Great Basin history, Jack Watson is still largely ignored. Jack lived a portion of his life in the vast sagebrush wilderness surrounding the Confusion and Wah Wah Mountains of western Utah, though at times he also lived in other parts of the state. In the stark sagebrush wilderness, he constructed a cabin and operated a lonely sheep "ranch" at a remote place called Ibex in the early 1900s. The area was so isolated in those days that no accurate and detailed maps yet existed, and Watson lived, at least for a time, in the midst of this solitary expanse. Older maps of the southern Confusion Mountains area show such names as "Jack Valley" and the "Ibex Hills," though these terms have now vanished from the topographical maps produced by the U.S. Geological Survey. I often have wondered what life was like for Jack, chasing his free-ranging stock through a land that was virtually unblemished by human contact. Whenever I travel the dusty dirt roads around the old Ibex area, I think about Jack and the beauty and the solitude of the area he inhabited. Though I probably would not have fared well in Jack's world, I still envy him. He was lucky to have such an exquisite and unspoiled place all to himself, a privilege that I, or anyone else now living, will never experience. Jack, of course, who had to confront the desolation and austerity of the region every day, may not have seen things quite that way.

CHAPTER SIX

THE SONOMA EVENT AND OTHER MESOZOIC CONVUL- SIONS

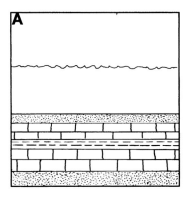

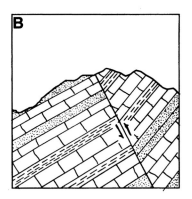

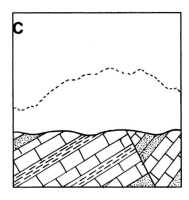

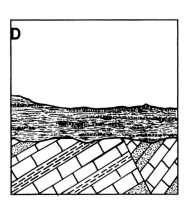

Figure 6.1 Development of an Angular Unconformity. A, deposition of horizontal layers of sedimentary rock; B, uplift and deformation of the initial layers; C, erosion of the deformed layers, creating a beveled surface; D, deposition of younger horizontal layers on the erosion surface carved on the tilted sequence. The angular unconformity represents a gap in the rock record equivalent to the time required for uplift and erosion of the initial strata. The angular unconformity in the southern Confusion Mountains represents a gap of approximately 450 million years, including the entire Mesozoic Era.

According to the prevalent theory, the Confusion Mountains were so named because their irregular topography, lacking a clearly defined north-south axis, was a source of constant confusion for anyone traveling between the House Range and the Snake Range, both of which are more clearly defined as north-south trending mountain systems. The "Ibex Hills," where the southern Confusion Mountains converge with the southern tip of the House Range, and where Jack Watson lived, appears on modern maps as the Barn Hills. The change in name no doubt reflects the realization that no real ibexes live the Great Basin, or ever did. The inspiration for the original name appears to be the common, though inaccurate, term used by early settlers for the pronghorns that still roam the region today. Even now, we still misidentify the pronghorns as "antelope"; in fact, they are the sole species of their own unique family (*Antilocapridae*). In any event, near the site of his cabin at "Ibex," up a small draw that is nameless on most modern maps, Watson was reputed to have constructed a cistern from rocks to catch and retain the runoff of desert downpours. Once, while exploring the region, I tried to locate the old

cistern, planning to camp there for a night to indulge my zeal for imaginary time traveling. I wanted to be Jack Watson for a night, and what better place to do it than his old domicile. After looping around the south end of the Barn Hills, past a crag of rock known as Warm Point, I veered north on a dirt road leading toward the old site of Ibex, the area where Watson's cabin once stood. A covey of chukars scrambled up the road and exploded into the air, veering off to the right of the truck as I raced along. Turning to watch them, I happened to glance to the west, across the sagebrush plain, toward the eastern escarpment of the Confusion Mountains. There, on the hillside and in all its glory, was one of the grandest **angular unconformities** in all the Great Basin. The lower strata exposed on the mountain face were tilted gently to the north, while the ridge was capped by dark-colored volcanic rocks arranged in sheets and layers that were nearly horizontal. The angular disjunction between the upper and the lower sets of strata slashed boldly across the escarpment. I never made it to Watson's cistern that night, but instead camped in the middle of the valley in order to examine the unconformity in the more direct light of morning. That night, while thinking about Jack Watson, I wondered if he ever happened to notice the angular unconformity. Even if Jack Watson knew no geology, this grand structure in the southern Confusion Mountains would have been hard to miss. If Jack did know rocks, it would have been a source of endless captivation for him. Since I first stumbled on it, I have shown the angular unconformity to many other geologists who, without exception, find it to be a thrilling sight. Why should geologists, and perhaps even pioneer ranchers, be so impressed with such a rock exposure? This magnificent structure is the icon of the Mesozoic Era in the eastern Great Basin; from it we can begin to glimpse a pivotal stage in the geologic history of the entire province. It offers a perfect starting point to begin an exploration of a fascinating slice of time in the Great Basin.

All angular unconformities develop in a similar four-step process (Figure 6.1) that explains the great significance attributed to them by geologists. The first step in the process is the deposition of the lower set of layers, one at a time, in a horizontal orientation. In the case of the Confusion Mountains unconformity, this sequence consists of limestone and shale strata of the Pogonip Group that accumulated some 485 million years ago on the shallow sea floor of the Cordilleran Geosyncline. After the Pogonip sediments accumulated and hardened, they were uplifted and tilted during some sort of geological upheaval. From the unconformity alone, it is impossible to determine the exact nature of this uplift, or its precise timing, but the inclination of the lower strata clearly reflects an event that occurred after they originated as horizontal layers. The third event in the sequence involves the erosion of the elevated and upturned edges of the Pogonip strata to create the beveled surface on which the horizontal layers rest. Without this step, the relatively flat and even boundary between the two sets of strata would not exist. Erosion is, of course, a destructive process, so no tangible record of this long period of rock removal exists—only the beveled surface separating the tilted Pogonip sediments and the overlying strata mark the passage of millions of years. Finally, the angular unconformity is completed when the surface of erosion is buried under horizontal layers of rock. The horizontal series in the Confusion Mountains is comprised of volcanic ash and lava flows of the Needles Range Group (we will learn more about these rocks later) that were erupted about 35 million years ago. Once these volcanic materials formed over the nearly flat erosion surface, the angular unconformity was complete.

Thus, a gap of approximately 450 million years exists between the tilted layers of the Ordovician Pogonip Group and the overlying Needles Range Group. This is the time required for the uplift of Ordovician sediments, the erosion of the tilted layers back down to a level surface, and the eruption of the volcanic materials that buried that surface. Angular unconformities of this magnitude provide evidence of a major period of geological unrest that created a persistent elevated terrain that yielded slowly to forces of erosion. It is an intriguing irony, I suppose, that angular unconformities record the existence

Figure 6.2 The angular unconformity in the southern Confusion Mountains. The dark lava flows in the Needles Range Group rest horizontally above the tilted layers of the Ordovician Pogonip Group. The boundary between the two sets of rock layers represents a gap of more than 450 million years.

and evolution of an entire mountain system through what is, in reality, nothing at all—just a surface separating rock sequences of differing orientations! Nonetheless, the evidence of major mountain building in such geologic structures is undeniable. Clearly, a substantial mountain system developed in the eastern Great Basin between the Paleozoic and Cenozoic Eras. The intervening era is, of course, the Mesozoic (248–65 million years ago), the great "Age of Reptiles," during which time dinosaurs and other reptiles dominated the global ecosystem. While we have seen evidence of other prehistoric upheavals in the Great Basin—the Proterozoic rifting, the Antler Orogeny, the Pennsylvanian unrest—none of these rivals the Mesozoic events that produced the Confusion Mountains unconformity. In fact, the disturbance that lifted the Pogonip strata in that area also permanently transformed the Great Basin region from a marine basin to rolling highland, in the process adding a considerable tract of land to the western edge of North America. But, local angular unconformities tell us nothing of the nature of the uplift, what its causes may have been, or exactly when it occurred. To explore this pivotal event further, we must look to the northwestern portion of the Great Basin, where geologists first glimpsed the fundamental event that triggered the Mesozoic chaos that ultimately affected the entire region.

THE SONOMA OROGENY

It has been more than forty years since geologists Norman Silberling and Ralph Roberts first noticed that the Pennsylvanian and Permian marine rocks in northwestern Nevada were folded and thrust faulted before they were buried under volcanic strata (the Koipato Group) of

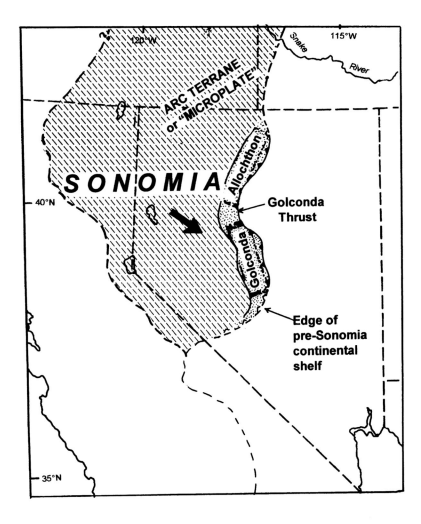

Figure 6.3 Sonomia and the Golconda Allochthon. The Sonoma Orogeny is related to the collision between a microcontinent known as Sonomia and the western edge of North America. The Golconda allochthon was driven over the edge of the continental shelf during this early Mesozoic collision.

Triassic age. The deformed rocks were part of the deep-water Havallah sequence (discussed in the preceding chapter), and consisted primarily of chert, silty shale and limestone of deep-water origin. By mapping the stack of deformed layers in the Tobin, Sonoma, Shoshone, and Humboldt Ranges, geologists subsequently demonstrated that the crumpled mass was positioned above shallow-water deposits of similar late Paleozoic age. In other words, the deep-water deposits had been driven eastward about 50 km (30 miles) out of the Havallah basin, up the continental slope, and over relatively undisturbed limestone and shale at the edge of the shallow continental shelf. If this sounds reminiscent of the Antler Orogeny, it is because the late Paleozoic event is a virtual replay of that disturbance some 100 million years later and slightly to the west.

Because the effects of this geological unrest are most noticeable in the Sonoma Range south of Winnemucca, Nevada, the period of moun-

tain building that it induced is known as the Sonoma Orogeny. Like the earlier Antler Orogeny, the Sonoma upheaval was generated by strong compression that crunched and folded rock layers and eventually generated several low-angle thrust faults. The most prominent thrust fault produced during the Sonoma Orogeny is the Golconda thrust, named for the Golconda Summit area of Edna Mountain, about thirty miles east of Winnemucca, Nevada. Above this nearly horizontal thrust, the mass of twisted strata that was transported eastward above it is known as the **Golconda allochthon**. The rocks of the Golconda allochthon comprise a heterogeneous mixture of argillite, chert, impure limestone, and volcanic sediments ranging in age from late Devonian to Permian. These materials were severely folded and faulted as they were transported east and thrust upward to create a north-south-trending mountain system just west of where the Antler highlands rose millions of years earlier (Figure 6.3). The youngest rocks under the Golconda thrust are of latest Permian age, so we know that the thrusting must have occurred after that time. It is much more difficult to determine when the deformation ended because, as we shall see, the Sonoma Orogeny was merely the first in a long series of geological disturbances of the Mesozoic Era. Thus the end of the Sonoma Orogeny is difficult to ascertain, because the Golconda allochthon was further crunched during later periods of deformation. However, because the transported rocks are buried under the late Triassic volcanic materials of the Koipato Group, it appears that the motion on the Golconda thrust system ceased, for all practical purposes, by the time of the Koipato eruptions. Therefore, geologists generally use the term Sonoma Orogeny to describe the mountain building in the central Great Basin that took place during the interval from late Permian (about 250 million years ago) to about mid-Triassic time (225 million years ago).

The Sonoma Orogeny, though it represents just the beginning of the geological chaos of the Mesozoic Era, is especially important because it signifies the early stages of a great geological collision (Figure 6.4). The volcanic materials in the Golconda allochthon include lava flows and granular rubble that was most likely derived from a volcanic island arc to the west (Figure 6.5). This volcanic chain was evidently developed above a subduction zone that dipped down to the west, under the approaching volcanic arc. Submarine landslides, possibly triggered by violent volcanic blasts, carried rubble off the slopes of the islands into the deep basin that separated the volcanic arc from North America, where it became interwoven with layers of deep-water sediment. When the volcanic arc collided with the continent, this volcanic detritus, along with the volcanoes themselves, was crushed against the western margin of the shelf and driven upward and eastward over the Golconda thrust. But it was not just the volcanic arc that appears to have collided with North America. To the west of the volcanoes, extending all the way to the Klamath Mountains region of northern California, was more rock. This block, mostly submerged beneath the sea, was an extensive slab of more arc-related materials of Permian or pre-Permian age. Rocks belonging to the mass have been identified in the Klamath Mountains and the northern Sierra Nevada, where they were deformed and/or penetrated by bodies of magma that migrated through them during later periods of the Mesozoic Era. In spite of the later complications, such Permian rocks as the McCloud Limestone in the Klamath Mountains or the Goodhue and Reeve Formations in the northern Sierra Nevada still provide evidence that a large mass of rock existed west of, and was contiguous to, the volcanic arc to the east. This mass has been referred to as **Sonomia**, which geologists have commonly described as a microcontinent or a "microplate" that moved from the west toward North America as a coherent slab of rock. Oceanic rocks of the North American Plate west of the old Antler belt were overrun by Sonomia as it approached, and this subduction generated the magma that erupted as the volcanic arc on the leading edge of the microcontinent (Figure 6.4A). Small, volcanically active microcontinents are especially abundant in the modern world in the southwest Pacific Ocean region; Japan, the Philippines, Java, and Sumatra might be modern analogs to ancient Sonomia.

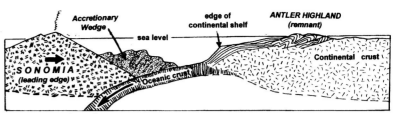

A: Mid-Late Permian

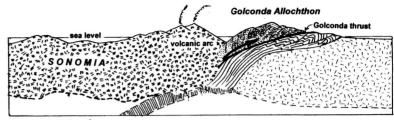

B: Early Triassic

Figure 6.4 The collision of Sonomia with the western edge of North America.

Sonomia eventually slammed into the western edge of North America, triggering the Sonoma Orogeny (Figure 6.4B) in late Permian–Triassic time. This great collision added as much as 400 kilometers of rock to the western margin of North America. After the collision of Sonomia, the western edge of the continent was located in the north-central California region, well beyond the modern Great Basin province. Thus, the Sonoma Orogeny of the Great Basin, and the Golconda allochthon that was deformed during it, are both local consequences of a larger event that dramatically transformed the western edge of North America. Such exotic masses of rock as Sonomia that become fused to the edge of a larger continental block through plate tectonic interactions are known in modern parlance as **accreted terranes**. Ever since the Sonoma Orogeny, the western margin of North America has grown steadily through the addition of numerous terranes, most of which were accreted during the Mesozoic Era. Such accreted terranes are

Figure 6.5 Outcrop of blocky volcanic rubble in the Golconda allochthon near Golconda Summit, northern Nevada. The notebook is about seven inches long.

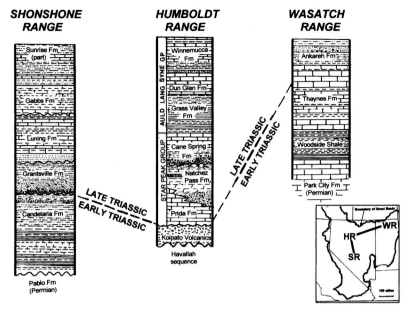

SHONSHONE RANGE

- Sunrise Fm (part)
- Gabbs Fm
- Luning Fm
- Grantsville Fm
- Candelaria Fm
- Pablo Fm (Permian)

LATE TRIASSIC
EARLY TRIASSIC

HUMBOLDT RANGE

AULD LANG SYNE GP
- Winnemucca Fm
- Dun Glen Fm
- Grass Valley Fm

STAR PEAK GROUP
- Cane Spring Fm
- Natchez Pass Fm
- Prida Fm
- Koipato Volcanics
Havallah sequence

WASATCH RANGE

- Ankareh Fm
- Thaynes Fm
- Woodside Shale
- Park City Fm (Permian)

LATE TRIASSIC
EARLY TRIASSIC

Figure 6.6 Triassic rocks of the Great Basin.

particularly prominent in Alaska and coastal British Columbia, both of which consist almost entirely of microcontinents that were jammed together at various times during the Mesozoic Era. Each time a fragment was accreted, a pulse of mountain building resulted. The net effect of this era of accretion along the western edge of North America was to create new land, and to lift the older sea floor, exposing it to the agents of erosion. While the accretion of Sonomia did not cause the angular unconformity in the Confusion Range, it was the first of several Mesozoic events that contributed to the elevation of the land in the Great Basin and the eventual destruction of the Cordilleran Geosyncline.

After the collision of Sonomia, a new subduction zone evolved along the reconstructed western edge of North America. The oceanic plate that delivered Sonomia to North America continued to move eastward but broke away from the accreted microcontinent and descended beneath it toward the east. Hence, an additional effect of the Sonoma Orogeny was to relocate the pre-collision subduction zone to the west, and to reverse the direction of subduction from westward (under the approaching arc) to eastward (beneath the accreted microcontinent). As we will review shortly, this change in the geological setting of western North America had profound consequences for the post-Triassic history of the Great Basin. But there was a more immediate response to the westward shift in the

position of the subduction zone: the volcanic activity in the region near the Golconda allochthon diminished as the zone of underground magma production moved into what is today central California. As the volcanic fires were extinguished, the accreted arc cooled, contracted, and eventually began to subside. In mid-Triassic time, the sinking land in western Nevada was submerged again, and marine sediments began to accumulate over the foundered surface of Sonomia. Eventually Triassic marine strata blanketed most of the northern Great Basin. To the west, however, a new volcanic arc was emerging from the sea and, by the end of the Triassic Period (about 208 million years ago), new forces were already elevating the sea floor again to permanently expel the oceans from the Great Basin. Before the seas vanished completely, though, during the relatively brief period of Triassic inundation, some amazing creatures evolved in this doomed ocean. Their remains can still be found in the high slopes of the Shoshone Mountains east of Gabbs, Nevada, scattered through the sediments that accumulated on the floor of the disappearing Cordilleran Geosyncline.

TRIASSIC MARINE ROCKS OF THE GREAT BASIN: MUD AND BONES

Triassic marine rocks are not widely distributed throughout the Great Basin, but good outcrops do exist in the Humboldt, Sonoma, and Trinity Ranges of northwest Nevada, the central Wasatch Mountains of Utah, and in the Shoshone and Tobin Ranges of central Nevada (see Figure 6.6). Marine Triassic strata in the central and western Great Basin, such as the Star Peak (Figure 6.7) and Auld Lang Syne Groups, comprise a complex sequence of limestone, shale, and sandstone layers shuffled with intervening strata of conglomerate and volcanic detritus. These rocks generally rest on the underlying Permian rocks, or on early Triassic volcanic rocks and their detritus (the Koipato Group and the upper Candelaria Formation), that marked the beginning of the Sonoma Orogeny. Collectively, the late Triassic rocks record the accumulation of sediment in several shallow marine basins isolated from each other by islands or volcanic highlands representing portions of Sonomia that rose above sea level. While these marine deposits accumulated in the Great Basin, far to the east, in what is now west Texas and New Mexico, a large river system flowed westward from a rugged highland. Small streams descended from this watershed to form larger rivers by the time they reached the Colorado Plateau region. There, near the eastern boundary of the modern Great Basin, a complex river system developed on a broad coastal plain. Throughout Triassic time, the rivers meandered back and forth across the coastal lowland, spreading out sand and mud that today comprise the colorful rocks of the Moenkopi and Chinle Formations of the Painted Desert region. Similar red rocks occur as far west as the Valley of Fire in southern Nevada, indicating that the river system must have extended into that area. To the northwest, the river continued across the coastal plain until it reached the shore of the shallow seas that covered the Great Basin region. As sediment accumulated along the edge of the ancient sea, the shoreline migrated westward from central Utah to a position in northern Nevada by the end of the Triassic Period (Figure 6.8). Thus, in the eastern Great Basin, the early Triassic strata consist of marine mud (e.g., the Thaynes Formation; Figure 6.6) that became buried under river-deposited sediment (the Ankareh Formation; Figure 6.6) in the late Triassic. In the central and western Great Basin, oceanic conditions generally persisted until the end of the Triassic Period, or even into the succeeding Jurassic. Some of the strata in the Auld Lang Syne Group of the Humboldt Range (Figure 6.6) consist of deltaic sediments that contain fragments of fossil wood. Such deposits signify the arrival of the ancient river system in central Nevada by the end of Triassic time. Areas to the west of the delta were still submerged at that time and, in places, remained so until some time in the following Jurassic Period. In west-central Nevada, offshore from the shoreline, a deep marine basin existed at the foot of a steep slope that plunged to the west from the shallow sea floor above. Occasionally, the powerful earthquakes that accompanied the ongoing collision of Sonomia

Figure 6.7 Gray Triassic limestone forms upper Star Peak in the Humboldt Range of northwest Nevada. The lower orange-colored slope marks outcrops of the Koipato volcanic rocks.

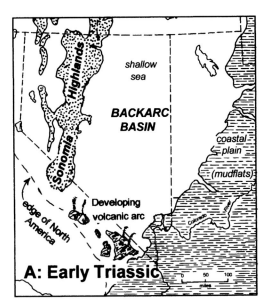

Figure 6.8 Triassic Paleogeography of the Great Basin.

A, in early Triassic time, much of the Great Basin was still submerged by a shallow sea;

B, following the Sonoma Orogeny and the early development of a volcanic arc to the west, the seas withdrew from the central Great Basin and a large river system flowed west across a broad coastal plain.

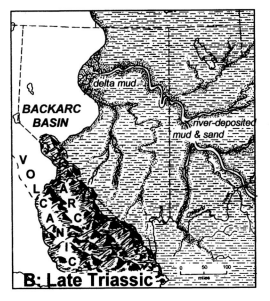

Figure 6.9 Limestone rubble in the Triassic deposits of the West Humboldt Range near Lovelock, Nevada. Such rocks resulted from great undersea "landslides" that carried fragments of self-deposited sediment into a deep marine trough.

loosened masses of rock and mud that slid down the steep slope as great undersea landslides. The rubble produced by such seismic jolts can still be seen in the Humboldt Range as the "megaconglomerate" that comprises part of the Dun Glen Formation (Figure 6.9).

Fossils have been collected from Triassic marine rocks in the Great Basin since the mid-1800s, when geologists first began to explore the region in detail. Such units as the Thaynes Formation of western Utah, the Prida Formation of the Humboldt Range, and the Grantsville and Luning Formations of the Shoshone Range consist primarily of silty limestone that yields many different species of marine invertebrates. These fossils, particularly the **ammonites** (Figure 6.10), have been used extensively in dating and correlating the various Triassic formations across the Great Basin. Compared to the array of invertebrates that populated the Paleozoic sea

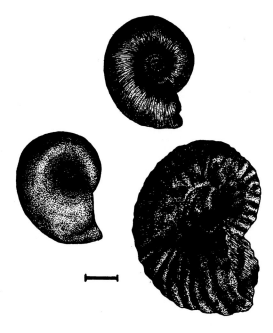

Figure 6.10 Triassic Ammonites from the Great Basin. Among the many types of ammonites that flourished in the shallow Triassic seas of the Great Basin were *Anasibrites* (top), *Meekoceras* (lower left), and *Acrochordiceras* (lower right). Scale bar equals 2 cm (0.75 inches), but the actual size of the ammonite fossils varies considerably depending on age of the organism at death, the environment it inhabited, and other factors.

floor, the Triassic assemblages are much richer in molluscs such as clams, snails, and ammonites. The striking differences between the Triassic marine invertebrates and those of earlier periods results from several factors. A severe worldwide extinction event at the end of the Permian Period resulted in the disappearance, or drastic reduction, of such sea-floor creatures as the trilobites, the solitary corals, the brachiopods, and the crinoids. In the Triassic Period, these creatures were replaced by new organisms, mostly molluscs, that either survived the extinction event or evolved afterward to fill the ecological niches left vacant by it. In addition, the Triassic seas in the Great Basin were probably much more turbid than the sparking tropical seas of the Paleozoic had been. The muddy character of the Triassic seas resulted from the sediment shed from the numerous islands created during the accretion of Sonomia, as well as from the fine silt washed from land by the Triassic rivers that drained into the Great Basin region from the east. The amount of fine sediment suspended in water is a limiting factor for organisms such as the crinoids and brachiopods that feed by filtering tiny planktonic organisms from seawater. Many molluscs, on the other hand, feed on organic-rich mud or, in the case of the ammonites, prey on other swimming creatures. Some of the Mesozoic molluscs were suspension-feeders, but the molluscan apparatus for straining food particles from seawater is much different from that of the brachiopods or crinoids and is generally less affected by muddy water. Thus, it was an extinction event, followed by a burst of evolution, coupled with a significant environmental change that resulted in the conspicuous shift in the character of marine invertebrates after the end of the Paleozoic Era. So notable was this biotic shift that, with a little practice, it is usually easy to distinguish Triassic limestone in the Great Basin from older limestone just by finding a few fossils.

Because of their abundance and nature, finding a rich deposit of Triassic invertebrate fossils in the Great Basin can be exciting experience. But no one ever could have felt a greater thrill of fossil discovery than Siemon Muller, a professor at Stanford University, who in 1929 stumbled upon thousands of large bone fragments weathering out of the Luning Formation in the Shoshone Mountains in central Nevada. The first Triassic vertebrate fossils discovered in the Great Basin were found decades earlier by miners in the Toiyabe Range northeast of Austin, Nevada. In 1868, the illustrious paleontologist Joseph Leidy identified these small pieces of fossil bone as the remains of a dolphinlike marine reptile known as an **ichthyosaur**. So, Muller's discovery was not the first time fossil bones had been found in the Triassic rocks of the Great Basin; however, the site he uncovered was different from anything that was known in his time, or has been encountered since. The mountain slopes of lower Union Canyon, where Muller found his treasure, were absolutely littered with fragments of fossil bone. And, when the fragments were pieced back together, they suggested bones of gigantic proportions. Some of the restored vertebrae were as large as 10 inches in diameter, indicating a behemoth animal of great dimensions. In all of the previously known sites, the bones were far less abundant and were dwarfed by the elements from the Shoshone Mountains. Muller had found one of the most spectacular fossil sites in the world, and he must have been giddy with excitement. However, the initial exhilaration of finding so many large fossils turned to disappointment when the source of the bone fragments on the surface was determined to be the extremely hard limestone layers of the Luning Formation that comprised the bedrock at that particular spot in the Shoshone Mountains. Such unyielding rock would make the large-scale excavation of the bone bed very difficult, if not downright impossible. It was not until University of California paleontologists Samuel Wells and Charles Camp returned to the site in 1953 that some of the fossils were found preserved in softer mudstone layers between the harder limestone strata. Encouraged by the soft matrix that surrounded these fossil bones, Wells and Camp returned to the site sporadically between the years 1954 and 1965. During this time, thousands of bones comprising many partial skeletons of ichthyosaurs were excavated from the extraordinarily rich fossil deposit in the Shoshone Mountains. As the excavations

Figure 6.11 Berlin-Ichthyosaur State Park in the Shoshone Range, central Nevada. Dark-colored fossil bones are scattered through the light-colored limestone.

proceeded, the locality began to attract global attention as one of the most important ichthyosaur fossil sites in the world. In recognition of the worldwide significance of Muller's fossil site, the State of Nevada established a state park at the locality and constructed a shelter over the bone bed to protect it and to facilitate its study and interpretation. Today, the site has been administratively combined with a nearby historic mining camp and is known as the Berlin-Ichthyosaur State Park (Figure 6.11). Fittingly, in 1977 the Nevada legislature designated the ichthyosaur *Shonisaurus popularis* (Figure 6.12) as the official Nevada State Fossil.

Ichthyosaurs are an amazing, and equally mysterious, group of marine reptiles that evolved in the sea at essentially the same time that the earliest dinosaurs were developing from their primitive reptilian ancestors on land. Ichthyosaurs possessed numerous adaptations to life as an active sea-going predator. Their streamlined, torpedo-shaped bodies were propelled through the water by a powerful tail that bore disparate fleshy fins (or "flukes") projecting up and down. Though there are no bones in the tail flukes of ichthyosaurs, the size and orientation of the fins are known from the outline of

Figure 6.12 Skeleton of *Shonisaurus popularis* from the Triassic Luning Formation of the Shonshone Mountains near Gabbs, Nevada. Scale bar = one meter (about three feet).

the entire tail that has been preserved in the bone-bearing rock of some localities. Ichthyosaurs also had two pairs of modified appendages, or "fins," along their flanks that were used to steer the body and control its position as it moved through the water. That these lateral "fins" are actually modified limbs is clear from the numerous internal bones that support them (Figure 6.12). These bones in the appendages of ichthyosaurs clearly indicate that they evolved from terrestrial, or land-living, ancestors. A large boneless dorsal fin, like those of sharks, extended upward from the back of most ichthyosaurs to help control rolling as they swam through the Mesozoic seas. The narrow snout tapered to a point toward the front, allowing it to slice through the water with ease. Though the teeth of ichthyosaurs are somewhat variable from one species to another, they are normally very sharp, conical in shape, and commonly set in a long groove in the jaw. Such teeth could effectively pierce shells, scales, and skin, and would be well suited for seizing swimming creatures as prey. Ichthyosaurs also had very large eyes, suggesting that they could see well in dimly lit water, such as the deep and muddy Triassic seas of the Great Basin. All of these traits seem to suggest that ichthyosaurs were speedy and agile swimmers that darted through the Mesozoic seas pursuing elusive prey such as ammonites, schools of fish, or even other marine reptiles. These voracious marine reptiles were relentless hunters, capable of chasing their prey into both deep and shallow water. The ichthyosaurs appear to be the approximate ecological equivalent to the modern dolphins and porpoises, mammals that have many similar adaptations and habits. Ichthyosaurs may even have swum in groups, or "pods," as do modern whales and dolphins. This may explain why the remains of no fewer than thirty-seven individual

ichthyosaurs, ranging from small babies to full-grown adults, have been found at Berlin-Ichthyosaur State Park. All of these animals appear to have lived, and died, together. No one knows exactly why so many ichthyosaurs died in central Nevada at the same time in the same place, but the most likely scenarios involve the stranding of the group in a pool of shallow water. Perhaps an unusually rapid tidal fluctuation doomed the pod of reptiles by leaving them marooned in a shallow basin with no outlet to the open sea. Maybe, on the other hand, a storm-generated surge washed them into some lagoon, or allowed them to swim over a reef barrier, where they became trapped after the seas calmed. Could the ichthyosaurs have been victims of their own appetites, perishing when they pursued a school of fish into water too shallow for their escape? Any of these events could have led to the mass mortality represented by the concentration of fossils in the Shoshone Mountains, and the geological evidence for each scenario is more or less equivocal.

Several different kinds of ichthyosaurs have been found in the Triassic marine rocks of the Great Basin. By far, the best known of these is *Shoshonius popularis* (see Figure 6.12), the animal that is represented by nearly all of the bones that Siemon Muller discovered. *Shoshonius*, when full grown, was a gigantic ichthyosaur, reaching a maximum length of 50 feet or more. Most of the *Shoshonius* individuals preserved at the state park are much smaller, though, and some are only a few feet long. Based on the measurements of the larger specimens, *Shoshonius* is by far the largest ichthyosaur that ever lived, comparable in size to the modern gray whales that live along the Pacific Coast. No other ichthyosaur even approaches the dimensions of this gargantuan beast, and a pod of *Shoshonius* in the late Triassic seas of the Great Basin must have been an astonishing sight. Elsewhere in the Great Basin, remains of other ichthyosaurs have been identified from rocks of about the same age as the *Shoshonius*-bearing strata of the Luning Formation. For example, two species of the genus *Cymbospondylus* were described in the early 1900s from bones recovered from the Prida Formation in the Humboldt

Range of northwest Nevada. *Cymbospondylus,* which also was identified in 1868 as the first ichthyosaur known from the Great Basin, was about 30 feet long with teeth set in sockets, but its detailed anatomy remains unknown due to the fragmentary nature of the fossils. *Omphalosaurus nevadanus,* also represented by incomplete fossil remains, was a smaller and rather bizarre ichthyosaur that had rows of button-like teeth that seem well designed to crush the shells of Triassic molluscs. Also, two additional species of *Shoshonius* (*S. silberlingi* and *S. mulleri*) have been added to the list of Great Basin ichthyosaurs. The Triassic seas of the Great Basin were literally awash in reptiles that mimicked such contemporary mammals as dolphins, whales, seals, and sea otters.

Where did all these sea-going reptiles come from? That question has perplexed generations of paleontologists and can still be regarded as the most mysterious aspect of the ichthyosaurs, as a group. From the bones in the shoulders, hips, skull, and paddles, it is clear that ichthyosaurs did not originate in the seas but evolved from some unknown land-living ancestor. Though many different groups of terrestrial reptiles from the Permian-Triassic interval are known, none of them shows the slightest evolutionary trend toward marine life prior to the emergence of the ichthyosaurs. The mystery of ichthyosaur origin is deepened by the fact that even the earliest members of this group, such as those found in Nevada, exhibit almost perfect adaptations to life in the seas. In other words, when they first appear in the fossil record, the ichthyosaurs are so completely adapted to the marine environment that we might expect them to have lived in the seas for million of years. But no fossils of anything even remotely similar to the ichthyosaurs have ever been found in pre-Triassic rocks, making it appear that these sea-going reptiles just magically appeared out of nowhere. If you like riddles, then consider this final ichthyosaur enigma: exhibiting a complete reversal of the normal pattern of evolution seen in most other vertebrate lineages, the earliest ichthyosaurs dwarf the later ones! *Shoshonius populfor* is not just the largest ichthyosaur that ever lived, it is also one of the oldest-known

species. The ichthyosaur lineage continued to flourish into the later periods of the Mesozoic Era, when hundreds of different species evolved, but none of them approached the size of their Triassic ancestors. We clearly still have much to learn about ichthyosaurs, and the solution to the mysteries will require many more fossils than are currently available for study. Given the abundance of ichthyosaur fossils in the Triassic marine rocks of the region, the Great Basin may be the place to look for the answers.

JURASSIC HISTORY

Around 208 million years ago, the Triassic Period came to an end and the Jurassic Period began. In the Great Basin, the upheavals caused by the Sonoma Orogeny had diminished, and the highlands it created were eroded to varying degrees in different places. However, no sooner had the rumblings quieted in the Sonoma belt than a new era of commotion began. The Jurassic heralded a series of tectonic events that would, by the end of the Mesozoic Era, complete the transformation of the entire Great Basin region from an ocean basin to a continental interior. In the process, some very dramatic changes occurred in the landscape as the land convulsed under multifarious forces arising from rapidly changing plate tectonic interactions. Because the geologic structures produced during each of the successive stages of unrest were commonly superimposed on one another, it has taken geologists several decades to work out the precise pattern of tectonic activity in the Jurassic and Cretaceous Periods. Though there is still considerable uncertainty about the timing, causes, and patterns of deformation in specific areas of the Great Basin, researchers have been able to piece together a general model for the widespread unrest that affected the region between about 200 and about 60 million years ago. It has become increasingly apparent in recent years that there were several discreet phases of Jurassic-Cretaceous upheaval that struck different areas, at different times, in different ways. Collectively, these mountain-building events were not merely extensions of the Sonoma Orogeny, but each signified the

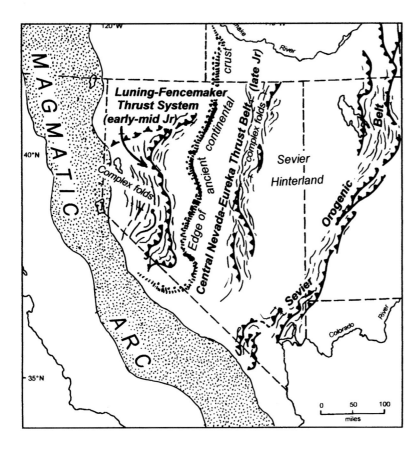

Figure 6.13 Jurassic and Cretaceous deformation in the Great Basin. Starting in the early Jurassic Period, a great wave of compression swept from west to east across the Great Basin. These forces resulted in several different episodes of mountain building, culminating in the late Cretaceous Sevier Orogeny in western Utah and southern Nevada. The compression that caused these disturbances was linked to the convergence of plates west of the magmatic arc in the Sierra-Klamath region.

evolution of a new domain of plate tectonic interactions along the west edge of North America. The structural complexity of rocks deformed by these multiple late Mesozoic events can be bewildering. In many places in the Great Basin, Jurassic and younger strata are mangled almost beyond belief and are commonly metamorphosed to one degree or another. In this brief review, it is simply impossible to describe in detail the complex patterns of deformation that occurred during this time. However, it is also impossible to overstate the importance of this geologic turmoil in setting the stage for later events in the region. Therefore, it is well worth our time to review the general pattern of late Mesozoic tectonic activity in the Great Basin, and to explore the significance and causes of it.

The complicated structure of Jurassic and Cretaceous rock sequences in the Great Basin provides evidence for several pulses of mountain building beginning about 200 million years ago and continuing to about 60 million years ago, just after the Cretaceous Period closed. In general, these numerous events can be grouped

into three broad categories: 1) those of early to middle Jurassic age that most strongly affected areas in the western Great Basin; 2) those that followed in late Jurassic to early Cretaceous time, which generally affected areas farther east, in the central and eastern Great Basin; and 3) the late Cretaceous Sevier Orogeny, which occurred along the eastern margin of the Great Basin (Figure 6.13). Each of these three broad episodes of deformation consists of multiple events that slightly overlap each other in both space and time. In spite of this overlap, the subdivision of the varied late Mesozoic disturbances into three phases allows us to recognize a general progression of geological disturbances from the west edge of the Great Basin to the eastern margin during Jurassic and Cretaceous time. It appears that a great wave of unrest swept eastward across the region over a period of about 130 million years. This overall pattern of Jurassic and Cretaceous mountain building in the Great Basin must reflect something about the interactions between North America and oceanic plates west of the volcanically active magmatic arc. We will return to these interactions later in this chapter, but first we must explore the way Great Basin rocks were deformed and lifted during each of the three phases of Mesozoic mayhem. The geological structures in the disturbed areas, along with their distribution and orientation, provide the keys to interpreting the fundamental plate tectonic mechanisms for such widespread and persistent turmoil.

The Luning-Fencemaker Belt

The earliest phase of Jurassic deformation was driven by a powerful wave of compression that penetrated inland from the magmatic arc into the western Great Basin. Recall that after the Sonoma Orogeny, a volcanic mountain range emerged along the ancient coast of North America as part of what geologists call the **magmatic arc**. Magmatic arcs develop on plates above subduction zones when one lithospheric plate descends beneath another slab of rock moving in the opposite direction. Offshore of the west coast of Jurassic North America, several different oceanic plates were moving eastward toward the leading edge of the continent, into which Sonomia had already been accreted. At the time, the North American Plate was moving to the west due to the initial opening and spreading of the northern Atlantic ocean basin. As the oceanic plate met North America head-on, immense compressive forces were generated where the two plates converged toward each other, like the jaws of a gigantic geological vice. By early Jurassic time, this compression was transmitted eastward through the leading edge of North America, into the Great Basin region. Eventually, the oceanic slabs that slammed into the edge of the continent were bent downward and slid beneath North America. In the hotter subsurface regions, the descending plates produced molten rock, or **magma**. The magma rose through overlying material either to erupt from a volcanic edifice on the surface or to solidify underground as a large mass of granite. We will discuss this process, and the rocks that resulted from it, in more detail in the next chapter, but for now it is important only to understand that the magmatic arc in the western Great Basin was well developed and active in early Jurassic time. The descending oceanic plates that dove eastward beneath the North American Plate were the ultimate source of the enormous compression that extended hundreds of miles inland from the ancient coast. In the early Jurassic, the rocks in the western Great Basin began to yield to these forces by breaking along low-angle thrust faults or by slowly bending to form complex patterns of folds. In western Nevada, numerous early Jurassic thrust faults occur in a belt that stretches almost 250 miles, from the Humboldt Range near Winnemucca southward to the Pilot Mountains, about 40 miles southeast of Hawthorne. Because the two most prominent faults in this belt are the Fencemaker Thrust in the north and Luning Thrust to the south, the entire belt of Jurassic deformation has been named the Luning-Fencemaker Thrust system.

Typically, the Jurassic thrust faults in the Luning-Fencemaker system separate numerous slabs of crumpled late Triassic marine strata, such as the Luning Formation, in a manner

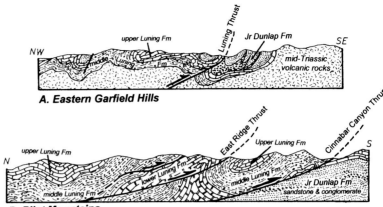

A. Eastern Garfield Hills

B. Pilot Mountains

reminiscent of the way shingles overlap each other on a roof (Figure 6.14B). As each slab of rock was transported to the east or southeast along an underlying thrust fault, it was driven over a wedge of similar strata below. The individual thrust plates appear to have piled up on each other about where the rifted edge of the ancient (Precambrian) continental crust ended, as indicated by the initial $^{87}Sr/^{86}Sr$ ratios (the 0.706 Line, discussed in Chapter 2). Evidently, the horizontal compression during the Jurassic drove the thrust plates eastward until the forces were directed upward by the strong subsurface buttress of thicker continental crust. This may be why no thrust faults related to the Luning-Fencemaker system extend east of the boundary between the ancient continental and oceanic crust (see Figure 6.13). In any case, the imbrication, or "piling up," of the thrust plates that comprise the Luning-Fencemaker system resulted in the contraction, or shortening, of the crust in the western Great Basin (Figure 6.15) and, at the same time, lifted the land surface and created a mountainous terrain in the region. The rising land expelled the last remnants of the Mesozoic seas from the area. After the mid-Jurassic, marine sediments were never again deposited anywhere in the Great Basin.

In addition to the eastward-directed thrusting in the Luning-Fencemaker belt, Jurassic deformation in the western Great Basin also involved complex folding. The older marine rocks that comprise most of the thrust plates were bent and crumpled as they were transported to the east (see Figure 6.14 and Figure 6.16). During the 40 million years of activity

Figure 6.14 Generalized structure of the Luning-Fencemaker thrust system in western Nevada. A, In the Garfield Hills, near the town of Luning, Nevada, late Triassic marine strata are transported southeast along the Luning Thrust. B, The imbricate nature of the multiple thrust faults in the Luning-Fencemaker system is evident in the geologic structure of the Pilot Mountains, southeast of Luning. Note that the Jurassic Dunlap Formation, which contains synorogenic sandstone and conglomerate deposits, is partially overridden by the thrust faults in both locations.

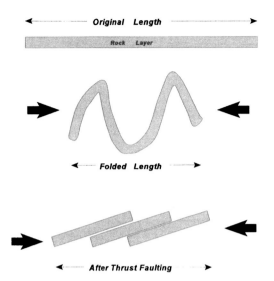

Figure 6.15 Shortening of a rock layer by folding and thrust faulting. Compression (black arrows) of rock strata can lead to either folding (middle), or thrust faulting (bottom), or both. In either case, shortening of the original rock layer occurs.

Original Length

Rock Layer

Folded Length

After Thrust Faulting

that geologists have estimated for the Luning-Fencemaker belt, some rock sequences were folded more than once, as numerous pulses of compression acted sequentially upon rock sequences. Such repetitive maiming of strata can produce extremely complex patterns of folding. This is because once they are contorted, rock sequences remain permanently folded, and later deformation is superimposed on earlier crumpling. Regardless of the complexity of folding in the Luning-Fencemaker belt, all folded rock sequences accommodate additional shortening of the crust (Figure 6.15). Whether we envision a sheet of paper or a layer of rock, it is impossible to fold either without shortening it. As a general rule, the more complex the folding, the greater the shortening. Through the

combined effects or folding and thrust faulting, the crust in the western Great Basin was shortened by at least 75 miles across the Luning-Fencemaker belt during the early part of the Jurassic Period.

As the early Jurassic deformation was underway in the western Great Basin, several small, isolated basins developed on the undulating surface of the rising land. These basins were commonly situated in low areas between thrust faults, in the vicinity of downward folds, or in places of localized down-faulting. Coarse sediment, eroded from the rising land adjacent to the basins, was washed into the lowlands, where it eventually hardened into sandstone and conglomerate. Such deposits are very spotty in the western Great Basin because the basins in which they accumulated were relatively small and isolated. Because they formed at the same time that mountain building (or orogeny) was underway, these sedimentary rocks are known as **synorogenic** deposits. In the Luning-Fencemaker belt, early to middle Jurassic synorogenic sandstone and conglomerate occurs primarily in the Dunlap Formation in the Luning area (see Figure 6.14) and in the Boyer Ranch Formation in the northern part of the belt. These land-deposited sediments generally rest upon deformed Triassic or older marine strata. Perched above the contorted layers of older limestone and shale, the synorogenic gravel deposits, now hardened into conglomerate layers, represent the earliest continental deposits in the western Great Basin. The fragments of rock comprising the conglomerates in the lower parts of the Dunlap and Boyer Ranch Formations are generally similar to the marine rocks they rest on, indicating a local source of the sediment. In addition to the coarse rubble shed from nearby highlands, however, the higher layers in the synorogenic sequences also contain some very fine grained sandstone that consists of tiny, well-rounded, and abraded quartz grains that probably originated from a more distant source. This finer-grained sandstone seems to be roughly contemporaneous with the widespread dune sandstones that blanketed the southwestern Colorado Plateau in the early Jurassic. In southern Utah, this thick mass of early Jurassic dune sand is known as the

Figure 6.16 Folded strata in the Luning-Fencemaker belt, southern Humboldt Range, Nevada.

Navajo Sandstone, the principal rock exposed in the spectacular cliffs of Zion National Park and in the sweeping slickrock domes surrounding Lake Powell. The Navajo Sandstone is incredibly widespread: it can be traced into northern Utah and southern Wyoming, where it is known as the Nugget Sandstone, and extends into the Mojave Desert region as the Aztec Sandstone (Figure 6.17). Such a widespread mass of sand suggests a Jurassic dune field of Saharan proportions, across which fierce winds blew sand and dust over thousands of square miles of western North America. Enormous dune fields of this type, literally "sand seas," are known as **ergs,** a word derived from the Hamitic term for a sandy desert. In addition to the size, shape, and abrasion of the grains comprising the Navajo and coeval sandstone units, the distinctive sweeping **cross-bedding** provides further evidence of wind deposition (Figure 6.18). The present western limit of Jurassic erg-deposited sandstone closely follows the eastern margin of the Great Basin, but such deposits do extend into the Mojave Region, where several hundred feet of dune-deposited sand comprise the main portion of the Aztec Sandstone. However, the present western limit of such sandstone deposits is due to the complete erosion of most Jurassic rocks from the eastern side of the Great Basin. Thus, the early Jurassic erg may have extended westward far beyond the present limits of the strata that record its existence. If so, it is certainly possible that the fine-grained sandstone in the upper parts of the Boyer Ranch and Dunlap Formations might conceivably have been blown into the basins by the powerful winds that spread sand across much of western North America. In any event, the presence of the Aztec Sandstone in the southern Great Basin clearly demonstrates that at least a part of the great early Jurassic sand sea spread westward all the way to the eastern foothills of the magmatic arc and the thrust belt immediately east of it.

The Central Nevada-Eureka Thrust Belt

In central and eastern Nevada, geologists have recognized Jurassic deformation in many different areas, stretching from the Ruby

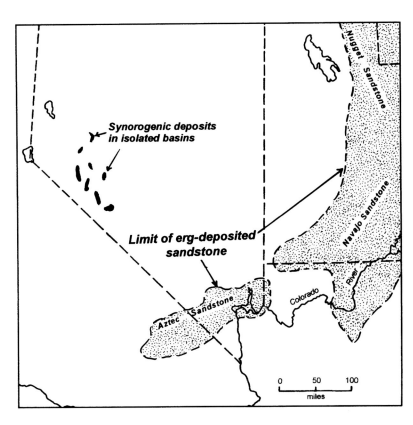

Mountains–Wood Hills–Toano Range area near Elko to the Grant and Quinn Canyon Ranges south of Eureka. This broad region of Jurassic deformation is less sharply defined than is the Luning-Fencemaker belt, and a variety of terms have been used to describe it, including the "Eureka Belt," the "Central Nevada Thrust Belt," and the "Elko Orogenic Belt." For simplicity, we will use the term Central Nevada–Eureka Thrust

Figure 6.17 Distribution of early Jurassic erg-deposited sandstones in the Great Basin region.

Figure 6.18 Cross-bedding in the Jurassic Aztec Sandstone, Spring Mountains, southern Nevada.

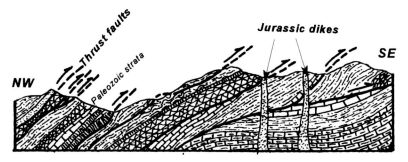

Figure 6.19 Simplified Geologic Structure of a Portion of the HD Range, northern Nevada. The numerous thrust faults are cut by tabular masses, or dikes, of late Jurassic igneous rock. Since the dikes are not offset by the faults, the faulting must have occurred prior to late Jurassic time.

Belt to encompass all of these areas, though it is important to remember that, in so doing, we are describing an area with heterogenous geologic structure and a complex, locally variable history. As the name suggests, Jurassic deformation in the central portion of the Great Basin was dominated by thrust faulting and associated folding. However, most of the unrest appears to have occurred somewhat later in the Jurassic Period than in the Luning-Fencemaker belt to the west. Nonetheless, because the geological structures in both regions are similar, the primary force that triggered the Jurassic unrest in both central and western Nevada was compression related to plate convergence. Therefore, it is tempting to envision a wave of deformation propagating eastward, across the Great Basin, as Jurassic time proceeds. While such a perception may be accurate in a very general way, detailed geological studies in central Nevada have revealed a much more complicated pattern of Jurassic orogenic events. For example, some of the Jurassic deformation in the central Great Basin began in the early part of the period, while in other places it persisted well into the Cretaceous Period. The timing of the various Jurassic disturbances in the central Great Basin does not appear to be synchronous or uniform. A further complication is the great variety of styles of deformation, different orientations of faults and folds, and the obscuring effects of metamorphism that occurs throughout the Central Nevada–Eureka belt. Remember also that, unlike the disturbances in the Luning-Fencemaker belt, the Jurassic deformation in this region was superimposed on rocks already contorted and faulted during the Antler and Sonoma Orogenies. This makes it sometimes very difficult to sort out the effects of the Mesozoic activity from those of the earlier events. No wonder that today, after

many decades of study, geologists are only beginning to unravel the Jurassic history of the central Great Basin.

In spite of the complicated pattern and imprecise timing of the unrest in the Central Nevada–Eureka belt, we can still grasp the overall effect of the geological ruckus by directing our attention to a few well-studied localities within the region. In the northern portion of the Central Nevada–Eureka belt, around the Jarbidge and Independence Ranges, Paleozoic and younger strata are strongly folded, broken into multiple thrust plates, and metamorphosed to varying degrees by igneous masses (Figure 6.19). The thrust faults in the HD Range, about twenty-five miles northeast of Wells, Nevada, are cut by unfaulted masses of late Jurassic igneous rocks, indicating that the faults were no longer active at that time. Farther south, in the area around Eureka, similar Mesozoic thrust faults are partially buried under layers of early Cretaceous sandstone and conglomerate (in the Newark Canyon Formation), further evincing Jurassic activity. Rock layers in many of the mountain ranges of north-central Nevada are conspicuously folded (Figure 6.20), and much of this deformation is probably of Jurassic age, though the geological evidence for the age is often disputable.

Because the severity and orientation of deformation varies significantly throughout the Central Nevada–Eureka Thrust Belt, it is not a simple matter to calculate the total amount of crustal shortening, or to compare it with the Luning-Fencemaker belt. However, given the widespread, and at least locally severe, nature of the Jurassic folding and faulting in the central Great Basin, it is inconceivable that this region did not experience substantial uplift at that time. Even before the advent of modern plate tectonic theory, and without knowing the extent of Jurassic deformation in the Great Basin, geologists recognized that the central and eastern Great Basin were elevated during the mid-Mesozoic Era. Older, and somewhat vague, terms such as the "Mesocordilleran High," "Mesocordilleran Geanticline," "Toiyabe Uplift," and "Nevadan Orogenic Belt" all have been used in the mid-twentieth century to describe a rather obscure region of the central and eastern Great Basin where no Jurassic sedimentary rocks existed. This suggested that erosion of an elevated tract, rather than deposition of sediment in a lowland basin, occurred in the Great Basin during that time. This perception is still valid, but through the analysis of the complex structures in the Central Nevada–Eureka Thrust Belt, we are now beginning to understand the forces that lifted the land and prevented sediment from accumulating in the mid-Great Basin during the Jurassic Period. There are still many gaps in our comprehension of the Jurassic tectonics of the Great Basin, but we can at least link the uplift and deformation to the compression generated between North America and converging oceanic plates along the ancient Pacific margin. We can be confident that at least some, and perhaps most, of the erosion that produced the grand angular unconformity in the Confusion Mountains (see Figure 6.2) occurred during the Jurassic Period.

The Sevier Orogenic Belt

The final, and arguably the greatest, Mesozoic mountain-building event in the Great Basin was the Sevier Orogeny. This disturbance was named in 1968 by geologist Richard Armstrong, who, after analyzing thrust faults and folded-rock sequences in the Sevier Desert region of western Utah, concluded that most of the deformation in that region occurred during the Cretaceous Period (144–65 million years ago). Armstrong also postulated that the deformation would have resulted in the uplift of an extensive mountain system, known as the Sevier Orogenic Belt, along the eastern edge of the modern Great Basin (see Figure 6.13). In the three decades since the term originated, the Sevier Orogenic Belt has been traced as far south as the eastern Mojave Desert, and at least as far north as the Yellowstone region. Many geologists consider the Sevier belt to be continuous well into Canada, where rocks have been deformed in the same manner and at about the same time as those along the eastern edge of the Great Basin. Because the geologic structures that developed during the Sevier Orogeny played a key role in

Figure 6.20 Folds in the rocks layers of the central Independence Range of northern Nevada probably originated in the Jurassic Period.

trapping petroleum resources in the "overthrust" region of northern Utah and Wyoming, the entire Sevier belt has been intensely studied by a small army of geologists. Consequently, our understanding of the style of deformation in the Sevier belt, its timing, and its regional significance surpasses that of any other Mesozoic mountain-building event in the Great Basin.

Almost everywhere along its length, the Sevier Orogenic Belt consists of several imbricate ("shingled") thrust plates which have been transported from the west to the east. The thrust faults are aligned nearly parallel to each other along the eastern and southern edge of the Great Basin, in a belt that varies in width from about 20 miles to more than 90 miles. Most of the principal thrust faults within this belt have been named for places where they are prominent: the Crawford Thrust of northern Utah, the Pahvant Range Thrust of west-central Utah, the Muddy Mountains Thrust of southern Nevada, and the Clark Mountain Thrust of the Mojave region are just a few examples. Within the Sevier belt, the earliest thrust faults typically developed toward the west, while later thrusts formed to the east, as the rocks responded to ongoing compression during the Cretaceous Period (Figure 6.21). In general, the thrust faults descend into the subsuface toward the west, where they commonly flatten and merge in a single, nearly horizontal surface known as a **décollement**, or "detachment," in French. Above the décollement, large but relatively thin plates

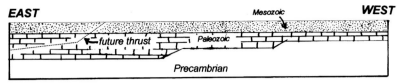

EAST   Mesozoic   **WEST**

future thrust — Paleozoic

Precambrian

A. Pre-Cretaceous

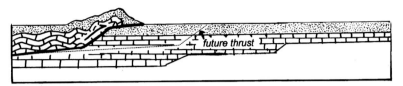

future thrust

B. Early Cretaceous-late Jurassic (?)

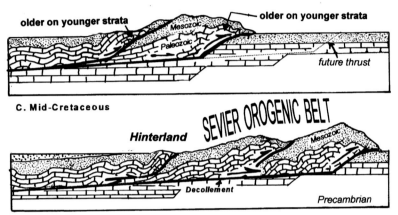

older on younger strata   older on younger strata

Mesozoic

Paleozoic

future thrust

C. Mid-Cretaceous

Hinterland   SEVIER OROGENIC BELT

Mesozoic

Decollement

Precambrian

D. Late Cretaceous-early Paleocene

Figure 6.21 Sequential development of the Sevier Orogenic Belt in the eastern Great Basin. Note the imbricate structure of the thrust plates and the manner in which low-angle thrust faults can place older rocks above younger strata (identified in panel C).

of mostly Paleozoic and Mesozoic rock were transported eastward during the Sevier Orogeny until they piled up, one at a time, in manner similar to what we have seen in the Luning-Fencemaker belt. In the case of the thrust plates in the Sevier belt, it appears that the pronounced eastward thinning of the great

wedge of Paleozoic strata caused the thrusts to cut upward toward the surface from the horizontal décollement. Much later, in the Cenozoic Era, the eastern margin of the modern Great Basin Province developed about where the Sevier thrust plates were forced upward by pronounced tapering of the thick underlying sequence of Paleozoic strata. The amount of displacement on the individual thrusts that comprise the Sevier belt varies from a few miles to scores of miles. In many places, the upper plate has moved enough to bring older (and thus originally deeper) layers of rock up and over younger strata (see Figure 6.21C). One such place is the Spring Mountains, west of Las Vegas, where limestone layers of the Cambrian Bonanza King Formation were placed *on top* of the Jurassic Aztec Sandstone by motion on the underlying Red Spring and Keystone thrust faults (see Plate 10). Each of the numerous thrust-bounded slabs of rock in the Sevier belt was internally deformed as it was transported to the east during the Cretaceous orogeny. Great folds developed, primarily in layers of Paleozoic limestone and shale, as the thrust plates were driven east. Such folding is sometimes displayed in spectacular fashion on the flanks and canyon walls of several different mountain ranges in western Utah (Figure 6.22). Together, the thrusting and folding that occurred during the Sevier Orogeny led to the uplift of a major mountain system that dwarfed the other elevated tracts that preceded it to the west. When

Figure 6.22 Folded layers of Pennsylvanian limestone in the walls of Provo Canyon in the Wasatch Mountains of Utah. This fold developed in the Cretaceous Period, when a thick slab of rock was driven eastward during the Sevier Orogeny.

they were fully developed, at the end of the Cretaceous Period, the mountains of the Sevier Orogenic Belt may have stood more than 15,000 feet above sea level. The western foothills of the Sevier mountains, known as the **hinterland** (see Figures 6.13 and 6.21) gently sloped for hundreds of miles into western Utah and eastern Nevada. The eastern flank of the Sevier mountain system was evidently a steep and bold escarpment that towered above the lowlands of central Utah and southwest Wyoming. Streams draining the summits of the Sevier peaks carried water and sediment in both directions, depositing great volumes of sand and gravel eroded from the rising mountains. Because the streams that drained the steep east flank of the Cretaceous mountains were much more vigorous than those flowing west, deposits of synorogenic rubble shed from the Sevier belt are much more widespread in the western Colorado Plateau than they are in the eastern Great Basin. In addition, whatever synorogenic sediments did accumulate in the hinterland of the Sevier Orogenic Belt were largely buried under younger material or eroded in the very recent geologic past. Nonetheless, there are enough small patches of Cretaceous river-deposited sediment in eastern Nevada to demonstrate that there was some drainage of streams from the Sevier belt in that direction. We will return to these deposits a little later in this chapter.

PLATE TECTONICS OF THE MESOZOIC

By the time the Sevier Orogeny ended about 65 million years ago, the crust had been shortened by about 100 miles across the entire Sevier Orogenic Belt. This crunching of the crust was the culmination of the great wave of compression that marched across the Great Basin during the Mesozoic Era. The total amount of crustal shortening caused by the compression (or "contraction," as many geologists now say) during the Mesozoic Era is difficult to estimate, but it must have been at least several hundred miles. Mesozoic tectonic forces in the Great Basin were generated at a new convergent-plate boundary that developed to the west following the accretion of Sonomia. Along the ancient west coast of

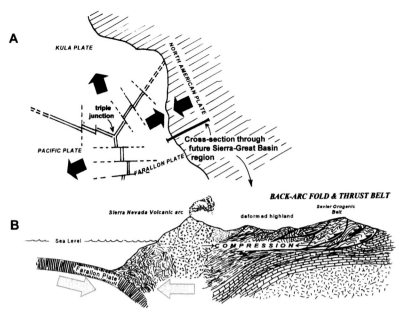

Figure 6.23 Mesozoic Plate Interactions of the Great Basin Region.
A, The Farallon Plate was moving east (black arrows) against the edge of North America, which was moving in the opposite direction during the Cretaceous and Jurassic Periods. Note the triple junction, where the Farallon, Kula, and Pacific Plates meet.
B, Convergence between the Farallon and North American Plates produced the compression that led to folding and thrusting in the Great Basin. It also resulted in the subduction of the oceanic slab, which is linked to volcanic activity in the Sierra Nevada (see Chapter 7).

North America, slabs of oceanic rock were being subducted toward the east beneath a continental crust that was moving in the opposite direction (Figure 6.23). This interaction produced immense compression forces behind the volcanic arc on the overriding plate, as the two great slabs of rock squeezed against each other even as subduction continued. Several times during the Jurassic and Cretaceous Periods, rocks in the Great Basin responded to this compression by bending into complex folds or rupturing along thrust faults. Though tension, or "stretching," forces were developed occasionally in the Mesozoic Great Basin (more on this in the next chapter), the major cause of uplift in the region throughout the era was the compression developed at the plate boundary.

As we have seen, the compressive deformation appears to have started first in the western Great Basin and then migrated east through Mesozoic time. Geologists refer to such areas of turmoil, associated with subduction-generated compression, as **back-arc** thrust and fold belts. The term "back-arc" refers to the interior location of the disturbed region, "behind" the volcanic arc that is also produced by the subduction process. The Great Basin, located behind the Sierra Nevada volcanic arc, was affected by several pulses of back-arc deformation during the Mesozoic Era. Because the phases of deformation outlined above overlap each other, there are many places in the Great Basin where rocks were mangled repeatedly during the Mesozoic Era. In such areas, it is almost impossible to tell exactly when the Mesozoic strata were mutilated, because folding and faulting both permanently disfigure rock whenever they occur, and their combined effects are cumulative. It is still clear, however, that the compression that affected the Great Basin came in pulses, or waves, during Jurassic and Cretaceous time: first to the Luning-Fencemaker belt, and last to the Sevier belt. What could have been responsible for such surges in deformation?

The magnitude of the compression in the back-arc region depends on several factors. The rate of convergence (or subduction), the inclination of the subducted plate, the angle of convergence, and the nature of the rock that transmits the force inland are all very important influences on the development of a back-arc basin. Geologists have developed ways, primarily using the magnetic properties frozen in rocks, along with certain chemical attributes of them, to reconstruct the likely position and movement of the ancient oceanic plates in the proto–Pacific Ocean. Though there is still some uncertainty about the precise pattern of plates and their exact trajectories, it seems clear that there were several different oceanic plates west of North America during the Jurassic Period. Three Mesozoic plates—the Kula, Farallon, and Pacific—all met at a common center, called a **triple junction** (see Figure 6.23A). This is a much more complicated arrangement than that of the modern Pacific basin, which is dominated by a single large oceanic plate. Sea-floor spreading between the Mesozoic plates caused each of them to slide away in different directions at rates of about an inch or so per year. One of the plates, the Farallon, was moving almost directly against the western edge of North America. It was the Farallon Plate that was eventually bent downward and subducted beneath the Great Basin during the Jurassic and Cretaceous Periods. The other two plates were traveling from the triple junction toward other continents, or other parts of North America, during most of the Mesozoic. Meanwhile, North America was being pushed to the west by sea-floor spreading that accompanied the development of the North

Atlantic Ocean. Recall that the supercontinent of Pangaea had been assembled by the end of the Triassic Period. In early Jurassic time, the fragmentation of this enormous landmass was already well underway. Linear rift valleys had opened between the various fragments of Pangaea, and the modern ocean basins were then in their early stages of development. Sea-floor spreading at the nascent oceanic ridge in the Atlantic basin was forcing North America west, away from Eurasia and Africa, causing it to override the Farallon Plate. Today, North America is still moving west from the mid-Atlantic ridge, albeit at only about 2 cm (a little less than one inch) per year. As we will see, the rate of westward motion was much higher in the Mesozoic Era.

At convergent-plate boundaries, the rate of convergence is the sum of the individual plate velocities. If two converging plates are each moving 2 cm/yr, then the convergence rate is 4 cm/yr. Obviously, the rate of convergence between the Farallon and North American Plates would directly influence the intensity of Mesozoic compression in the Great Basin: when the convergence rate was high, compression would be intensified; when it was low, compression would be relaxed. Geological evidence suggests that the convergence rate between these two plates varied between a few centimeters per year to as much as 15 cm (or about 6 inches) per year in Jurassic and Cretaceous time. This variable convergence reflects continuous changes in the rates of sea-floor spreading at both the Pacific triple junction and the mid-Atlantic ridge, as well as variations in other factors. For example, the angle of convergence is important because, if the plates meet head-on, inland compression is intensified. If, on the other hand, they meet obliquely, delivering a glancing blow to each other, then the convergence rate, as well as the intensity of compression, is somewhat diminished because the plates will slip past one another slightly as they simultaneously squeeze together. In fact, the force generated by this type of compression, produced by the oblique convergence of plates, is commonly called **transpression** by geologists. Transpression can result in folding and faulting, but the intensity of deformation is generally less than it is when the plates are converging more directly. In addition, faults that result from transpression will have a lateral, or sideways, component of displacement that "normal" thrust faults do not possess. Fnally, the angle of subduction is also important because it affects the area over which the compression is applied. If the Farallon Plate descends steeply beneath the continental slab, comprssion is focused on a narrow zone on the western side of the overriding plate. If the inclination, or "dip," of the subduction zone is low, such that the Farallon Plate scrapes against the bottom of the North American Plate, forces can be transmitted much farther to the east and the resulting deformation would be dispersed over a broader area. So, whenever changes occurred in the angle of convergence, in the rate of sea-floor spreading in the Atlantic and Pacific Oceans, or in the angle of subduction, the style, intensity, and extent of deformation in the Great Basin varied accordingly.

During Jurassic and Cretaceous time, the mountain-building forces in the Great Basin came and went in rhythm with the changing interactions between the North American and Farallon Plates. Currently, it is thought that several different reorganizations of the plates, each fostering a new epoch of deformation, appear to have occurred in western North America since the beginning of Triassic time. The Luning-Fencemaker belt, for example, seems to coincide with a strong westward "push" of North America from the mid-Atlantic ridge that intensified compression along the western margin. The Central Nevada–Eureka belt is more difficult to correlate with any discreet plate tectonic event, but most of the deformation in this area occurs during a very complex transpressional event from 165 to about 140 million years ago. Finally, after a lull in mountain building from about 140 to 120 million years ago, the Sevier Orogeny represents an extremely strong surge in compression that resulted from an increase in the rate of sea-floor spreading in the Atlantic Ocean from 2 cm/yr (less than one inch/year) to as much as 15–20 cm /yr (up to 7 inches/year). Thus, the Sevier Orogeny appears to coincide with the time when North America was "cata-pulted" west by this relatively sudden acceleration. This jolt would have intensified compression in the back-arc region, which resulted in the most prominent fold and thrust belt of them all. The tectonic chaos in the Great Basin during the Mesozoic Era is ultimately the consequence of a runaway continent plowing into the oncoming traffic of oceanic plates. Some of the collisions were head-on, some were glancing. Some were high-speed, some were slow-motion. At the close of the Mesozoic Era, what was left was a continent with a badly crumpled front end, but one that was hidden beneath a magnificent landscape.

MESOZOIC LAND AND LIFE IN THE GREAT BASIN: WHERE ARE THE DINOSAURS?

The complicated cavalcade of Mesozoic mountain-building events in the Great Basin raised land everywhere in the region. The older seas were forever expelled by the Jurassic uplift and, by the time the Sevier Orogenic Belt was fully developed in late Cretaceous time, a rugged highland stretched from California to central Utah. In the western Great Basin, this elevated terrain was capped with majestic, snow-clad volcanic cones, while to the east stood the magnificent Sevier mountains, composed mostly of badly gnarled layers of sedimentary rock. In between, in the central Great Basin, a hilly, undulating terrain separated the high regions on either side. Throughout this Mesozoic highland, erosion was the dominant geological process. The intensity of erosion was no doubt greatest on the higher peaks due to the greater abundance of water and the extreme climatic conditions that always prevail in high-mountain environments. It is difficult, however, to estimate the amount of rock worn away during the long period of erosion that followed the Mesozoic unrest in the Great Basin. Almost certainly, it varied from place to place as a function of elevation, climate, and other factors. However, we do know that the unconformities resulting from post-uplift erosion sometimes indicate the removal of thousands of feet of sedimentary rock. This is certainly true in the case of the angular unconformity in the Confusion Range

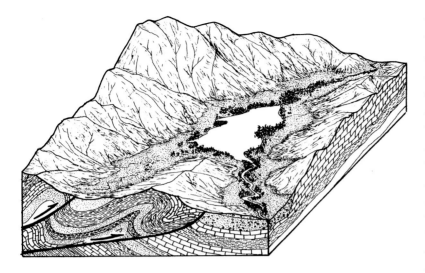

Figure 6.24 The Cretaceous Landscape of the Central Great Basin. Uplift resulting from Jurassic and Cretaceous deformation (front subsurface panel) created a rugged highland with numerous small, isolated basins. Forests of redwoods, cycads, and ferns lined the well-watered stream channels and lakeshores.

(Figure 6.2), where the entire sequence of late Paleozoic and early Mesozoic strata was stripped away. Such deep erosion following Mesozoic uplift is the rule, not the exception, in the Great Basin. Consequently, little sediment was deposited anywhere in the Great Basin during this late Mesozoic interval of erosion, and sedimentary rocks of this age are almost, but not entirely, nonexistent in the region.

During the Jurassic and Cretaceous Periods, western North America was situated between 25° and 35° north latitude, a little closer to the tropics than where the region is located today. With elevations that may have exceeded 15,000 feet in places, the Mesozoic peaks in the Great Basin must have received abundant water from passing storms. Because the Great Basin was not yet a region of internal drainage, the Mesozoic rivers draining the mountain slopes flowed primarily toward adjacent lowlands to the east and west. In the west, the rivers drained into the proto–Pacific Ocean from the coastal volcanoes of California. To the east, the streams emerging from the Sevier Orogenic Belt gathered on a broad plain that sloped toward a vast inland sea that covered the Rocky Mountain region in the Cretaceous Period. Enormous amounts of sediment, derived from the deep erosion of the Great Basin highland, were flushed into the oceans that bordered the region to the west and east. However, not all of the water in the Mesozoic highland drained away completely; at least some of it cascaded into relatively small basins nestled between the hills and peaks in the cen-

tral portion of the province (Figure 6.24). In these basins, sediments were deposited in river channels and floodplains, on lakebeds, and in swamps. This sediment is represented by the conglomerate, sandstone, and shale comprising such rock units as the Newark Canyon Formation (early Cretaceous; Cortez-Diamond Range area of central Nevada), the King Lear Formation (early Cretaceous; Jackson Mountains of northwest Nevada), and the Willow Tank Formation–Baseline Sandstone (early–late Cretaceous; Muddy Mountains of southern Nevada). Each of these sequences records somewhat different conditions in the places where the sediment was deposited, but they collectively demonstrate that there were many small basins and valleys scattered among the hills and mountains of the Mesozoic Great Basin. Exposures of such terrestrial deposits are never very widespread in the Great Basin, suggesting that sediment only accumulated in small, isolated areas in a larger terrain that everywhere else was subject to erosion.

The small and isolated exposures of late Mesozoic terrestrial strata in the central Great Basin are easy to overlook. Not only are the outcrops usually relatively small, but the soft rocks typically erode to form a smooth slope covered with pebbles and dirt, and masked with an overgrowth of vegetation. Moreover, the drab neutral colors of these deposits commonly blend into the dusky desert environment so well that the outcrops virtually disappear to the eye. The paltry exposures of late Mesozoic sediments, so utterly overshadowed by the spectacular cliffs of Paleozoic rocks and panoramic expanses of Cenozoic strata, are well concealed in the midst of such a geological wonderland. And yet, in 1882, Charles D. Walcott, one the first geologists to explore the Great Basin, not only found these rocks but also discovered fossils in them. Walcott's fossils included the remains of land plants and freshwater molluscs that were found in the shale and sandstone of the Newark Canyon Formation, exposed in patches around the Eureka Mining District of central Nevada. Since the late nineteenth century, additional fossils have been found in this formation, including the remains of small fish and teeth from a ratlike primitive mammal. Additional Cretaceous plant fossils are now known from the King Lear Formation; and the Willow Tank Formation in the Muddy Mountains area has produced plant, fish, and mollusc remains similar to those that Walcott found in the Eureka District. Collectively, these fossils add an intriguing blush of life to our reconstruction of the rugged Cretaceous terrain of the Great Basin (Figure 6.25).

The Cretaceous plant fossils in the Great Basin suggest that the undulating mountains and valleys supported forests dominated by trees very similar to the modern sequoia or redwoods, as well as to sego "palms," which are not palms at all, but belong a primitive group of plants known as cycads. The Cretaceous cycads and redwoods were amazingly similar to their modern counterparts, so it is not difficult to envision shadowy groves of such trees dotting the Great Basin landscape 100 million years ago. There were also some bizarre elements in the Cretaceous flora of the Great Basin. For example, the remains of *Tempskya*, a tree-sized fern with a thick trunk and dense wood, have been identified in the Willow Tank Formation. This monster fern is much different from the delicate, shrub-sized ferns of the modern world. Several other types of plants, mostly conifers, existed in the Cretaceous forests of the Great Basin, but their fossil remains are either too fragmentary or too poorly preserved to be identified with precision. These forests were undoubtedly thickest in places where the soil was deep and fertile and climatic conditions were optimal. Geological evidence from areas adjacent to the Great Basin suggests that the climate of the early Cretaceous was strongly seasonal across the entire western part of North America. If so, then the redwood-cycad-fern forests of the Great Basin were probably concentrated along river courses or along the edges of lowland lakes, where water generally would be more plentiful and would tend to linger during the dry season. Because it was also in these same low-lying regions that sediment accumulated in lakes and streams, most of the Cretaceous plant fossils we have discovered probably represent the lowland flora. What types of trees and shrubs grew on the higher flanks and summits

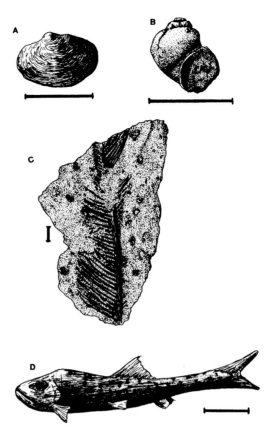

Figure 6.25
Early Cretaceous Fossils from the Great Basin. A, *Musculiopsis russelli*, a freshwater clam; B, *Scalez powelli*, a tiny snail; C, a cycad frond; D, *Leptolepis nevadensis*, a small minnowlike fish. A, B, and D are from the Newark Canyon Formation near Eureka, Nevada. C is from the King Lear Formation in the Jackson Mountains of northern Nevada. Bar in all sketches = 1 cm (0.3 inches).

of the Cretaceous mountains, where erosion prevailed, remains a mystery.

The Mesozoic plants of the Great Basin must have provided food for animals that lived in and near the lowland forests. Fossils documenting about a dozen different species of aquatic snails and clams have been collected from Cretaceous lakebeds and stream deposits in the Great Basin (Figure 6.25A and B). Schools of small minnowlike fish (Figure 6.25D) swam in the isolated lakes that developed in the basins. Almost certainly, these fish moved upstream into the rivers that brought water into the lakes. The small fish must have been pursued by larger predators, but no fossils documenting fish larger than about 4 inches long have yet been recovered from Cretaceous deposits. Thus, it is clear that at least the lakes and rivers were teeming with life, and there must have been nonaquatic creatures crawling through the nearby forests as well. Unfortunately, there is very little fossil evidence pertaining to the Cretaceous land fauna of the Great Basin. Worse yet, the few fossils of terrestrial organisms that have been found have not been studied in detail or are just too poorly preserved to yield much useful information. Nonetheless, there are tantalizing hints that some very interesting creatures probably roamed the Great Basin in the late Mesozoic Era. For example, tiny teeth of primitive mammals have been found in the Newark Canyon Formation, suggesting that small, rodentlike animals scurried through the redwood forests. Even more intriguing are the persistent, but mostly undocumented and unanalyzed, reports of dinosaur bone fragments in Cretaceous strata of the Cortez Range and the Muddy Mountains. Even though such fossils are extremely rare, and none have ever been identified with precision, they suggest the presence of dinosaurs in the Great Basin. Having spent many days in the Great Basin, unsuccessfully scouring Cretaceous strata for vertebrate fossils, I can personally attest to their dearth. But, regardless of the rarity of dinosaur fossils in the Great Basin, I think that we can safely go beyond implying their presence: dinosaurs *had* to live in the central Great Basin during the late Mesozoic Era. To understand why, consider the following.

Though no dinosaur skeletons have ever been found in the Mesozoic rocks of the Great Basin, there is indisputable proof that these reptiles existed in the southern part of the region during early Jurassic time. In the Mescal Range of the eastern Mojave Desert, dozens of three-toed dinosaur footprints, generally less than 6 inches long, occur in the Aztec Sandstone. Most of these tracks were probably left by small, bipedal dinosaurs that scurried about the dunes while the sand was moistened enough by dew or a light drizzle to preserve their footprints. Accompanying the dinosaur tracks in the Mescal Mountains are some smaller three-toed tracks that could have been made by birds or bird-reptiles. In addition, some of the footprints in the Aztec Sandstone appear in a quadrupedal pattern, are only about an inch long, and have four-toe impressions. Such tracks likely were made by lizards. The assortment of footprints and trackways in the Aztec Sandstone clearly proves that a rather rich reptile (and bird?) fauna populated the edges of the early Jurassic erg in the southern Great Basin. While the erg environment was obviously not biologically in-

hospitable, it was still an austere habitat for dinosaurs compared to the relatively lush Cretaceous forests that would develop to the north millions of years later. If dinosaurs could flourish in a bleak Jurassic dune field, then why would they not be equally successful in a verdant Cretaceous highland?

Moreover, we know that the alluvial plain of Utah and Colorado, immediately east of the Sevier Orogenic Belt, was populated by a diverse array of dinosaurs during the early Cretaceous. The Cedar Mountain Formation of Utah, the Cloverly Formation of Wyoming, and the Burro Canyon Formation of Colorado have produced abundant fossils of many different types of dinosaurs. These rock units are all roughly coeval with the Newark Canyon, King Lear, and Willow Tank Formations of the Great Basin. In the past few years, several new discoveries of Cretaceous dinosaur fossils have been made on the Pacific Coast, most notably in the marine sediments of the California coast ranges. These dinosaur remains almost certainly signify the drift of dinosaur carcasses out to sea following the death of the animals along the Cretaceous shore, which was located 80 miles or so farther east than the modern West Coast. Particularly well represented by the California fossils are the duck-billed dinosaurs and the quadrupedal armored dinosaurs known as ankylosaurs. Even large bipedal carnivores are represented by a few scraps of bone. Overall, the new fossil information from California evinces a rather rich array of dinosaurs living on the seaward flanks of the volcanic chain that bordered the Great Basin during the late Mesozoic. So, if the dinosaurs were such adaptable animals, and if there were herds of them to the east, and a gaggle of them to the west, what would have prevented them from living in the Great Basin? Nothing; and that is why it is reasonable to conclude that dinosaurs *must* have prowled the interior basins and hilly terrain of Mesozoic Nevada. The scarcity of their fossils in this region simply reflects the dominance of erosion, rather than the deposition of sediment, during the time that the great reptiles were roaming through the ferns and redwoods. The dinosaurs were there, but they rarely died in just the right place and at the right time for their remains to be fossilized. Someday, we'll know more about Great Basin dinosaurs, but first someone must find the rare fossils they left behind. Those fossils are probably lying on a slope somewhere in the Great Basin right now, just waiting to be discovered. Never pass up a chance to explore one of those bland and boring outcrops of Cretaceous rock. In them, there are treasures yet to be found!

The quest to find dinosaur fossils in the Great Basin is more compelling than a similar expedition would be almost anywhere else in North America. This is because those dinosaurs, whatever types they prove to be, are very likely to comprise an assortment much different from the communities already known from either the Pacific Coast or the Rocky Mountain region. This is because the physical and biotic conditions in the Great Basin, particularly during the Cretaceous Period, were much different from the habitats to the east and west. The California dinosaurs lived amidst lagoons and bays along the volcanically active coast of North America, while the reptile faunas of Utah and Colorado flourished on a broad coastal plain laced with rivers, ponds, and swamps. In the Great Basin there were no coastal lagoons or swamps. Most of the rivers followed short courses into the isolated basins, and probably withered entirely during the dry season. The climate of the Cretaceous highlands was probably much cooler, and perhaps more arid, than the maritime coast to the west or the steamy plain to the east. These environmental differences probably resulted in a unique and distinctive flora in the Great Basin, which means that the food resources for herbivorous dinosaurs would have been different from those to the east and west. Moreover, long-range dinosaur migrations, and therefore genetic mixing with distant populations, may have been limited in the Great Basin by insurmountable mountains and the relatively impenetrable terrain. Perhaps there were unique local populations of dinosaurs, living in near-isolation from other such groups, scattered across the Great Basin. If so, these "lost tribes" of dinosaurs would probably have included types that did not exist, or were very uncommon, in the faunas of the adjacent regions. Dinosaur fossils from the Great Basin, once they are located, are likely to reveal some pretty fascinating creatures, potentially unlike anything already known in North America

In our review of the Mesozoic history of the Great Basin, we have emphasized how plate tectonic forces deformed rock and lifted the land from a sea floor to a highland. We also have briefly explored how these events changed the terrain, and how living creatures adapted to those changes. But, there were also important geological events taking place *under* the Great Basin throughout the Mesozoic Era. The volcanic activity to which we have frequently alluded in this chapter is the outward expression of the subsurface turmoil that existed while the land in the Great Basin was being crunched and elevated by plate tectonic forces. All the events described in this chapter took place under a veil of continuous volcanic eruptions. We have neglected this igneous activity not because it was a trivial ruffle in the progression of momentous Mesozoic events but because it was sufficiently significant in the deep history of the Great Basin to merit our full attention. It is to this underground realm that we now shift our focus.

Plate 1. The Humboldt River in northern Nevada.

Plate 2. Western escarpment of the House Range in western Utah.

A

Plate 3. Pyramid Lake in northwestern Nevada.

Plate 4. Cambrian strata in the Marble Mountains, southern Mojave Desert.

B

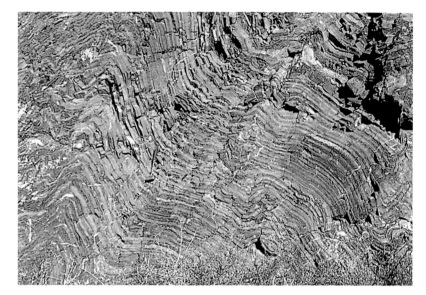

Plate 5. Interbedded Cambrian chert and shale in the Hot Creek Range of central Nevada. Such sediments accumulated in the deep-water setting of the miogeocline ramp and eugeocline basin.

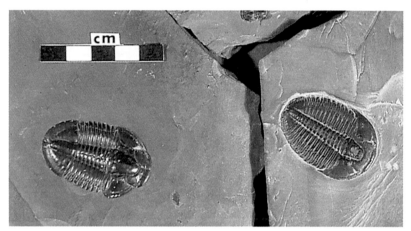

Plate 6. Cambrian trilobite fossils from the House Range, western Utah.

Plate 7. The threefold Ordovician succession in the Arrow Canyon Range of south-central Nevada. The thick dark band in the middle slope is the upper dolomite sequence, below which rest the light tan quartzite and light gray limestone components. Strata above the dark band are post-Ordovician in age.

C

Plate 8. The Roberts Mountains: Heartland of the Antler Orogeny.

Plate 9. A prominent unconformity near the Carlin tunnel on Interstate-80 in northern Nevada. The angular discordance between the steeply tilted Mississippian rocks (Diamond Peak Formation of the Antler Overlap sequence) and the gently inclined Strathern Formation (Pennsylvanian age) clearly indicates that the geological disturbances in the region continued beyond the main phase of the Antler Orogeny.

D

Plate 10. In the Spring Mountains of southern Nevada, gray Cambrian limestone was thrust over red Jurassic sandstone during the Sevier Orogeny. The boundary between the older gray and younger (but lower) red rocks is known here as the Red Spring Thrust.

Plate 11. Granite exposures beneath Notch Peak in the House Range of western Utah.

E

Plate 12. Layers of Miocene sediments and volcanic ash (light greenish-gray) at The Sump in Esmeralda County, western Nevada.

Plate 13. The U-shaped profile of upper Lamoille Canyon, in the Ruby Mountains of central Nevada, was carved during at least two periods of late Pleistocene glaciation.

F

Plate 14. Owens River Gorge near Bishop, California.

Plate 15. Reconstruction of the Ice Age flora and fauna in the hills above Lake Bonneville. Photo courtesy of the Utah Museum of Natural History.

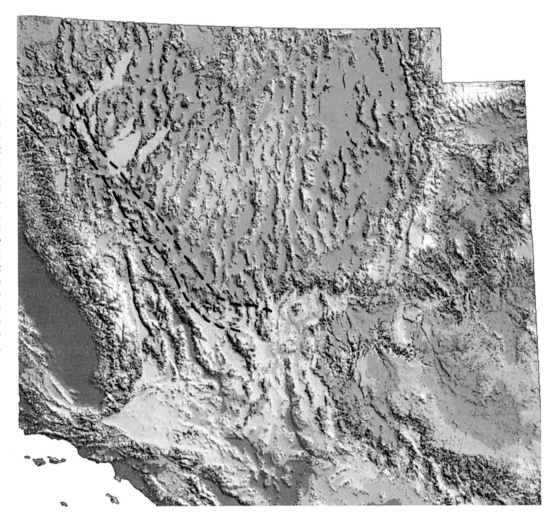

Plate 16. Modern topography of the Great Basin and adjacent regions. On this map from the U.S. Geological Survey, the highest elevations appear in white, lower elevations in deepening shades of green, and the lowest terrains are represented by brown colors. Note that the southern Great Basin is significantly lower than the northern part of the province. Short dashed lines indicate the Walker Lane zone, where the orientation of mountain systems changes from north-south to northwest-southeast.

Plate 17. Hot Creek, in the Long Valley caldera of eastern California.

"How's that for a hang-gliding spot?" I offered the question to the school-teachers in the Suburban as we waited at the intersection of two dirt roads in Tule Valley, a broad dusty basin separating the House and Confusion Ranges of western Utah. I meant to direct their attention to the magnificent cliff, rising to an elevation of nearly 9,600 feet, that formed the sheer western face of the House Range to the east of us. The massive wall of limestone plunged headlong more than 4,600 feet, from the top of Notch Peak to the apex of the alluvial fan that sloped toward us. In reality, the cliff is not quite vertical, but it's close enough. I remembered from a previous hike, the knee-shaking thrill, or more accurately the terror, of standing atop that wall, looking over the edge into the depths of the rocky canyon cut into the escarpment of the mountains. The plunge of Notch Peak is so impressive that many people, when examining it from Tule Valley, fail to even notice the light tan-colored mass of granite that punches through the limestone along the base of the cliff (see Plate 11). With their nearly vertical columns and ragged pinnacles, those granite exposures would merit considerable attention anywhere else. But, juxta-posed beneath such a magnificent cliff, they attract little attention from visitors and passersby. Unless, of course, those people happen to be geologists.

I didn't hear anyone in the truck answer my question, but it didn't matter, since it was more or less rhetorical. Besides, it was late in the day and I had been dragging the poor teachers all over the desert since daybreak, exploring the wonderland of Utah's West Desert. I'm sure they were beginning to view the Teachers Workshop on Great Basin Geology as something of an

ordeal. Our numerous uphill sprints earlier in the day seemed to have left them with too little energy to respond to anything with much vigor. They were obviously tired and anxious to get to our intended camp site, which was nestled among those unnoticed granite crags, just a mile or two away. At long last, the other vehicles in our caravan rumbled into sight from the north, accompanied by a great cloud of swirling yellow dust. With our cortege re-formed, we continued on our way beneath the monumental escarpment of Notch Peak. We turned east, toward the House Range, bouncing along a rough road leading through the sagebrush toward our destination: Painter Spring.

I had chosen the spring for our camp because, aside from their beauty, the rocks at this site are especially interesting. Painter Spring is in the middle of the granite exposure in the House Range and is surrounded by excellent outcrops of the 170-million-year-old rock. These rocks formed from magma that oozed into a mile-thick succession of Cambrian limestone, then cooled and hardened deep underground, more than 145 million years before the House Range was lifted to its present height. Due to the extensive erosion that accompanied the Mesozoic uplift we explored in the last chapter, the thick succession of rock that originally covered the granite has been stripped away at Painter Spring, revealing the body of crystalline rock entombed within the limestone. Since erosion was the dominant process in the region during the Mesozoic Era, Jurassic rocks of any type are rare in the eastern Great Basin. Granite, it so happens, is surprisingly rare everywhere. A lot of what is called granite is actually some other variety of igneous rock such as monzonite, quartz diorite, or granodiorite. The

CHAPTER SEVEN

THE FIRES BELOW

distinctions between these related rock types, to which we will return later, have to do with subtle differences in the fractions and chemistry of the specific minerals that comprise them. Even in the Sierra Nevada, a "granitic" mountain range, true granite is actually rather uncommon. But not at Painter Spring: this rock is real, textbook, 200-proof, authentic granite, and there's a lot of it! It is the combined rarity of both granite and Jurassic rocks in the Great Basin that makes Painter Spring a place of special significance to a geologist. The site is therefore a mandatory stop on a geological tour such as ours, and, as a scenic and instructive campsite, it is unsurpassed.

As we approached the spring, the rough road faded into two faint threads of bare dirt as it swung around a small knoll, then curved back into the mountains. In the angled light of early evening, the pinkish tan granite flashed brilliant sparkles as we drove by a low outcrop. Finally, the spring came into view: a fifteen-foot-high cascade of clear water, spilling over a low wall of granite. The sparkle of the granite combined with the bright flashes of rushing water to produce a glittering display of light. A small rivulet gathered at the base of the waterfall, flowed through a little gully into a pipe that, in turn, delivered it to a stock pond out in the valley, now 500 feet below us. Between the pooling water and the pipe it drained into, decomposed granite had been washed out onto a level surface about 150 feet wide: a perfect campsite. At the top of the waterfall was a narrow slot through which the water tumbled. The cleft was surrounded by boulders and slabs of granite, stacked haphazardly in a chaotic mound. High above the waterfall, the walls of the canyon rose almost vertically to the jagged skyline that blazed in the early evening light. The granite pillars standing above us were much more impressive from this near vantage point than they had been from a distance. As the teachers disembarked from the trucks, all were astonished by the unsuspected scenery that surrounded them. None of them was overlooking the granite outcrops now.

We were unpacking the trucks and setting up camp when the sun touched the western horizon, igniting a brilliant display of orange and yellow streaks that arched across the sky. The purple shadows on the canyon floor gradually deepened as the higher peaks of the House Range dimmed to a rosy glow. When our camp was deployed, and as Happy Hour commenced, I took a seat on the tailgate to admire the sunset colors splashed across the mountain walls above Painter Spring. One of the teachers approached and asked where the large blocks of rock at the top of the waterfall came from. "Did they fall from those cliffs up there?" she asked, pointing to the knobs and crags of granite rising hundreds of feet above us. I nodded, explaining how rockslides, combined with flash floods, might have transported the boulders down to the lip of the waterfall. Of course, millions of years before those blocks of granite fell to the water's edge they began their journey by moving *up* from their original position deep underground. The granite at Painter Spring literally rose out of the Jurassic underworld to create the enchanting alcove surrounding the gurgling waterfall. Though they are rarely exposed with such grandeur as in the House Range, similar Mesozoic granitic rocks are scattered throughout the Great Basin. Such rocks add an intriguing subsurface dimension to the story of evolving landscapes that we began in the last chapter. In addition, these rocks have played a critical role in the development of some of the richest mineral deposits in the Great Basin. Just what is the nature of these Mesozoic granitic rocks, and what do they tell us about the ancient underworld?

SUBDUCTION, MAGMA, AND MINERALS

As a starting point, let's briefly reconsider granite. Granite is a rock characterized by large (that is, easily visible) intergrown crystals of a specific assemblage of minerals. This network of interlocking crystals is what gives granite its characteristic sparkle and glitter; the flat, reflective surfaces of the randomly oriented crystals flash light to the observer in a twinkling pattern as a sample is rotated. The actual size of the crystals in granite varies from barely visible to several inches, which explains why some granites appear more crystalline than others. The crystals in granite formed when magma cooled very slowly, over a period of perhaps thousands of years, deep underground. The intergrowth of relatively large crystals in granite and similar igneous rocks is described by geologists as **phaneritic** texture (Figure 7.1). In contrast, all volcanic igneous rocks represent magma that reaches the surface in the fluid form and cools relatively quickly. Volcanic rocks may contain crystals, but because the magma cooled rapidly, those crystals are either microscopic in size or are not completely intergrown. Volcanic rocks do not exhibit phaneritic texture. Geologists use the term **plutonic**, from Pluto, the Greek god of the underworld, to describe the underground origin of granite and related igneous rocks. The term **intrusive**, reflecting an origin *within* the earth, is sometimes used as a synonym for plutonic. A mass of granite thus represents a body of liquid magma that cooled, crystallized, and became solid deep underground. Accordingly, such a solid mass of rock is called a **pluton**, of which there are many types, depending on the size and shape of the rock body. Because a pluton can only form underground, its exposure on the surface is always an indication of uplift and erosion into the deeper levels of the crust.

The plutonic igneous rocks, all of which exhibit phaneritic texture, are subdivided on the basis of mineral composition. That is, granite contains a specific assortment of certain mineral crystals. Other varieties of plutonic igneous rocks are texturally similar to granite, and may look much the same, but have different assemblages of minerals. For example, the term granite specifically describes a phaneritic-textured igneous rock that contains about 20 percent quartz, 60 percent feldspar (most of which is a potassium-rich variety), 5 percent muscovite, (a silvery, flaky mica), with the remaining fraction composed of dark-colored biotite and/or amphibole (see Figure 7.1). Because the predominant constituents of granite—quartz, potassium feldspar, and muscovite—are light-colored minerals, granite typically has an overall light shade, though its actual color may be gray, tan, or pinkish. **Monzonite** is a phaneritic-textured igneous rock that has less quartz and more calcium or sodium feldspar than does granite.

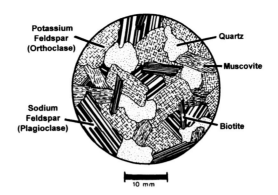

Figure 7.1 Microscopic View of Granite. Granite is a plutonic igneous rock consisting of large intergrown crystals of feldspar, mica, and quartz. Such crystalline igneous rocks can only form deep underground where magma cools very slowly.

Diorite has no quartz and more amphibole than does monzonite, while **gabbro** consists almost entirely of calcium feldspar, pyroxene, and olivine. The ratio between dark-colored minerals (biotite mica, amphibole, and pyroxene) and light-colored minerals varies in each of these rocks. Therefore, the overall shade of the crystalline igneous rocks can be a useful approximation of their identity: gabbro is typically greenish black, diorite is medium gray, granite is light gray or tan.

When magma cools slowly deep underground, the various minerals that crystallize from the fluid do not form simultaneously. Instead, they each crystallize over a specific range of temperatures. Though the sequence of crystallization can be very complicated, it does follow a temperature-dependent pattern that was first discerned by geologists in the early 1900s. This is the main reason why quartz, muscovite, and potassium feldspar are almost always found together in granite. Those three minerals crystallize at similar temperatures and, as a set, they constitute a natural and recurring association in igneous rocks. The same is true of the pyroxene-calcium feldspar-olivine assemblage in gabbro. As in most classification systems used in natural science, the names mentioned above represent our attempt to subdivide a continuum of variation in the mineral composition and texture of the plutonic igneous rocks. The distinction between granite and diorite, for example, is defined more or less arbitrarily for human convenience, not because any natural process prevents other mineral assemblages from existing. Many rocks have mineral assemblages that place them astride the subjective boundaries we have superimposed on the natural continuum of variation.

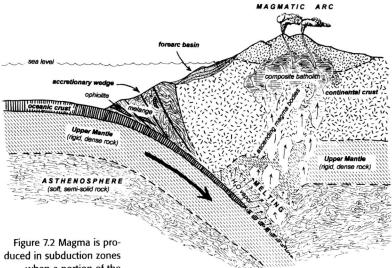

MAGMATIC ARC

forearc basin

sea level

accretionary wedge

ophiolite

Oceanic crust

melange

composite batholith

continental crust

descending magma bodies

Upper Mantle
(rigid, dense rock)

Upper Mantle
(rigid, dense rock)

ASTHENOSPHERE
(soft, semi-solid rock)

MELTING

HO vapor is subduction

Figure 7.2 Magma is produced in subduction zones when a portion of the descending plate is melted and water vapor is released in the lower crust and mantle. The magma bodies then migrate upward to erupt on the surface or solidify underground as part of a composite batholith. The magmatic arc is composed of both the batholith and the overlying volcanic edifice. To the west, materials scraped off the descending slab are piled against the continent as an accretionary wedge containing deformed strata (melange) and fragments of oceanic crust (ophiolites).

Such terms as **granodiorite** and **monzogranite** have emerged for these transitional rocks, expanding the lexicon of igneous rock names considerably in the process. Sometimes, the magma that crystallizes has unusual chemical properties that result in uncommon mineral associations. In this case, the classification of plutonic igneous rocks is further complicated by additional names for the odd-ball rocks, such as **syenite** (potassium feldspar, but no quartz) or **anorthosite** (almost all calcium feldspar, and little else). These are just a few examples of the many different names that have been applied to the Mesozoic plutonic rocks of the Great Basin. This proliferation of terms can be frustrating to the nongeologist, but it reflects the broad variation in the compositional and textural details of such igneous rocks. Though the plutonic rocks of the Great Basin go by many names, most of them can still be collectively referred as the "granitic" and "granitoid" rocks. This means that they all are of a fundamentally similar origin, even if the mineral compositions and/or textural features are not identical. Ironically, granite, as mentioned earlier, is actually relatively rare among these rocks; it is much less abundant than monzonite or granodiorite in the Great Basin. But all the plutonic rocks are "granitic" in that they represent magma that cooled slowly, deep underground, millions of years ago. But, where did the magma in the Great Basin come from, and why was it produced in such impressive quantity during the Mesozoic Era?

We have already mentioned that most magma is generated in subduction zones where one lithospheric plate descends beneath another. In understanding the origin of magma it is important to remember that the plates are actually composed of two slabs of rock: an upper layer of low-density crust and a lower mantle layer, composed of relatively dense rock enriched with iron and magnesium (Figure 7.2). Some plates carry only relatively thin and dense oceanic crust above the mantle slab, others may also have some thick and buoyant continental crust as well. The boundary between the crust and mantle, known as the moho, is located *within* the plate and is characterized by a sharp increase in density from the "lighter," or less dense, oceanic and/or continental crust to the "heavier", or more dense, mantle rocks. The two slabs of rocks, mantle and crust, are sutured together at the moho and comprise a solid plate that moves as a unit over a soft and weak zone, known as the **asthenosphere**. The asthenosphere, located between about 100 km (60 miles) and 350 km (210 miles) beneath the surface, is soft because the temperature at this depth is high enough to melt about 10–15 percent of the minerals in the mantle rock. This partial melting weakens the rock by converting it to a semi-solid "sludge" that flows very slowly in response to heat emanating from deeper zones of the earth's interior. It is commonly written in geology textbooks that a lithospheric plate descending into a subduction zone melts to produce bodies of magma when it encounters temperatures that are sufficiently high. In fact, some magma may be produced this way, but it's not just the descending plate that liquefies. Deep under the surface of the overriding plate, near the crust-mantle boundary, the rock is still rigid, but its temperature is very close to the melting point (on average, about 1000°C). When an oceanic slab descends into this zone of hot (but still solid) rock, it can induce melting by introducing a substance that is otherwise very rare in this deep subterranean realm: water. What does underground water have to do with magma, and how is it related to the subduction process?

Though lithospheric plates may carry either type of crust, only those that have oceanic crust can be subducted. For plates that have both types of crust, such as the North American Plate, only the portion with oceanic crust is prone to subduction. This is because continental crust is too thick and buoyant to be forced downward into the asthenosphere. It is oceanic plates, with their thinner and more dense crustal component, that more readily descend into the deeper interior. As they are subducted, the oceanic slabs carry a great amount of water, in several different forms. In our brief review of sea-floor spreading in Chapter 2, we learned that oceanic lithosphere is created by submarine volcanic eruptions at the axes of mid-ocean ridge systems. Once the oceanic crust forms, and before of the process of sea-floor spreading moves it away from the volcanically active region of the ocean floor, it may be affected by highly reactive fluids associated with hydrothermal vents along the ridge axes. Stewed in the vicinity of such submarine hot springs, several different alteration minerals may develop in the oceanic crust. Many of these minerals are hydrated, meaning that they contain water in their molecular lattices. Such water remains in the rocks as the oceanic crust spreads away from the undersea ridge. In addition, as the oceanic plate moves slowly away from the mid-ocean ridge, sediment accumulates on it. Depending on the length of time the plate travels before it reaches the subduction zone, the blanket of sediment that settles onto it may be several kilometers thick. This thick sedimentary cover also contains large amounts of water in the tiny spaces between the individual sediment grains. When the oceanic plate meets an opposing plate, some of this wet sediment is scraped off the top of the descending slab, folded against the leading edge of the overriding plate, and pushed upward in what geologists call an **accretionary wedge** (see Figure 7.2). However, a portion of the sediment, and the water it contains, remains affixed to the oceanic slab as it descends toward the zone of melting. Finally, the subducted oceanic lithosphere also carries seawater in tiny fractures that open when the plate bends and cracks as it begins to plunge beneath the overriding slab. After it enters a subduction zone, the oceanic lithosphere descends into the subsurface zone of high temperature, carrying with it a substantial store of water. Eventually, when the temperature of the descending plates rises to several hundred degrees (C), the water it carries is liberated as vapor or as a "superheated" fluid from the hydrated minerals, from the wet sediments, and from the water-filled cracks. This water is under extremely high pressure and penetrates along cracks and voids in the surrounding rock. After the water vapor escapes from the subducted lithosphere, the dehydrated slab continues to sink deeper into the subduction zone, into regions of even higher temperature. The melting of this slab begins at a depth of roughly 75 km and continues to about 250 km below the surface. Parts of the oceanic lithosphere may remain solid to a depth as great as 700 km, but the liquid formed from melting at such as great depth in the earth rarely escapes to the surface. It remains in the mushy asthenosphere, the molten residue of a vanished plate.

But what about the escaped water? Remember that when it was forced from the subducted plate, it penetrated into rocks of the upper mantle and lower crust adjacent to the descending plate. Those rocks, near the base of the overriding plate, are extremely hot but not yet molten. Under the conditions that exist deep in the earth, the water vapor has an interesting effect on the hot rock it migrates through: it causes melting! This effect is somewhat counterintuitive, since we most often think of water as a cooling medium that would be unlikely to cause melting. But, deep in the earth, water reacts with the hot rock much the same way that salt reacts with ice on a sidewalk on a cold winter day. Sprinkling salt on the sidewalk lowers the melting point of the ice, and melting begins even when the temperature remains constant. Water vapor reacting with the hot subsurface rocks has a similar effect: it reduces the melting point and the rocks begin to liquefy without any increase in temperature. In the case of both the icy sidewalk and the subterranean rocks, the temperature of the environment remains constant while the melting points of the materials present are reduced. The melting of both the ice and the rock is a response to the introduction of an exotic substance, salt or water, respectively.

To illustrate how effectively subsurface water can reduce the melting point of rock, consider the following: dry granite, at the surface of the earth, begins to melt at about 1,000°C, but deep underground in the presence of water vapor, this melting point is reduced by as much as 400°C. So, it is not just the descending slab that melts to produce magma in a subduction zone; in fact, the melting of the upper mantle and lower crust, triggered by the dehydration of the consumed plate, probably produces most of the molten rock associated with the process of subduction. Because both mechanisms of melting rock operate simultaneously whenever oceanic lithosphere descends beneath an overriding plate, the production of great quantities of magma is a primary consequence of the subduction process. The magma produced in this way usually contains large amounts of dissolved water vapor and other gases. Under the high pressures that exist deep underground, these gases remain absorbed into the magma, rather like the carbon dioxide that is present in carbonated beverages. However, these gases separate from the fluid at lower pressures, and can become an important influence on the way magma moves underground and the potential style of volcanic activity it may produce at the surface. During most of the Mesozoic Era, as the oceanic Farallon Plate was subducted under the western edge of the North American Plate, massive bodies of magma originated beneath the Great Basin and adjacent regions. Eventually, these magma bodies became the plutonic rocks we observe today at the surface. But, how?

Bodies of magma that form in a subduction zone can be thought of as "bubbles" of liquid rock, surrounded by hot rock of nearly the same temperature, positioned some 50–250 km (30–150 miles) underground. The cooler rock above the zone of melting is hard and rigid, forming what may seem to be an impenetrable barrier between the magma bodies below and the cold surface above. No granite, or any other type of igneous rock, would ever form unless the magma somehow moves into a cooler realm where it can lose heat to the surrounding environment. As long as the "bubble" of magma remains in the hot zone where it originated, it cannot cool sufficiently to crystallize and solidify. Magma must move *up*, toward the cold surface of the earth, to encounter the lower temperatures necessary for cooling and solidification to occur. Even though this upward movement is resisted by miles of overlying rock, there are several mechanisms that allow magma to penetrate the overlying barrier. First, remember that the magma bodies are fluids under high pressures. The pressure deep underground results from the weight of the overlying materials pushing down under the influence of gravity. In this manner, the pressure within the earth is similar to the pressure you may have felt in your sinuses at the bottom of a swimming pool: the water of the pool squeezes your head the same way that rocks in the earth squeeze materials buried deep underground. Underground pressure, commonly called **lithostatic** pressure, intensifies with depth due to the increasing weight of the overlying rock. On the surface of the earth, there is no lithostatic pressure. The only pressure that exists on the surface is the "weight" of the atmosphere, about 15 pounds per square inch. But 100 km (60 miles) underground, in the realm where magma is produced, the lithostatic pressure is 35,000 times greater than the atmospheric pressure at the surface. Because all liquids move toward lower pressures, magma bodies are actually "squeezed" toward the earth's surface. The pressures within a magma chamber may propel the molten fluid into cracks in the overlying rock with enough force to widen those cracks, resulting in an upward surge of additional magma. Recall, also, that the magma contains dissolved gases that will separate from the liquid at low pressures. As the magma migrates upward, gas pressure within the underground fluid increases, even as lithostatic pressure decreases. This internal pressure can help drive the magma even closer to the surface.

In addition to the pressure-driven movement of magma, the molten rock is also slightly less dense than the surrounding rock, at least where it initially formed. The density contrast between the magma and its surroundings causes to the magma to "float" upward against the ceiling of overlying rock at the same time that the external and internal pressures are also driving it in that direction. Furthermore, as the magma body migrates into the overlying rock, it passes into cooler and cooler material as it ascends toward the surface. This will eventually result in cooling of the magma, but at least initially the fluid is hot enough to melt some of the surrounding rock, assimilating the resulting liquid in the process. This means that the volume and chemistry of the magma may be affected to some degree by the rocks it penetrates en route to the surface. It also suggests another mechanism of upward movement: magma can actually "melt" its way to the surface.

For three reasons then, magma always tends to move up from its deep source: it is "squeezed" toward the surface at the same time that it tends to "float" in that direction, all the while "welding" holes in the overlying rock ceiling. In reality, the advancing front of the magma mass is probably highly irregular and uneven. Fingers of pressurized magma may protrude from the main body as internal pressures force the fiery fluid along cracks in the surrounding rock. Chunks and slabs of the solid wall rock may fall into the molten magma and become partially liquified in the hot solution. As a single magma body ascends, it may separate into several smaller masses that fill separate pockets and voids in the enclosing rock. The magma may squirt toward the surface rapidly along some zone of weakness, then remain motionless in an underground puddle until sufficient internal pressure develops to propel it upward again. This erratic, but relentless, upward movement sets the stage for the formation of igneous rocks because it carries the magma into cooler zones where it will begin to crystallize. If the magma rises through the overlying rock fast enough to arrive at the surface while still molten, volcanic eruptions occur. If, on the other hand, it ascends more slowly, it may cool and solidify as a pluton embedded in the subsurface rock. The largest plutons are known as **batholiths**, great irregular masses of granite that represent, in essence, an entire magma chamber that cooled and solidified deep underground. In the Great Basin, both volcanic and plutonic igneous activity resulted from the magma that rose through the crust in

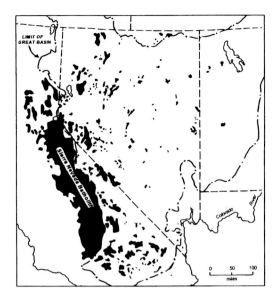

Figure 7.3 Distribution of Mesozoic Plutonic Rocks in the Great Basin. Granitic rocks are scat-tered across the entire Great Basin, but they are most abundant in the Sierra Nevada and the western part of the region.

the Mesozoic Era. Starting in the Triassic Period, magma produced by the subduction process ignited volcanic eruptions on the surface while, at the same time, numerous batholiths were forming deep underground. For more than 100 million years the magma continued to gush upward from the subduction zone, resulting in a massive swarm of plutons that coalesced beneath an extensive volcanic edifice. This area of Mesozoic igneous activity is known as the **Cordilleran magmatic arc**.

ROCKS OF THE CORDILLERAN MAGMATIC ARC

Even though the Cordilleran magmatic arc has been inactive for at least 80 million years, and the area it occupied in the Great Basin has now been stretched and highly fractured, geologists can still reconstruct its former extent on the basis of existing outcrops of Mesozoic igneous rocks (Figure 7.3). Because the entire region has been uplifted since the Mesozoic Era closed, most of the volcanic rocks related to the arc have been eroded and are exposed only in isolated places or as inclusions imbedded in the underlying plutonic rocks. This same erosion, however, has exhumed much of the plutonic rock that formed the foundation of the Mesozoic volcanoes. These plutonic rocks are widespread in western North America, including such impressive exposures as those of the Sierra Nevada and Klamath Mountains in California, the rugged ranges of Baja California, the

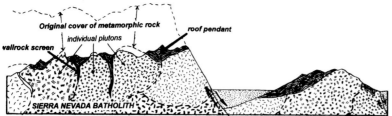

Original cover of metamorphic rock   roof pendant
individual plutons
wallrock screen

SIERRA NEVADA BATHOLITH

Figure 7.4 Schematic Diagram of the Sierra Nevada Batholith. The plutonic rocks of the Sierra Nevada comprise a composite batholith consisting of many smaller plutons. Roof pendants and wallrock screens embedded in the granitic rocks represent older materials, or portions of the overlying Mesozoic volcanic edifice, that have not been completely removed by erosion. Regardless of their original character, rocks in the pendants and screens have been highly metamorphosed. Note that plutonic rocks associated with the Sierra Nevada batholith extend east into faulted terrain of the western Great Basin, where they are exposed in scattered outcrops.

majesty peaks of central Idaho and western Montana, and the British Columbia coast ranges. In the Great Basin, exposures of Mesozoic plutonic rocks are concentrated along the western margin, but are scattered across the entire province (Figure 7.3). From these outcrops, it is clear that the Cordilleran magmatic arc extended from Canada to Mexico, following a general northwest-southeast trend. The original width of the arc is difficult to determine because, as we have already seen, the Great Basin has been ruptured and extended in an east-west direction in post-Mesozoic time. However, plutonic rocks that can be related confidently to the arc are now distributed from California to western Utah, a distance of more than 450 miles.

In the Great Basin, scattered remnants of the rocks erupted from the volcanoes of the Cordilleran arc can be recognized from the Sierra Nevada to the Jackson Mountains in the Black Rock Desert region of northwestern Nevada. In addition, some of the rocks comprising the Jurassic Pony Trail Group in Eureka County, Nevada, consist of volcanic ash and rubble that may have originated in the Cordilleran magmatic arc. In the Sierra Nevada, the volcanic rocks are present in several different **roof pendants**, which represent small remnants of older rock, including some that comprised Mesozoic volcanoes, that were later engulfed by ascending magma. In some places, the volcanic sequences are preserved as **wallrock screens**, partitions of older rock between adjacent plutons (Figure 7.4). East of the Sierra Nevada, some of the largest remnants of the Mesozoic volcanic sequence occur in the Alabama Hills and the Inyo Mountains of eastern California, in the Jackson Mountains and Pine Forest Range of northeast Nevada, and in the Gold Range area of west-central Nevada. These volcanic rocks all have been metamorphosed to one degree or another by the heat and fluids associated with the

magma that formed the plutonic rocks beneath them. For this reason, they are commonly referred to as **metavolcanic** rocks. Such rocks in the western Great Basin are generally rather drab, dark-colored, splintery rocks that appear to have been originally composed of volcanic ash and broken rubble, the product of violent and explosive eruptions. In addition, detailed studies of the metavolcanic sequences in the pendants of the Minarets (Cretaceous), Ritter Range (Jurassic), and Tioga Pass (Triassic) regions of the Sierra Nevada have revealed good evidence that some of the Mesozoic eruptions created large **calderas,** depressions formed when magma literally explodes from a subsurface reservoir. Crater Lake, in the modern Cascade Range of Oregon, is a famous example of a very youthful (6,600-year-old) caldera that offers a glimpse of the kind of eruptions that must have occurred in the western Great Basin during the Mesozoic. The six-mile-wide depression at Crater Lake resulted from a blast so powerful that at least ten cubic miles of magma was blown from the ancient volcano as a fine mist. The droplets of magma sprayed out of the Crater Lake volcano cooled in the atmosphere to form tiny grains of ash that traveled as far away as Saskatchewan before they settled out of the air! After the eruption, several thousand feet of the volcano collapsed into the evacuated magma chamber beneath the summit, creating the present flooded caldera. Though the products of their eruptions largely have been erased by erosion, it appears that the volcanoes of the Cordilleran arc were equally, if not more, violent. The ash from the Mesozoic volcanoes of the Great Basin fell as far away as Montana. It is very difficult to establish the size of the Mesozoic calderas, because later faulting, metamorphism, and erosion have obscured and/or obliterated much of the evidence. Nonetheless, the extent of the ash deposits they produced suggests that they were at least as large as Crater Lake.

Exposures of the plutonic rocks below the volcanic remnants are particularly abundant in the western Great Basin (see Figure 7.3). By far, the single most impressive exposure is in the Sierra Nevada, where granitic rock underlies roughly 15,000 square miles of beautiful alpine scenery. The crystalline rocks of the Sierra Nevada do not reside within a single pluton, but instead comprise an immense **composite pluton**, known as the Sierra Nevada batholith (see Figure 7.4). The Sierra Nevada batholith, 380 miles long and about 60 miles wide, is one of the largest masses of granitic rock in the world. Within this mass, scores of individual plutons are packed so close to one another that they have become fused into a coherent block of speckled rock. Geologists commonly refer to the structure of the Sierra Nevada batholith as similar to a bag of marshmallows. According to this analogy, the individual marshmallows in a package represent the internal plutons of the Sierra Nevada. Both the plutons and the marshmallows are bound together as a coherent mass: the marshmallows within a plastic bag, the plutons by geologic fusion to adjacent masses of rock. No one knows for sure how many individual plutons (or "marshmallows") make up the Sierra Nevada batholith, because some of them have already been erased by erosion, while others are still buried deep in the subterranean core of the mountains. Potentially, there may be hundreds of plutons in the Sierra Nevada batholith, as is suggested by the fact that just west of Bishop, California, geologists have identified no fewer than twenty-three major plutons (along with several smaller ones) exposed over an area of less than 3,000 square miles. Screens of metamorphic rocks sometimes separate the plutons of granite in the Sierra Nevada, while pendants of older rock are often embedded in them. East of the Sierra Nevada, across the Great Basin proper, exposures of Mesozoic plutonic rocks are much smaller and are widely scattered among the mountain ranges of eastern California, Nevada, and Utah (Figure 7.5). In particular, extensive outcrops of granitic rocks can be seen in the Toyiabe Range near Austin, Nevada; in the southern Wassuk Range (west of Walker Lake), Nevada; in the Pine Forest Range and Jackson Mountain of northwest Nevada; in the White-Inyo Range of eastern California; and in many places in the Mojave Desert. In general, the smaller plutons east of the Sierra Nevada represent individual bodies of magma that in-

Figure 7.5 Outcrops of weathered Mesozoic granite in the Toiyabe Range of central Nevada.

truded into deformed Mesozoic and Paleozoic rocks, crystallizing within tangled masses of older marine sediments or volcanic materials. Though they are minuscule by Sierran standards, the isolated plutons, like the granite at Painter Spring, are all related to the Cordilleran magmatic arc.

Over the past several decades geologists interested in the history of the Cordilleran magmatic arc have used radiometric dating techniques to establish the age of the Mesozoic plutonic rocks of the Sierra Nevada and the Great Basin. For plutonic igneous rocks, this sophisticated technique provides the most accurate and reliable age determination because it reveals the time that has elapsed since the crystals formed from the cooling magma. Of course, the magma that the rocks originated from was created earlier from older rock in the subduction zone. However, when old rock melts to form magma, the accumulated decay products are released while the residual radioactive elements, such as uranium and certain isotopes of strontium and potassium, tend to be reincorporated into the next generation of igneous rock. In effect, when the magma resolidifies through cooling and crystallization, the internal "atomic clocks" are reset. Because the products of radioactive decay will begin to accumulate in individual mineral crystals once they have formed, the atomic clocks start running as soon as the magma cools and solidifies. Thus, the radiometric dating technique results in ages that represent the elapsed time since the magma crystallized, regardless of how old the preexisting rock that melted to produce it might have been.

The radiometric ages of the Mesozoic

plutonic rocks in the Great Basin have revealed some interesting details about the history of igneous activity in the Cordilleran arc. Though all the Mesozoic plutons in the region share a common origin in the ancient subduction zone beneath western North America, they are not all the same age. A few of them crystallized in the Triassic Period, a few more in the Jurassic Period, but the vast majority are of Cretaceous age. This is particularly true in western Nevada and eastern California. In fact, 90 percent of the plutons that comprise the Sierra Nevada batholith are of late Cretaceous age, between about 120 and 80 million years old. Along the crest of the central Sierra Nevada, nearly all of the granitic plutons formed during an even shorter interval—from about 98 to 86 million years ago. Across the Great Basin region there is no precise and well-defined relationship between age and location of the Mesozoic plutons; however, younger (late Cretaceous) plutons generally are more common in the western Great Basin and the Mojave region, whereas older plutons (Jurassic or early Cretaceous, such as the Painter Spring body in western Utah) appear be somewhat more common to the east. Aside from this very general pattern, plutons of various Mesozoic age are scattered around the Great Basin in an almost random pattern. In any case, the radiometric dates clearly suggest that magma was not produced at a steady rate throughout the Mesozoic Era. About 100 million years ago in the late Cretaceous there was an upward surge of magma so impressive that to describe it as a "flood" is no overstatement. At that time, hundreds of magma bodies swarmed up from the subduction zone at virtually the same time, rather like the beginning of a hot-air balloon race. The volcanoes of the Cordilleran arc, sporadically active for 150 million years, exploded into fiery fury at this time. About twenty miles beneath the volcanoes, magma pooled in pockets and cavities, where it slowly crystallized into the plutons now exposed at the surface in the Great Basin region. Then, about 80 million years ago, the surge of magma stopped dead in its tracks. The volcanoes became quiet, dying as abruptly as they had been ignited 20 million years earlier. What could have caused such a dramatic spurt of magma from the subduction zone beneath western North America? Once again, the answer lies with the arrangement and interactions of lithospheric plates along the western margin of ancient North America.

LATE MESOZOIC PLATE TECTONICS

We have already seen that the subduction of oceanic plates under western North America began after the Sonoma Orogeny, which was related to the accretion of the Sonomia landmass. Throughout the Mesozoic Era, the principal plate that descended under the west-moving continent was the Farallon Plate, though there were probably other plates in the Pacific basin (Figure 7.6). The amount of magma produced in this subduction zone varied during the Mesozoic Era, depending on the speed of the Farallon Plate, the angle of its convergence with the North American Plate, the tilt of the subduction zone, the thickness and density of the descending plate, and the amount of water it delivered to the subsurface melting zone. None of these factors remained constant through the 180 million years of Mesozoic time, and therefore the amount of magma rising upward varied as well. The late Cretaceous "flood" of magma seems to have resulted from several factors that accelerated the melting of rock about 75 miles under the continent. Most geologists agree that there was a rather sudden increase in the rate of seafloor spreading at the Mesozoic triple junction in the Pacific basin (see Figure 7.6) in the late Cretaceous. Some estimates place this rate as high as 20 cm/yr (about 8 inches/year), compared to the current global average of about 5 cm/yr. If the Mesozoic sea floor in the Pacific basin was spreading four times faster than it is today, then the Farallon Plate would have descended into the subduction zone at a similarly high rate. More rock moving into the zone of melting means more magma will be produced deep underground. The extreme concentration of Cretaceous plutonic rocks in the Sierra Nevada and western Great Basin also implies a relative steep angle of subduction that would have localized the zone of magma intrusion as a linear belt running along the western margin of the continent. Earlier in the Mesozoic Era, in the Triassic and Jurassic Periods, the Farallon Plate was evidently moving slower, and descending beneath North America at a lesser angle. These conditions would have led to the production of less magma than in Cretaceous time, and the plutons would have been emplaced into the overlying crust over a broader area. This is precisely what the age distribution of Mesozoic plutonic rocks in the Great Basin suggests.

In addition to the flood of magma into the crust beneath the Great Basin, the convergence between the Farallon Plate and the North American Plate also produced forces that ruptured the Cordilleran magmatic arc along extensive northwest-trending faults. Most of these faults were eventually engulfed by the magma that was rising from the subduction zone at the same time that the faults were active. However, careful studies of the altered rocks in the pendants and screens of the Sierra Nevada, which represent older rock that was not completely consumed by the magma, have disclosed evidence for lateral displacement along the ancient faults. The blocks of rock west of the faults were generally shifted to the northwest, relative to the rocks on the east side of the faults (see Figure 7.6). In this type of lateral motion, rocks on either side of the fault would appear to have moved to the right, if we were to look across the fault from the opposite side. This "right-handed" lateral movement is known to geologists as **dextral slip**. Though much of the evidence for dextral-slip faults in the Cordilleran arc has been obscured by later events, their development is clearly linked to the transpression generated by the oblique convergence between the Farallon and North American Plates along the continental borderland. The Farallon Plate was, at the time, sliding toward the northeast toward westbound North America. As it slipped beneath the continent at an angle, the Farallon Plate produced forces that sheared slivers of rock from the overriding plate and shuffled them to the northwest along the plate boundary. This activity was probably most active during the Cretaceous Period, when the rates of convergence between the two plates was greatest. At this time, several dextral-slip faults probably developed in

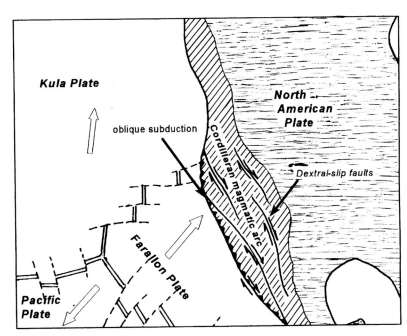

Figure 7.6 Late Mesozoic plate tectonic interactions along the west coast of North America. During Cretaceous time, the Farallon Plate was moving toward North America from the southwest to meet the continental margin at an oblique angle. As the Farallon Plate was subducted beneath the continent, the oblique convergence generated numerous dextral-slip faults in the Cordilleran magmatic arc.

the Cordilleran arc, breaking the region into elongated slivers of rock displaced to the northwest, relative to each other. Geologists Richard Schweikert and Mary Lahren have suggested that the block west of one ancient fault, the Mojave–Snow Lake Fault, moved northward by about 300 miles. This estimate was based on the displacement of pre-Cretaceous rocks presumed to have been on opposite sides of the cryptic fault. Other dextral-slip faults in the Cordilleran arc, such as those in the Pine Nut Range region (northwest Nevada) and the Lake Tahoe region, may have had similar displacements; but the amount of movement along these old faults is difficult to establish with precision because much of the evidence has been masked by magma intrusion, metamorphism, or later faulting. The famous San Andreas Fault system in California is also a dextral-slip fault, albeit one that developed much later in time and in a different plate tectonic setting from the Mesozoic faults of the Cordilleran arc (see Chapter 8). Nonetheless, the modern San Andreas system and the ancient faults of the western Great Basin surely had something else in common: earthquakes!

Even as the great volcanoes were exploding throughout the Cordilleran arc, powerful earthquakes associated with the dextral-slip faults were rocking the entire Great Basin region. These earthquakes must have been at least as violent as those that occur along the San Andreas system, and they probably were more frequent because of the numerous faults that were active during Mesozoic time. In the modern world, most of the earthquakes associated with major lateral-slip faults occur at relatively shallow depths. This is because there is no mechanism to transmit stress deep underground when blocks of rock slide past each other laterally. When the geological forces overcome the friction between the blocks along the fault, rocks shatter at shallow depths to produce strong seismic vibrations. Such vibrations travel a short distance upward to the earth's surface, and are not significantly weakened when they arrive there. Consider, for example, that even though the crust is nearly 20 miles thick along the modern San Andreas system, almost all of the earthquakes it generates result from underground ruptures that occur within 8 miles of the surface. Because the seismic energy is released so near the surface, the ground vibrations can be strong enough to pose a serious risk of catastrophic destruction. Though the ancient dextral-slip faults of the Great Basin developed in a different plate tectonic setting, they probably produced a pattern of earthquake activity similar to that of the modern San Andreas system. The earthquakes that shook the Great Basin during the Mesozoic Era must have been thrilling accessories to the periodic volcanic blasts that roared from the calderas and darkened the sky with ash. The West has never been more wild than it was in the Cretaceous Period!

RICHES FROM THE MAGMA

The Great Basin is world renowned for its incredibly rich mineral deposits. Any casual exploration of the back country of Nevada and western Utah is certain to reveal at least one old mining camp "ghost town" reflecting the endless cycle of boom-and-bust that has always typified mining activity. In some places in the Great Basin, several such graveyards of former prosperity exist within a few miles of each other, clustered near what was formerly an immense body of mineralized rock. So rich and extensive are the mineral resources of the Great Basin that the role of the mining industry in the historical development of towns and cities across the region cannot be overstated. Nor is it possible to dismiss the political, economic, social, and environmental influences of the modern high-tech mining operations that exist today in even the most remote corners of the region. After more than 160 years of gouging, scraping, drilling, blasting, and boring, there is still enough quality ore in the Great Basin to currently sustain no fewer than 150 active mining operations. In 1999, the total value of mineral products produced in Nevada alone exceeded $2.9 billion, 78 percent of which was attributable to the 8.26 million ounces of gold that was produced that year. If we include the value of mineral products produced from western Utah, eastern California, and the Mojave region, these numbers become even more impressive. In addition to gold, the mines of the Great Basin have produced great quantities of silver, copper, lead, zinc, iron, molybdenum, mercury, uranium, tungsten, beryllium, and other metals. A geology student, after learning about the geological history and mineralization in the Great Basin, once remarked to me that she felt as if nature had compensated for so severely mangling the rocks of the region by sprinkling them with glittering metal. She may have been right: some of the richest ore bodies are located in the most highly deformed and altered host rock. Moreover, in many of the mining centers of the Great Basin, the events of the Mesozoic Era played a critical role in the localization of mineral riches.

As we have already learned, the Mesozoic Era was a turbulent time in the geologic history of the Great Basin. The combination of compressive deformation and magmatic invasion during this era subjected the older rocks to elevated temperatures, crushing forces, and corrosive vapors and fluids. Under such conditions, rocks experienced some degree of metamorphism almost everywhere in the region. In addition, as we will see in the next chapter, the volcanic and tectonic unrest continued into the Cenozoic Era, so that any rock that escaped metamorphism in the Mesozoic might have been affected by similar unrest million of years later. It is small wonder, then, that most outcrops of Mesozoic rocks in the Great Basin exhibit some evidence of alteration. With so much magma moving through the crust during that time, and with the tectonic forces so extreme, metamorphism was almost guaranteed. Although the development of the rich ore bodies in the Great Basin continued into the Cenozoic Era, the Mesozoic maiming set the stage for the extensive mineralization of rock bodies across the region.

The association between mineralization and Mesozoic plutonic rocks can be illustrated in many places in the Great Basin, but it is especially well documented in the Robinson and Hamilton Mining Districts near Ely, in the Eureka Mining District in central Nevada, and in the Yerington District of western Nevada. While each of the ore bodies related to the Mesozoic plutons are unique, with different histories resulting in varying mineral assemblages, there are some general features that apply to all of them. In all of these areas, the first step in forming ore minerals involved the **intrusion** of magma during the Mesozoic Era, usually during the Jurassic or Cretaceous Periods. As the molten rock ascended toward the surface, the heat associated with it began to transform the surrounding older rock. A narrow zone of metamorphic rock developed around the periphery of the magma body as it slowly cooled deep underground. After the magma completely solidified, the resulting granitic pluton was surrounded by a metamorphic halo that may have contained minerals richer in metals than the crystals that comprise the granite. This is because the minerals in the metamorphic zone, unlike the quartz and feldspar in the granite, developed from the transformation of older rock that differed chemically from the magma. In some forms of metamorphism, elements are chemically repackaged into new combinations that may have elevated concentrations of valuable metals. One good example of this process is the formation of **skarns**, a Swedish term for calcium-rich metamorphic rock assemblages that result from the interaction between magma and impure limestone (Figure 7.7).

Recall that thick sequences of silty limestone and dolomite formed in the Great Basin during

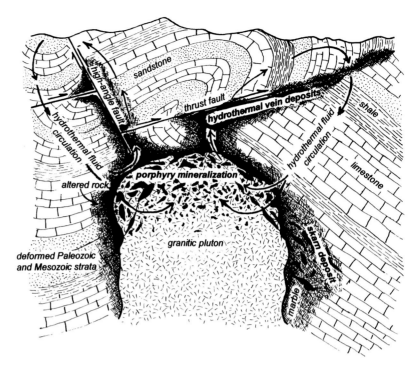

Figure 7.7 Ore deposits associated with plutons. The intrusion of a mass of magma may result in several different types of ore deposits, including skarns, hydrothermal veins, and porphyry deposits.

the Paleozoic and Mesozoic Eras. After these strata were deformed during multiple orogenic events (Antler, Sonoma, and late Mesozoic disturbances), they were penetrated by the numerous bodies of magma that rose through the crust during the late Mesozoic Era. Responding to the invading magma, the older limestone sequences were commonly transformed by metamorphism into skarns near the periphery of the granitic plutons. Such skarn assemblages are dominated by marble, the metamorphic equivalent of limestone. Marble is composed primarily of ordinary crystalline calcite (calcium carbonate), and is not particularly valuable. But in skarn deposits, the marble is commonly accompanied by a variety of other minerals, such as calcium-bearing garnet, diopside and wollastonite (a calcium pyroxene), and vesuvianite (a complex calcium silicate mineral). Less commonly, but often enough to stimulate significant mining activity, concentrations of minerals such as scheelite and wolframite, both excellent sources of tungsten, occur within skarns. In the Great Basin, tungsten skarn deposits include those mined at Mill City, Nevada, and at the Pine Creek mine along the east side of the Sierra Nevada. In the Mount Hamilton area, about forty-five miles west of Ely, Nevada, skarn mineralization associated with Cretaceous

Figure 7.8 Altered and mineralized rock in the Singatse Range of western Nevada, several miles from the Yerington batholith.

Figure 7.8 Altered and mineralized rock in the Singatse Range of western Nevada, several miles from the Yerington batholith.

intrusions has produced ore bodies enriched in molybdenum, bismuth, tungsten, gold, tellurium, copper, and zinc.

In addition to the formation of skarns along the margins of plutons, magma intruded during the Mesozoic commonly had a more extended effect on the surrounding rock. As the magma migrated upward into lower-pressure environments, the gases dissolved in it began to escape into the surrounding rock. At some distance from the rising magma body, these gases would condense into briny fluids laced with reactive chemicals such as sulfurous compounds and corrosive acids. The hot fluids and gases could move great distances from the magma along faults, fractures, and bedding planes in the enclosing rock. Such migration of reactive fluids extended the zone of alteration far beyond the border zone surrounding the pluton (Figure 7.8). Reactions between the older rock surrounding the magma and the fluids migrating from it caused the precipitation of many different minerals in veins or pods dispersed through the host rock. Sulfide minerals (those consisting of metals combined with sulfur) dominate the mineralized veins due to the high concentration of sulfur in the gases and fluids that typically emanate from magma. Examples of such minerals include bornite (Cu_5FeS_4), chalcocite (Cu_2S), and chalcopyrite ($CuFeS$), all of which are very important sources of copper (Cu). Because the fluids and vapors would migrate most effectively

along planes of weakness in the rock surrounding the magma, such ore deposits are often aligned along ancient thrust faults or along boundaries between layered rock successions (see Figure 7.7).

In addition to the vein and skarn ore deposits, sometimes the granitic rock itself became mineralized. This internal mineralization can happen when the magma migrated upward far enough to come into contact with water percolating downward from the surface. This surface water, along with other water condensed from the magmatic vapors, was heated and driven back toward the surface. Near the surface, the rising hot water could cool and sink back toward the subterranean mass of molten rock, only to be reheated and forced upward again. In this manner a complex **hydrothermal** system involving cyclic circulation of hot underground water would develop as the magma rose toward the surface (see Figure 7.7). As the magma below slowly solidified, the fluids might begin to circulate through the upper, solid portion of the pluton. There, in the newly formed granite, exotic minerals might be introduced into the normal assemblage crystallized directly from the magma. Depending on what sort of rock the hydrothermal fluids circulated through, the exact temperature of the fluid, and the concentration of various elements in it, a great variety of different minerals could be precipitated from such a system. Some of the minerals might form where hydrothermal fluids moved up or down through the rock surrounding the cooling pluton, while others might precipitate as mineral masses dispersed through the granite. In fact, the hydrothermal ore deposits of the Great Basin are among the richest and most heterogeneous types known in the region. In the Great Basin, very distinctive types of hydrothermal ore deposits that are especially important are the **porphyry copper** systems. In such deposits, the circulation of hydrothermal fluids through a portion of a pluton leads to the formation of copper sulfide minerals within the granitic mass. This process usually leads to the mineralization of the upper portion, or sometimes the sides, of the pluton where large crystals of normal feldspar and mica have already formed. The term **porphyry** is used to describe a plutonic igneous rock with larger than average crystals. The ore from a porphyry copper deposit is typically a coarse-grained granitic rock sprinkled with glittering copper-bearing sulfide minerals such as chalcopyrite. Usually, several zones of mineralization surround the unaltered core of an igneous intrusion, each with a unique set of minerals. The metal-bearing sulfides in the granite generally comprise less than about 20 percent of the minerals in the rock. However, in places where the mineralized granite has been exposed at the surface, the weathering process can leach (dissolve and remove downward by percolation) some of the minerals from the surface exposures and carry them deeper underground, where they may eventually be re-precipitated. This process of concentration, known as **supergene enrichment,** can lead to the development of ore bodies much richer in metals than was the original porphyry ore. In many of the porphyry copper deposits of the Great Basin, molybdenum, platinum, gold, or silver also can be extracted from the primary copper ore. Though the accessory metals are always much less abundant than copper in porphyry systems, these commodities are so valuable that they can sustain mining operations that would otherwise be unprofitable. In the Great Basin, large porphyry copper deposits include those at Bingham, Utah; Battle Mountain, Nevada; and the Robinson District near Ely, Nevada (Figure 7.9).

Because several different types of mineralization are associated with magma bodies migrating into the crust, it is easy to understand the extreme richness and geologic complexity of the mining districts in the Great Basin that are centered around Mesozoic (and younger) plutons. The Yerington District, on the flanks of the Singatse Range in western Nevada, is just one of many examples in the region of an area that experienced multiple episodes and styles of mineralization, primarily during the Mesozoic Era. Mining began in the Yerington area in 1865, when copper-rich veins were discovered in garnet and epidote-bearing skarn deposits. The skarn deposits were associated with two Jurassic plutons in the Singatse Range, about 168 mil-

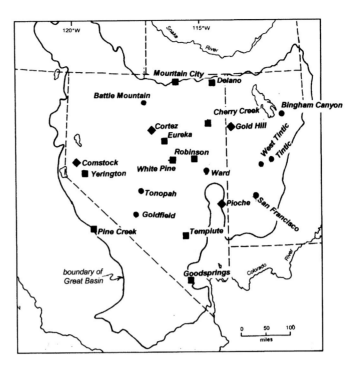

lion years old, that intruded into older (Triassic) marine sedimentary and volcanic rocks and are capped by roughly coeval volcanic materials. Some of the veins were related to fingerlike projections from the plutons, called **dikes,** that extend in a northeast-southwest direction across the region. Initial mining in the Yerington area was directed toward the copper vein deposits, but there was only minimal production until 1912. By 1930, the vein deposits had been depleted. However, a porphyry copper deposit was discovered in the 1940s, and Anaconda Mining Company began to develop this deposit in the 1950s. Between 1952 and 1978, more than 160 million tons of ore were mined from an immense open pit (Figure 7.10). The richest ore from the Yerington porphyry copper deposit averaged about 0.6 percent copper, contained mostly in the minerals chalcopyrite, chrysocolla, bornite, and covellite dispersed through the plutonic rock. The relatively low-grade and dispersed ore at Yerington required the mining of an enormous volume of rock via open-pit techniques, an operation that is typical for all porphyry copper deposits situated at or near the surface. Some of the largest holes that humans have ever excavated in the earth are open-pit mines developed to exploit porphyry copper systems, and there are several classic examples in

Figure 7.9 Major mining districts of the Great Basin associated with plutonic rocks. Mineralization in these areas resulted from intrusion of magma during the Mesozoic Era (■), the Cenozoic Era (●), or both (◆).

Figure 7.10 The Yerington open pit, currently inactive and partially flooded.

the Great Basin. In spite of the low grade of the ore, more than 800 million pounds of copper was produced from the Yerington pit between 1953 and 1965. Today, with copper prices declining and the best ore depleted, the Yerington mine and the milling and processing facilities at nearby Weed Heights are inactive. But it took decades of feverish mining activity at Yerington to extract the metallic wealth that geologic events of the Mesozoic Era localized there. Across the Great Basin, there are many other examples of such rich and varied ore deposits clustered around Mesozoic plutons that were emplaced under the Cordilleran magmatic arc and adjacent regions.

THE MESOZOIC-CENOZOIC TRANSITION IN THE GREAT BASIN: THE STORM SUBSIDES

The igneous activity in the Cordilleran magmatic arc appears to have concluded around 80 million years ago, the approximate age of the youngest Mesozoic granitic rocks in the region. A few million years after the upward flood of magma in the Great Basin diminished, the Rocky Mountain region to the east began to experience deep-seated compressive forces. Deformation and uplift related to these forces initiated the development of the modern Rocky Mountains during an event known as the Laramide Orogeny. While the Great Basin was unaffected by the Laramide disturbance, the

cessation of igneous activity during the earliest Cenozoic Era may have resulted from the same plate tectonic mechanism as the mountain building to the east. It appears that the Farallon Plate, which had been descending under western North America rapidly and at a high angle during the Cretaceous Period, flattened out under the continent in early Cenozoic time (Figure 7.11). This flattening of the subduction zone may have been the consequence of North America approaching the spreading ridges in the proto-Pacific basin, as it rapidly overran the Farallon Plate sliding beneath it. As the continent drew closer to the source of the Farallon Plate, the subducted oceanic slab became younger, hotter, thicker, and more buoyant. Such an oceanic slab would be less likely to bend downward at a high angle than the older, thinner, and less buoyant part that was subducted earlier in the Mesozoic Era. In any case, if the subduction zone flattened, then the oceanic plate would not descend deep enough under the Great Basin region to generate magma. This concept may also explain the compression that resulted in the Laramide Orogeny: as the Farallon Plate scraped along the base of the continent at a low angle, it could have transmitted compressive forces to the Rocky Mountain region, far inland from the actual plate boundary. So, the end of magmatic activity in the Great Basin and the initial rise of the Rocky Mountains may be related to a single plate tectonic event—a shift to low-angle subduction of the Farallon Plate.

By the time the final Mesozoic magma body hardened underground, scores of plutons had been jammed into the crust beneath the Great Basin, most of them comprising the composite batholiths beneath the western volcanic highlands. Beyond the magmatic arc itself, in eastern Nevada and western Utah, the last few magma bodies generated by subduction rose into the crust, baking subterranean rocks as they migrated upward. The rocks metamorphosed at this time would surface millions of years later when the metamorphic core complexes developed during early to mid-Cenozoic time (we will explore these features more thoroughly in the next chapter). Along the eastern margin of the Great Basin, deformation related to the Sevier Orogeny was still in progress as the plutonic fires flickered, but it would only persist for another 10–15 million years before the wave of mountain building passed eastward into the Rockies. When the Cenozoic Era began in the Great Basin about 65 million years ago, the geological commotion of the late Mesozoic had given way to an epoch of relative tranquility.

For the first 15–20 million years of the Cenozoic Era, erosion was the dominant process in the Great Basin region. Great volumes of rock were worn away from the summits of the mountains that had been built during the late Mesozoic, and the resulting debris was washed into low areas adjacent to, or within, the dwindling highlands. The sediment shed during the early Tertiary Period (the first period of the Cenozoic Era) was very similar to the sand and gravel worn from the highlands during the preceding Cretaceous period, and it commonly accumulated in the same isolated basins. For this reason, it can be very difficult to separate early Cenozoic sediments from the underlying late Cretaceous strata. In the Pinyon Range near Carlin, Nevada, for example, late Cretaceous and early Tertiary sediments are grouped together within the Newark Canyon Formation. Cretaceous and Tertiary sediments probably also occur together within the Sheep Pass Formation in the Egan Range of Nevada, though most of this formation is composed of early Tertiary lake beds. In the Diamond Mountains northeast of Eureka, Nevada, coarse, river-deposited gravel that appears to merge with the upper part of the Newark Canyon Formation (Figure 7.12) is also probably early Tertiary in age. In the El Paso Mountains, along the southwest edge of the modern Great Basin, the conglomerate and sandstone of the Goler Formation was deposited by west-flowing rivers during the earliest part of the Tertiary. Though exposures of early Tertiary sediments are rare and widely scattered in the Great Basin, they record the local accumulation of significant quantities of sand and gravel during a period of widespread erosion in the region. The scarcity of early Cenozoic deposits in the Great Basin suggests that either much of the geological litter produced by erosion at that

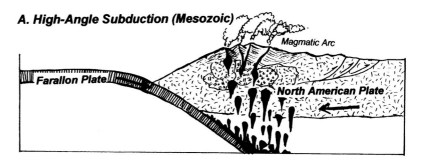

A. High-Angle Subduction (Mesozoic)

Magmatic Arc

Farallon Plate

North American Plate

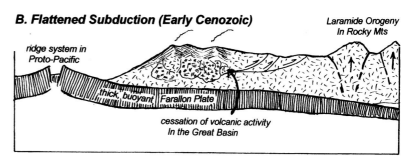

B. Flattened Subduction (Early Cenozoic)

Laramide Orogeny In Rocky Mts

ridge system in Proto-Pacific

thick, buoyant Farallon Plate

cessation of volcanic activity In the Great Basin

time was washed completely out of the Great Basin region or it is currently buried under younger deposits. Along the western side of the region, recent studies have identified the ancient channels, or "paleovalleys," of early Cenozoic rivers that flowed from the Walker Lake area of western Nevada to central California, directly across the present western boundary of the Great Basin. These paleovalleys clearly demonstrate that the Great Basin was not an area of internal drainage 45 million years ago; instead, it was a rolling highland, with a few local basins, and with river valleys that extended well beyond the present limits of the province. Coincidentally, the western paleovalleys were the conduit through which much of the gold-bearing gravel in California was transported and deposited in early Cenozoic time. The famous "auriferous gravels" that touched off the California gold rush actually consist largely of sediments washed out of Nevada millions of years before their discovery by miners and sawmill operators. Early Tertiary paleovalleys draining east, into eastern Utah and Colorado, have not yet been clearly identified. However, the widespread river and lake sediments of the Flagstaff Formation and the upper North Horn Formation of central Utah were probably washed into place during early Tertiary time by rivers draining the eastern Great Basin highlands. Throughout the early Cenozoic Era, the topographic relief across

Figure 7.11 Early Cenozoic flattening of the Farallon Plate subduction zone. The decrease in the angle of subduction under western North America may explain the end of igneous activity in the Great Basin and the coeval uplift in the Rocky Mountain region.

Figure 7.12 Early Tertiary conglomerate in the northern Diamond Range of central Nevada.

the Great Basin became increasingly subdued as the mountains were planed down to lower elevations, while sediment piled up in the small, isolated basins. In spite of the vigorous erosion in the region, the Great Basin remained an elevated tract for the first 15–20 million years of the Cenozoic Era.

During the quiet post-Mesozoic interlude, a pervasive erosion surface was developed across most of the Great Basin. Almost everywhere in the central and eastern parts of the region, that erosion surface is now represented by a major unconformity that separates deformed pre-Tertiary rocks from middle Tertiary and younger strata. These erosion surfaces, such as the Confusion Mountains unconformity discussed at the beginning of the last chapter, started to develop in the late Mesozoic Era but were en-hanced and deepened by additional wearing down in early Tertiary time. About 45 million years ago, the geological tranquility of the Great Basin region was shattered by a new wave of fiery havoc that swept across the region. The rolling plain created by the long interval of quiet erosion was again jolted by powerful geological forces, beginning in mid-Cenozoic time. Across the region, mountains heaved, valleys collapsed, and volcanoes detonated. These convulsions heralded a new phase of geological evolution in the history of the Great Basin, during which the earliest outlines of the modern province began to emerge. The tumultuous beginnings of the modern Great Basin must be considered one of the most violent phases in the turbulent geological history of the region. The middle Cenozoic Era was a time when, quite literally, all hell broke loose in the Great Basin.

More than one hundred billion days have elapsed since the ancient supercontinent of Rodinia was ripped apart and the progression of geologic events leading to the modern Great Basin began. One hundred billion times stars twinkled against the inky black night sky, just as they do today in the dark corners of the remote desert. The configurations of the stars in prehistoric constellations were undoubtedly different from the familiar modern patterns, but stars still gleamed overhead in the primeval darkness. Sometimes during the deep geological past those stars glittered above tropical seas, sometimes they were veiled by clouds of volcanic ash, and sometimes they rose and set over majestic mountain peaks. Almost certainly, primordial eyes of some sort must have noticed the stars that floated above the prehistoric landscapes. Perhaps the ichthyosaurs of the Triassic seas, with their large eyes well adapted to darkness, looked upward at the cosmic canopy as they bobbed in the undulating swells. Maybe the dinosaurs, like many migratory animals in the modern world, were guided by the stars when they crossed the rolling hills of the Great Basin uplands during the Cretaceous Period. Could the gleaming eyes of the primitive reptiles in the Mississippian forests have captured the ancient starlight? Could the trilobites have glimpsed the glow of the Milky Way through the crystal clear tropical waters of the Cambrian seas? Maybe. But, no celestial sight beheld by any eyes, anywhere, any time could have surpassed the stunning spectacle that arched across the Great Basin sky on the night of March 23, 1997.

On that night, I was leading a group of students on a geology field trip to Death Valley, as I had done every spring for the preceding twenty-five years. Spring field trips to desert regions are long-standing traditions in the geology departments of most western colleges and universities. It is easy to understand the appeal of a weeklong jaunt in the warm, sunny deserts of the American West after several months of winter sequestration in gloomy classrooms, libraries, and labs. For geology students and faculty, the post-winter field trip is even more refreshing than the traditional sprees that have made the collegiate "spring break" infamous. A field trip to the Great Basin desert releases just as much pent-up tension as does an excursion to Florida, but it does so in a stunningly beautiful wilderness rather than in a dismal, overpopulated, and boisterous swamp. I don't recall ever having an empty seat on any of the more than thirty spring field trips that I have led or attended. Such is the allure of the Great Basin desert in the spring.

Our initial destination in 1997 was Death Valley, which had been the starting point for our spring desert junkets for years. It took two days to reach Death Valley from our northern California campus because there is so much scintillating geology to view along the way. While there are many places in the western Great Basin to pause en route to Death Valley, my favorite midway campsite is an area known as The Sump, tucked away in the hills adjacent to the northern Fish Lake Valley of western Nevada. The Sump was not just conveniently located near our route, but it is also one of the best places in the Great Basin to glimpse a piece of mid-Cenozoic history. In this bowllike pocket are beautiful exposures of tilted strata of fossiliferous sedimentary rocks, interlayered with pale greenish gray beds of volcanic ash (Plate 12). These rocks accumulated during the middle Cenozoic Era in a down-faulted depression as

CHAPTER EIGHT

THE ARMY OF CATER-PILLARS AND THE CONFLA-GRATION

the modern basin-and-range topography was beginning to take shape. The area was especially significant on our spring field trip because one of our primary themes in exploring Death Valley, and of this chapter as well, is the evolution of the Cenozoic basins. We were, after all, headed for the grandest and deepest basin in the entire region, and The Sump offers a glimpse of the early events that resulted in such features. We were chatting about those basins, the faults that bound them, and the forces that created them as we casually crawled over the sandstone layers looking for fossils during the final hour of daylight on March 23, 1997.

There was a chilly breeze in The Sump that night. We circled the vehicles as a windbreak and built a large fire in the center for warmth as the evening meal was prepared. Some of the students had found a fragment of a jawbone in the sandstone and it was being passed along for examination of the group in the dim firelight. From the shape of the teeth, I thought it was the remains of a dog of some sort, but reserved final judgement for the bright light of morning. The layers of rock at The Sump are famous for the remains of Cenozoic mammals that have been collected there over the past fifty years. In addition to dog remains, fossils representing camels, pigs, cats, horses, and other primitive mammals have been identified at the site. Though most of the best fossils were collected by scientists years ago, we always seem to find a scrap or two of bone on every visit to The Sump. The fragmental fossils always provide for entertaining speculations on their identity and stimulate lots of discussion about the ancient fauna that lived in western Nevada. In addition to all its other attributes, The Sump is a mid-Cenozoic graveyard. Which, of course, was another good reason for spending a night there during our spring excursions.

Soon after the sun set below the western rim of The Sump, I suggested a hike up the nose of a gentle ridge to view Comet Hale-Bopp. This celestial visitor had been putting on a spectacular show throughout the month of March, and I was anxious to see how it looked against the clear, dark sky of the remote Great Basin desert.

Most of the students followed me up the ridge, and when we neared the crest of it, all faces turned west to find the comet. From our vantage point on the ridge, the jagged profile of the White Mountains of eastern California framed the western horizon. Hanging just above the angular profile of the distant mountain crest, Comet Hale-Bopp blazed like a great luminous feather, its forked tail arcing upward across the western sky. Though most of us had seen the comet before, none had viewed it through skies so lucid as those above The Sump that night. In the chilly desert darkness, the comet seemed to be almost alive with color and light. All eyes were riveted to it for several minutes before we noticed something almost equally spectacular over the opposite horizon. To the east, the moon had just risen as a thin crescent peeking above the dark outline of the Silver Peak Range. Lost in the beauty of the moonrise, it took me a minute or two to realize that something was amiss: a crescent moon cannot rise in the east at sunset. Only near *sunrise* does the proper sun-moon geometry exist for crescent moons to ascend in the eastern sky. Unless, of course, it is the shadow of the earth that darkens the disc of the moon. I mentioned the dilemma to the students, and we watched for another few minutes for evidence of a lunar eclipse. Sure enough, as the moon gradually lifted above the horizon, the ruddy orange color, typical of the earth's shadow on the lunar surface, became obvious. For the next hour, we all sat on the ridge and watched the thin crescent widen as the moon passed out of the planetary shadow. Huddled against the cold wind on that night, we were treated to the most spectacular celestial display any of us had ever seen: an eclipsed moon hanging like a rusty beach ball over the eastern horizon, while Comet Hale-Bopp glittered to the west. When the eclipse ended, the restored brightness of the full moon made it more difficult to see the comet, the head of which had dipped below the horizon. With the cosmic display over, we all descended the ridge back to camp, a mood of solemn appreciation for the sights we had witnessed overtaking all of us. Such scenes cannot be observed every day, and

the beauty of that celestial vision is still permanently burned into my memory. I am sure that
none of those who climbed that ridge on the
night of March 23, 1997, will ever forget what
they saw.

The next morning, we packed up to continue
the journey to Death Valley. In preparing to
leave, I found the jaw fragment discovered the
previous evening in my shirt pocket. I stuck
with my initial identification of the fossil as the
remains of a primitive doglike creature, and
passed it around for final inspection, before returning it to the ground where it was found. I
was asked about the frequency of comets and
eclipses by a student, obviously still charmed by
the events witnessed on the ridge the night before. I didn't have an answer to the question of
how often comets accompany lunar eclipses,
but I suggested that the probabilities of a co-
occurrence of such fleeting celestial phenomena
must be very low. Pondering the fossil in my
hand, I wondered: had we been born on the day
that the dog died and could somehow have lived
for the millions of years that have since elapsed,
how many times would an eclipsed moon and a
bright comet have simultaneously appeared in
the night sky? Once? Twice? Never? Might it be
millions of years before such a conjunction re-
occurs? How lucky were we all to have witnessed
such a rare cosmic confluence? The rock layers
surrounding us at The Sump, millions of years
old, seemed to provide the perfect foundation
for grasping our good fortune. As we discussed
the incredible history that those strata record,
the eons of change in land and life embedded in
them, the singularity of our evening became
somehow more comprehensible. But, appreciating our experience in the modern world is only
one of the fringe benefits of contemplating the
middle Cenozoic rocks of the Great Basin. There
are fascinating stories embedded in those rocks,
tales that make the ground under our feet as enchanting as the rare events in the sky overhead.
The rocks, of course, are more dependable than
transitory comets and eclipses. They can be experienced anytime by anyone who ventures into
places like The Sump.

C E N O Z O I C E R A	QUATERNARY PERIOD	Holocene Epoch	0.01 m.y.
		Pleistocene Epoch	1.8 m.y.
	TERTIARY PERIOD	Pliocene Epoch	5.5 m.y.
		Miocene Epoch	24 m.y.
		Oligocene Epoch	34 m.y.
		Eocene Epoch	55 m.y.
		Paleocene Epoch	65 m.y.

Figure 8.1 Time scale for
the Cenozoic Era
(m.y. = million years).

MIDDLE CENOZOIC ROCKS
OF THE GREAT BASIN

As a general rule, exposures of Cenozoic rocks
are generally more widespread and less deformed than those of older strata, everywhere in
the world. Not surprisingly, geologists have uti-
lized the wealth of data derived from Cenozoic
rock sequences to develop a relatively refined
time scale for the past 65 million years (Figure
8.1). In this time scale, the Cenozoic Era is divided into two dramatically unequal periods:
the older Tertiary Period (65–1.8 million years
ago), and the later Quaternary Period (1.8 million years ago to the present). The Tertiary comprises 97 percent of Cenozoic time, while the
Quaternary encompasses only the most recent
ice ages and post-glacial time. Because these
subdivisions are so highly disproportionate, the
terms "middle Cenozoic" and "middle Tertiary"
are nearly synonymous. Both periods of the
Cenozoic Era are further subdivided into formal
epochs. The Tertiary Period is comprised of five
epochs, while the Quaternary Period is subdivided into two. The first two epochs of the
Tertiary Period, the Paleocene and Eocene, constitute what we will refer to as the "early" Tertiary, or the early Cenozoic, and it was briefly
discussed in the previous chapter. Our present
focus is the middle Tertiary (or middle Cenozoic), which includes the Oligocene, Miocene,
and perhaps a part of the Pliocene Epochs. Our

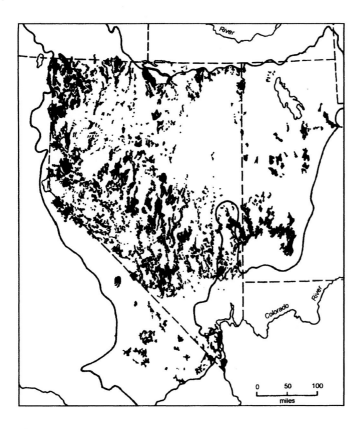

Figure 8.2 Distribution of Oligocene and Miocene igneous rocks in the Great Basin. Areas in black indicate exposures of mid-Tertiary (43–6 million years old) igneous rocks, including some plutonic rocks. The map is not complete for the Oregon portion of the Great Basin.

processes—explosive volcanism and active faulting—that began at about the same time, and for similar reasons. We will return to the connection between middle Tertiary volcanic activity and faulting, and the plate tectonic mechanisms for both, later in this chapter. For now, let us simply acknowledge that these two geological phenomena represent the beginning of a chain of events that would ultimately lead to the evolution of the Great Basin province as we know it today. This was a pivotal time in the history of the region, one that is worth exploring in a little more detail.

Middle Tertiary Volcanic Rocks of the Great Basin

Following the erosional interlude of the early Tertiary, igneous activity flared up again in the Great Basin about 43 million years ago in the late Eocene Epoch. The eruptions intensified dramatically during the Oligocene and Miocene Epochs, and waned in late Cenozoic time, though some areas have been active as recently as a few thousand years ago. This prolonged and widespread volcanic episode produced an amazing variety of rocks that cover more than 20 percent of the Great Basin (Figure 8.2). Sometimes, sticky lava oozed onto the surface to harden into dome-like masses of **rhyolite** or **dacite**, two of the most common volcanic rock types in the Great Basin. At other times, particularly later in the Cenozoic volcanic episode, torrents of very fluid lava gushed out of elongated fissures to inundate hundreds of square miles, leaving black **basalt** flows as evidence of the incandescent flood. As we shall soon see, there were also times of immense volcanic explosions that blew droplets of magma and small bits of rock hundreds of miles from their source, where the hot rubble coalesced into a fragmental volcanic rock known as **tuff**. Tuffs, basalts, rhyolites, dacites: the great variety of volcanic rock types in the Great Basin is a reflection of the varied style of eruptions that shook the Great Basin during middle Tertiary time.

The oldest Cenozoic volcanic rocks in the Great Basin occur in the northern portion of the province, in the area around the Tuscarora, Independence, and Jarbidge Mountains of Elko

discussion of the late Cenozoic history of the Great Basin in subsequent chapters will focus primarily on the events of the late Pliocene Epoch to modern times, including the entire Quaternary Period.

As defined above, the middle Cenozoic is chiefly the interval from about 35 million years ago to about 5 million years ago, from the Oligocene to the early Pliocene Epoch. Rocks of this age are widespread across the Great Basin, and can be divided into two broad categories. One group consists of fragmental volcanic rocks and lava flows that bespeak the return of volcanic activity to the region after the calm respite of the early Cenozoic. In some areas, the middle Tertiary volcanic rocks are associated with coeval plutonic masses, but this is not true everywhere in the Great Basin. The second broad category of middle Tertiary rocks is coarse granular sedimentary deposits, such as conglomerate, breccia, and sandstone that accumulated in down-faulted basins. The faults began to develop in the Great Basin about 35 million years ago, and many of them are still active today. These two types of middle Tertiary rocks reflect two contemporaneous and related

County, Nevada. These late Eocene–early Oligocene rocks are a mixed assemblage of lava flows and tuffs representing the earliest phase of the Cenozoic renewal of volcanic activity. Following these initial eruptions, a very interesting pattern in the subsequent volcanic activity developed across the Great Basin. While the older volcanoes in the north remained active, new volcanoes began to erupt farther and farther south through middle Tertiary time. Though each of the volcanic centers remained active for millions of years after they flared up, the activity spread progressively south during Oligocene and Miocene time (Figure 8.3). For example, the earliest Cenozoic volcanoes in southern Oregon and northern Nevada erupted in the late Eocene Epoch; those in central Nevada and western Utah commenced their activity from Oligocene to early Miocene time; and the eruptions in the Mojave–southern Nevada region began during the mid-to-late Miocene Epoch and continued into the Pliocene. Like a wave of fire marching southward across the Great Basin, hundreds of mid-Tertiary volcanoes burst into activity in sequence. By the end of the Miocene Epoch, the Great Basin was virtually covered by a thick succession of volcanic rocks, the refuse of a 30-million-year volcanic rampage.

A fascinating shift in the character of the mid-Tertiary eruptions accompanied the general southward migration of volcanic activity across the Great Basin. The earliest eruptions appear to have been relatively quiet effusions of thick lava, punctuated by small blasts from diminutive volcanic cones. This early activity produced stubby lava flows and layers of volcanic ash that are more or less localized and cannot be easily traced for great distances across the Great Basin. Starting in the Oligocene Epoch, and usually to the south of older volcanic areas, things began to change. Instead of quietly oozing onto the surface, the magma literally exploded from the ground with extraordinary violence. As was true of the Mesozoic volcanoes of the Cordilleran arc, these violent eruptions produced great quantities of volcanic ash and rubble that were expelled across vast areas surrounding the blast zone. Such volcanic events primarily produce great sheets of **pyroclastic** ("broken by fire")

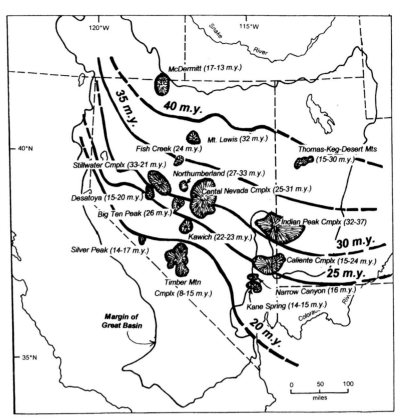

igneous rocks, such as fine-grained tuff (consisting of the tiny particles of ash; see Figure 8.5) and coarse-textured **volcanic breccia** (composed of larger, angular fragments of rock). In such violent eruptions, little lava flows out of the exploding volcano; most of it is blown out as a spray consisting of minute droplets of magma that cool while they are airborne to form the microscopic ash particles. Powerful pyroclastic eruptions can eject ash as high as 100,000 feet above the earth's surface. Such ash, as it settles out from the atmosphere, can be carried great distances by the wind and eventually accumulate on the ground hundreds, or even thousands, of miles from its source. In addition, pyroclastic eruptions sometimes occur on the flanks of volcanoes, rather than at the summit. When this happens, the lateral blast can send clouds of ash racing across the landscape at speeds of several hundred miles per hour, scorching everything in their path. Thus, even flank eruptions can scatter ash great distances from the detonating volcanic cone. For these reasons, unlike older volcanic sequences, many of the Oligocene and Miocene ash blankets produced during the catastrophic volcanic

Figure 8.3 Major mid-Tertiary caldera complexes of the Great Basin. Only the larger calderas are shown, and the indicated size and shape of each is approximate. Bold lines indicate the timing of the earliest explosive eruptions in millions of years (m.y.) before the present. Note that volcanic activity starts earliest in the northern Great Basin and migrates south during the Oligocene and Miocene Epochs.

Figure 8.4 (above) Crystal Peak, Millard County, Utah. The white rock is the Tunnel Spring Tuff, erupted about 33 million years ago during the Oligocene Epoch.

Figure 8.5 The pyroclastic texture of the Tunnel Spring Tuff at Crystal Peak. The larger rock fragments are mostly Paleozoic limestone that was blown out during the same eruption that produced the finer, light-colored crumbly ash.

eruptions in the Great Basin can be traced for hundreds of miles over extensive areas. Geologists estimate that Oligocene and Miocene ash flows once covered about 80,000 square miles of the Great Basin before they were disrupted by faulting or buried under younger material. Some of the *individual* layers of ash, each representing a single volcanic blast, can be traced over areas of nearly 15,000 square miles!

Hundreds of names have been applied to the mid-Tertiary volcanic rocks of the Great Basin, each term reflecting a local exposure of such rocks. But, because the ash sheets tend to be rather uniform in character throughout the areas they cover, many of them can be traced with little difficulty from mountain to mountain across the south-central Great Basin (Figure 8.6). The explosive nature of the eruptions in the Great Basin is reflected in the dominance of tuff among the various formations that geologists have described. For example, the late Oligocene Quichapa Group (also known as the

Condor Canyon Formation) of southwest Utah consists of at least four major tuff units: the Leach Canyon Tuff, the Swett Tuff, the Bauers Tuff, and the Harmony Hills Tuff. Likewise the Needles Range Group of western Utah consists of, in ascending order, the Cottonwood Wash Tuff, the Wah Wah Springs Tuff, the Ryan Spring Tuff, and the tuff-rich Lund Formation. In the vicinity of the Nevada Test Site of the southwestern Great Basin, no fewer than twelve different tuff formations were erupted in the Miocene Epoch, between about 14 and 8 million years ago. With so many different layers of mid-Tertiary tuff, and with so many geologists studying and naming them in different places, the nomenclature applied to these rocks can be a confusing tangle of names. The widespread nature of the ash sheets exacerbates this problem because different scientists studying tuff layers in widely separated locations may actually be examining (and naming!) the same deposit. In recent years, some of the confusion has been eliminated by the radiometric dating and correlation of the various layers of ash. But, old names die hard, and many invalidated terms still linger in the geological literature of the Great Basin.

It is not so much the names of the tuff formations as it is their nature that reveals details about the middle Tertiary events in the Great Basin. The various layers of tuff are actually quite heterogeneous in their textural and compositional details. For example, some of the tuff contains broken rock fragments embedded in the matrix of volcanic ash. Such rock is commonly called **lithic tuff** (Figure 8.7B), and probably represents a deep-seated blast from within an existing volcano. The fragments of rock in a lithic tuff are the remains of the dismembered volcano. Sometimes, as in the case of the Tunnel Spring Tuff (see Figure 8.5), powerful explosions even shattered the Paleozoic strata deep under the volcanoes, blasting limestone and dolomite fragments high into the atmosphere. Such rock fragments could have returned to the surface hundreds of miles from their source in a downpour of rock and ash. **Crystal tuff** (Figure 8.7A) contains tiny geometric crystals of such minerals as quartz and feldspar within the ashy matrix, which commonly causes a sparkling appearance in bright sunlight. This type of tuff represents magma that had partially crystallized prior to the eruption that ejected it from its subterranean quarters. After the crystal-magma mush was blown into the sky, the crystals accumulated with the hot ash to become fused into the solid tuff. Usually, the mineral crystals in such tuff are no more than a few millimeters (1 mm=1/25 inch) long, and they are commonly broken during the violent eruption. Sometimes, however, the crystals can be much larger, complete, and well formed. The beautiful topaz and red beryl crystals from western Utah, prized among mineral collectors, both occur in mid-Tertiary crystal tuff and rhyolite sequences. Another collectors' favorite, the famed "thunder eggs" (geodes) found in many places across the Great Basin, also originate primarily in mid-Tertiary volcanic rocks. Geodes represent open pockets that develop within either a lava flow or a layer of hot ash as gases escape during, or soon after, an eruption. While the ash or lava is hardening into solid rock, vapors and fluids may continue to move through the open pockets depositing various minerals in thin layers on the inside of the roughly spherical void. Even long after the rock mass cools, water and volcanic fumes from later eruptions may continue to migrate through the open "bubbles," depositing additional minerals inside the cavity. If this process continues long enough, the hollow becomes completely filled with a solid mass of quartz and similar minerals. Because quartz is very hard and durable, the filled bubble is much more resistant to weathering than is the ash or solidified lava that surrounds it. Eventually it weathers free as a more or less egg-shaped rock that may be hollow, if the original pocket was not completely filled, or completely solid, if it was. Geodes, therefore, represent the in-fillings of gas bubbles and pockets in volcanic rocks. Not all geodes found in the Great Basin occur in mid-Tertiary rocks, but most of them do.

A third type of tuff that is extremely common in the Great Basin is particularly captivating because it evinces terrifying volcanic events. **Vitric tuffs**, sometimes also referred to as **vitrophyres**, are those that have become fused into a

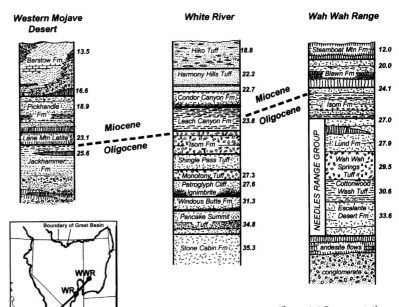

Figure 8.6 Representative mid-Tertiary volcanic rocks of the Great Basin. Oligocene and Miocene rocks in the Great Basin are predominantly tuffs and ignimbrites produced by violent pyroclastic eruptions. Sedimentary rocks, rich in volcanic ash and rubble, also accumulated in localized basins. Small numbers in bold type are average radiometric dates for specific rock units, in millions of years before the present.

glasslike material that can easily be mistaken for black obsidian. This dense welding can only happen when ash accumulates while it is still glowing hot. Under such conditions, the microscopic ash particles can be literally welded to each other, forming a gooey mass that quickly cools into a glasslike material. Most of the tuffs in the Great Basin are "welded" to one degree or another, but the vitrophyres represent the extreme in consolidation of volcanic ash. Vitrophyres are usually associated with other kinds of tuff in crudely layered rock sequences that geologists call **ignimbrites** (Figure 8.8, Figure 8.9). The term ignimbrite literally means "fire cloud rock," and that phrase is an excellent description of the frightening origin of such rocks. When sudden and violent flank eruptions discharge a great mass of lava laterally, it can roll over the

Figure 8.7 Microscopic views of
A, a crystal-rich tuff, and
B, a lithic tuff. Both types of tuff consist mostly of fused volcanic ash.
Scale bar = 3 mm (about 1/8 inch).

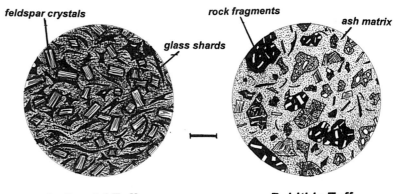

A. Crystal Tuff   **B. Lithic Tuff**

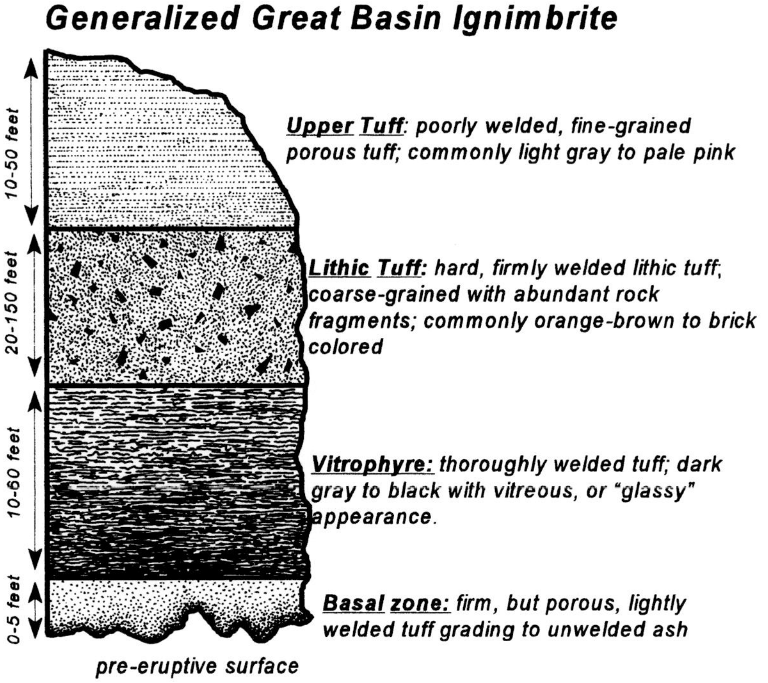

Generalized Great Basin Ignimbrite

Upper Tuff: *poorly welded, fine-grained porous tuff; commonly light gray to pale pink*

Lithic Tuff: *hard, firmly welded lithic tuff; coarse-grained with abundant rock fragments; commonly orange-brown to brick colored*

Vitrophyre: *thoroughly welded tuff; dark gray to black with vitreous, or "glassy" appearance.*

Basal zone: *firm, but porous, lightly welded tuff grading to unwelded ash*

10-50 feet

20-150 feet

10-60 feet

0-5 feet

pre-eruptive surface

Figure 8.8 A generalized ignimbrite sequence. In the Great Basin, Oligocene and Miocene ignimbrite sequences commonly exhibit an upward decrease in the degree of welding. Some of the ignimbrite zones illustrated above may be missing or poorly developed in individual outcrops.

surface at temperatures above 1,000 degrees. At this temperature, the ash particles will be incandescent, glowing with a white-yellow color as they are blasted from the volcano. Lubricated by escaping gases, such clouds of glowing ash can travel at supersonic speeds across the land, scorching everything in their path. The French term *nuée ardente* ("glowing cloud") has been applied to such masses of ash that can bury the land under an avalanche of fire. This is precisely what happed in 1902 at Mount Pelée on the Caribbean island of Martinique. On the morning of May 8 of that year, a *nuée ardente* descended the flanks of Mount Pelée, engulfed the port city of St. Pierre, incinerated the entire town, and killed nearly all of its 30,000 inhabitants. There were four survivors.

The numerous ignimbrite sequences in the Great Basin suggest that the region was swept by *nuées ardente* repeatedly during the Oligocene and Miocene Epochs. The glassy vitrophyres that formed during such volcanic cataclysms usually occur near the base of the tuff sequence, where the heat, aided by the weight of overlying ash, promoted thorough welding. A thin zone of lightly welded material, representing ash cooled by contact with the ground, is typically present just below the vitrophyre (see Figure 8.8). Above

the vitrophyre, a layer of welded lithic tuff represents a flow of ash and rock fragments from the same eruption that caught up with the fast-moving *nuée ardente* and accumulated on top of it. Most ignimbrite sequences are capped by fine-grained, moderately welded tuff that records the accumulation of fine airborne ash onto the pile as the eruption waned. The entire sequence, normally between 30 and 250 feet thick, records a single violent event. Some of the best-known ignimbrite sequences in the Great Basin include those in the middle Oligocene Needles Range Group of western Utah, the late Oligocene Bauers and Swett tuffs of east-central Nevada, the Petroglyph Cliff tuff of south-central Nevada, and the Shoshone Volcanics of the Death Valley region. A complete listing of all the Great Basin ignimbrites would require much more space than is available here. In any case, such rocks are sufficiently common to indicate that deadly *nuées ardente* were regular occurrences throughout the Great Basin during mid-Tertiary time. Some geologists have even referred to this interval as the "great ignimbrite flare-up." It was a hellish time in the Great Basin during the great mid-Tertiary conflagrations.

Many times when *nuées ardente* were blasted from mid-Tertiary Great Basin volcanoes, huge calderas were created across the region (see Figure 8.3). As we learned in the preceding chapter, calderas are great volcanic depressions that originate when large volumes of magma are expelled from subsurface reservoirs. Some of the pyroclastic eruptions of the Oligocene and Miocene Epochs left calderas of staggering dimensions. For example, the Kane Spring caldera in southern Nevada, measuring roughly 30 by 13 km (18 by 8 miles), originated in the mid-Miocene (about 14 million years ago), while the older Kawich caldera of central Nevada collapsed about 23 million years ago to form a depression more than 50 km (30 miles) long. For comparison, Crater Lake in Oregon is about 10 km (6 miles) in diameter. However, some of the mid-Tertiary blast holes in the Great Basin are even more impressive. In many places, geologists have identified **caldera complexes**, composed of several individual depressions that have coalesced over an extended period of volcanic activity. The Timber Mountain caldera complex of southwest Nevada, for example, is comprised of at least seven intermingled calderas. One of the best-known caldera complexes in the Great Basin is the Indian Peak caldera, astride the modern Nevada-Utah boundary. This oblong depression measured approximately 100 by 60 km (60 by 36 miles) when it was created by a series of powerful eruptions between about 33 and 27 million years ago. About thirty miles south of the Indian Peak complex is the somewhat younger (24–15 million years old) Caliente caldera complex (see Figure 8.3), measuring about 80 by 35 km (50 by 20 miles). In addition to these large compound cavities, there are dozens of other caldera complexes in the Great Basin that affirm the notion of widespread, prolonged, and explosive volcanic activity in the Oligocene and Miocene Epochs.

It is difficult to imagine the volcanic ferocity that must have accompanied the formation of the caldera complexes and ignimbrites in the Great Basin. Humans have witnessed several tragic volcanic calamities in historic times, such as the eruptions of Mount Vesuvius in 79 (CE), the island of Krakatoa in 1883, and, of course, Mount St. Helens in 1980. Though thousands of people perished and whole villages disappeared during these volcanic disasters, they all pale in comparison to the immense eruptions that ravaged the Great Basin in mid-Tertiary time. None of the catastrophic eruptions that humans have encountered created anything like a caldera complex. The geological scars of even the most powerful historic eruptions will most likely vanish in a few thousand years. The mid-Tertiary blasts that shook the Great Basin, on the other hand, have left discernable signs of wreckage that have persisted for tens of millions of years. However, do not expect to see any colossal pits on a jaunt through the central Great Basin today. Without exception, the great calderas have all been at least partially dismembered by later faulting, obscured by post-collapse erosion, or filled and buried by younger rock and sediment. In some cases, rubble created by the volcanic explosions filled the calderas almost as soon as they collapsed. In other calderas, continuing eruptions built dome-like masses of lava, known

as **resurgent domes** on the floor of craters, partially filling the great depressions (Figure 8.10). Because the rock that accumulated within the calderas was commonly affected by later volcanic activity, it is usually harder, and more resistant to erosion, than the surrounding sheets of tuff that formed during the initial outburst. After millions of years of erosion, the harder caldera-filling material is now higher than the surrounding terrain. What was a huge depression millions of years ago may now be a high peak, hill, or plateau surrounded by lower ground. Geologists describe this situation, when the ancient topography becomes reversed by erosion, as **inverted topography**. This is another reason why no gaping chasms remain in the Great Basin from the mid-Tertiary paroxysms. Nonetheless, geologists have been able to reconstruct the mid-Tertiary caldera complexes by finding the remnants of them scattered among the mountains of the Great Basin. Usually, just a small part of a caldera wall, or perhaps a mass of ash that avalanched from the rim, is all that is revealed in a single outcrop of Oligocene or Miocene volcanic rock (see Figure 8.9). Reconstructing the ancient calderas and volcanoes is a time-consuming process of assembling such subtle hints and clues from many localities. Though the evidence of extreme volcanic violence is often disjointed and fragmentary, it is also undeniable. If we could somehow revisit the mid-Tertiary Great Basin, we would experience a land ablaze with terrifying volcanic chaos.

Figure 8.9 A portion of the caldera wall preserved in the Caliente Caldera Complex, southern Nevada. The light-colored mass of rock on the right is a hardened ash flow, while the layers to the left represent volcanic rubble that accumulated within a large caldera.

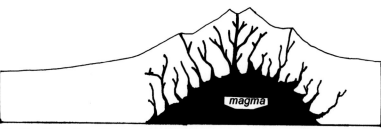

A. Pre-Eruption Stratovolcano

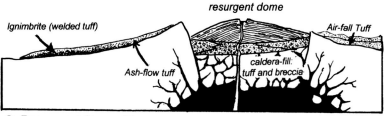

B. Caldera-forming Pyroclastic Eruption

C. Resurgent Dome Phase

Figure 8.10 Explosive origin of mid-Tertiary calderas in the Great Basin.

Magma and Minerals, The Sequel

The surge of magma into the crust of the Great Basin during the middle Tertiary Period helped induce a wave of mineralization that was, in many ways, a replay of the ore-forming events of the Mesozoic Era. The lavas that reached the surface in Tertiary time were relatively rich in **silica** (a compound comprised of silicon and oxygen), and commonly cooled to form tuffs or flow rocks that are classified as rhyolite, dacite, or latite. Such igneous rocks contain 60–70 percent silica, as opposed to **mafic** igneous rocks, like the younger black basalt of the region, which normally have about 50 percent silica or less. Abundant silica in magma tends to increase the viscosity, or thickness, of the molten rock, making it very thick and sticky. The mafic magma that forms basalt is, in comparison to silicic magma, very fluid and flows readily, like warm honey. The highly silicic mid-Tertiary magma

was more similar to gooey taffy in consistency. This, of course, is a partial explanation of the volcanic violence of the time: viscous magma is more likely to block its own path to the surface until pressures in the magma chamber beneath blow the obstruction free. Such sluggish magma would also tend to migrate upward through the crust very slowly and, therefore, would be in contact with older rocks for protracted intervals of time. Consequently, during the time that the mid-Tertiary volcanoes were exploding on the surface, magma was also pooling underground throughout the Great Basin. Slowly oozing through the subsurface rock en route to the surface, the magma reacted with the surrounding rock, metamorphosing whatever materials happened to be present, and sometimes mineralizing them as well.

Some of the underground magma never made it to the surface, crystallizing into plutons of granitic rock similar to the Mesozoic masses we explored in the last chapter. In general, Oligocene and Miocene plutons are smaller than the great Mesozoic batholiths and are commonly referred to as **stocks**, which can be thought of as miniature batholiths. Such small masses of mid-Tertiary plutonic rock are exposed in many scattered localities in the east-central Great Basin, from the Ruby Mountains of Nevada to the Wasatch Mountains of Utah. Mid-Tertiary plutonic rocks are much more common in the eastern part of the Great Basin than they are to the west, a distribution that is the opposite of the pattern for Mesozoic plutons. For example, in the central Wasatch Mountains along the eastern margin of the Great Basin, four different Tertiary plutons (the Little Cottonwood, Alta, Clayton Peak, and Pine Creek stocks) were emplaced from 24 to 36 million years ago. West of the Wasatch Mountains, mid-Tertiary granitic rocks are present in an east-west-trending zone that includes the numerous stocks in the central Oquirrh Mountains (37–39 million years old), the Deep Creek Mountains (37–42 million years), and the Ruby Mountains of Nevada (34–40 million years old). This zone of mid-Tertiary plutons has been called the "Uinta trend" by geologists because it is in line with the east-west axis of the Uinta

Mountains of northeast Utah. Elsewhere, Oligocene and Miocene granitic rocks are widely scattered in western Utah: from the Tintic Mountains (three plutons, 31–34 million years old), to the Mineral Mountains (a compound pluton about 25 million years old), to the Iron Springs Mining District of southwest Utah (three plutons, each about 20 million years old). In eastern Nevada, mid-Tertiary plutonic rocks are less abundant, and such rocks are very uncommon in the western part of the Great Basin.

Through mechanisms similar to those related to the Mesozoic intrusions, the mid-Tertiary igneous activity in the Great Basin led to the development of varied types of rich ore bodies in many locations. Hydrothermal, skarn, and porphyry ore deposits that developed during the Oligocene and Miocene Epochs augmented the considerable riches already localized during the Mesozoic Era, transforming the Great Basin into one of the most heavily mineralized regions in the world. Some of the areas that were affected by ore-forming events during the Jurassic and Cretaceous Periods were further enriched by mid-Tertiary igneous activity. The Pioche District in eastern Nevada is one example of such a fortunate region. The silver, gold, and zinc ores of this area resulted from two major ore-forming events; the second related to the intrusion of several Oligocene stocks millions of years after the rocks in that area were initially mineralized by magma emplaced during the Cretaceous Period. In scores of other localities, the mid-Tertiary volcanic and plutonic rocks were solely responsible for the genesis of prolific ore bodies. The numerous and diverse silver-lead-copper-zinc-gold ore bodies in the Tintic Mining District of western Utah, for example, were primarily formed through the interaction between the Paleozoic host rock and Oligocene igneous rocks. The deformed Paleozoic strata in this district were subject to caldera-forming volcanic events 31–33 million years ago and then intruded by the Silver City monzonite stock a short time later. By the time the igneous activity in the Tintic Mountains ended about 17 million years ago, millions of tons of rich ore had formed adjacent to the igneous masses. Some 20 million tons of this ore, worth more than a half-billion dollars, has been mined in the Tintic District since operations began there in 1869.

There are many other places in the Great Basin where extensive ore bodies developed during mid-Tertiary time. In western Utah, the Bingham porphyry copper deposit, the Mercur District, and the Ophir District in the Oquirrh Mountains; the Iron Springs District west of Cedar City; and the Marysvale and Frisco Districts in Beaver County all experienced mineralization during the Oligocene or Miocene Epochs. Likewise, in Nevada, mid-Tertiary igneous rocks played a key role in the origin of the rich ore bodies in the Getchell, Battle Mountain, and Carlin mining districts of northern Nevada. In the west-central part of the Great Basin, the fabulous gold ores that made Goldfield, Tonopah, and Rhyolite famous in the early 1900s are all associated with Miocene igneous rocks ranging in age from about 20 million years to less than 10 million years old. Along the western margin of the Great Basin, the extraordinarily rich silver ore in the historic Comstock lode was primary located in the Alta Formation, a sequence of Miocene volcanic rocks. The Comstock ore was so rich that between 1859 and 1900 more than 15 million pounds of silver was produced. Mineralization also spread into the Mojave region in mid-Tertiary time, as illustrated by the $20 million worth of silver ore extracted between 1880 and 1940 from the Miocene Pickhandle Volcanics and Barstow Formation in the Calico Mining District. Clearly, while the igneous activity of the Oligocene and Miocene Epochs may have ravaged the surface of the Great Basin, it also concentrated precious metals in the rocks below ground throughout the region.

In the mid-Miocene Epoch a major shift occurred in the character of the volcanic activity in the Great Basin. About 15 million years ago, the explosive activity began to decline in the region, though it would persist in some areas of the western Great Basin to the end of the Miocene Epoch, about 6 million years ago. Volcanoes continued to erupt in the Great Basin, some until very recently, but the magma that poured from them became less silicic (or more mafic) in composition after mid-Miocene time.

Figure 8.11 Small late Tertiary cinder cones in the Fish Creek Mountains of western Nevada

This low-silica magma was much more fluid than the mid-Tertiary lava, and most often it seeped from erupting volcanoes with little violence. The runny lava flooded large areas of the Great Basin, forming extensive sheets of black basalt after it had cooled sufficiently to harden. The great pyroclastic explosions, and the thick layers of welded tuff and ignimbrites they created, became less common after mid-Miocene time. Instead, vast tracts of black lava, dotted with relatively small **cinder cones** (Figure 8.11), became the dominant products of late Tertiary-Quaternary volcanic activity. Occasionally, during and after the shift to more mafic volcanism, relatively small blobs of thick silicic magma would reach the surface to form rhyolite domes or stubby flows. Thus, the volcanic activity in the Great Basin for the past 15 million years or so produced two general types of igneous products: 1) extensive flows and cinder cones of low-silica basalt, and 2) dome-like effusions of high-silica rhyolite. Basalt and rhyolite represent two ends of the compositional spectrum of volcanic rocks, with dacite, latite, and andesite reflecting intermediate compositions. For this reason, geologists commonly describe the late Tertiary and Quaternary volcanic activity in the Great Basin as **bimodal**, meaning that it produced rocks (rhyolite and basalt) of two distinct modes of composition. The shift from explosive silicic volcanism to more passive bimodal eruptions began about 15 million years ago, but it was a gradual transition that spanned several million

years. Caldera-forming eruptions continued well into the Pliocene Epoch but were never again as common or violent as they were in mid-Tertiary time. As the violent eruptions diminished, bimodal activity spread across the Great Basin and has persisted to modern time. Though contemporary humans have not observed any historic eruptions in the Great Basin, some of the cinder cones, basalt flows, and rhyolite domes are less than a thousand years old. Young volcanoes in such areas as the central Mojave Desert, the Long Valley Caldera along the foot of the Sierra Nevada, or the Black Rock Desert of western Utah could potentially awaken in the near future. We will return to such recent volcanic features in Chapter 10, but there is one other aspect of the mid-Tertiary history of the Great Basin that must be explored before we can completely understand this pivotal stage in the development of the region. Even before the age of bimodal volcanism began, near the peak of the mid-Tertiary conflagration, new forces begin to stretch the crust under the Great Basin. Eventually, these extensional forces would fracture the crust into blocks that would shift up or down relative to each other to produce the modern landforms. It was in mid-Tertiary time, in the midst of the volcanic maelstrom, that the Broken Land, as we know it today, began to break.

EXTENSIONAL FAULTING: THE CATERPILLARS HATCH

As we reviewed in the introductory chapter, when Clarence Dutton alluded to the "army of caterpillars" in describing the terrain of the Great Basin, he was fancifully portraying isolated mountains and valleys bounded by normal faults that are produced by extensional, or "stretching," forces (see Figure 1.5). Until mid-Miocene time, such faults were relatively rare in the Great Basin because the convergence of lithospheric plates along the western edge of North America generated mostly compressive stress, the opposite of that required for normal faults to develop. Some minor normal faulting and extension during the Mesozoic and Paleozoic Eras has been postulated by some geologists to explain certain geologic anomalies, but until the mid-Cenozoic Era extensional deformation was localized and short-lived. Starting in the mid-Tertiary interval, about the same time that the explosive volcanic eruptions began, the crust in the Great Basin began to experience widespread and persistent normal faulting. Throughout the remainder of Cenozoic time, great blocks of rock isolated by the normal faults shifted up and down to create the alternating pattern of basin and ranges that typify the region today. It was in just the past 30 million years or so, following more than 2 billion years of earlier geological evolution, that the physiology of the modern Great Basin began to materialize.

The association between normal faults and extensional stress is based on the distinctive and consistent geometry of fractures that are generated when rocks are stretched. Geologists define normal faults as fractures in which the block of rock above the fault plane, the **hanging wall**, moves down relative to the mass below the fault, known as the **foot wall** (Figure 8.12). If the fault is perfectly vertical, such that neither side can be designated as resting above or below the fault plane, then it is difficult to classify the fault. Fortunately, this is rarely the case, and most faults are tilted slightly from a vertical orientation such that the fracture passes above one rock mass (the foot wall) and below another (the hanging wall), as diagramed on the right side of Figure 8.12. The important point is that rocks cannot fracture with this geometry under compressional ("squeezing") or shear ("tearing") stresses. Only if rocks are extended, or "stretched," can the hanging wall move down relative to the foot wall to create a normal fault. Compression produces **reverse** faults, such as the thrust faults we have discussed in preceding chapters. Shear forces will cause a lateral shift of rock masses, either **dextral slip** (to the right) or **sinistral slip** (to the left), with minimal vertical offset. However, normal faults can only form by extensional stresses. Moreover, all the faults that separate the major mountain ranges of the Great Basin from the adjacent basins are normal faults, the consequence of extensional deformation in the recent geologic past. Therefore, the rugged mountains and level valleys of the mod-

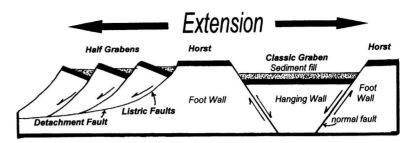

ern Great Basin are all relatively youthful features; none of them existed prior to the era of late Cenozoic normal faulting.

Many geologists have referred to the mountain-and-basin landscape of the Great Basin, produced by repetitive normal faulting, as **horst-and-graben** topography (Figure 8.12). These terms originate from the German words for an eyrie, or high place ("horsts"), and for a ditch or trench ("graben"). The mountain blocks represent horsts in this terminology, while the valley floors signify the down-faulted grabens. The normal faults between the horsts and grabens are generally assumed to descend into the subsurface at a high angle, until the extreme pressures miles underground prevent the movement of rock along the fracture. Many of the mountain ranges of the Great Basin are, in fact, simple horsts that represent the elevated footwall of a high-angle normal fault. Such horst-block mountains can be identified by their characteristic symmetry: the opposite flanks tend to slope at about the same angle. However, some of the most prominent normal faults in the Great Basin exhibit a somewhat different geometry. Instead of descending into the subsurface at a high angle indefinitely, many normal faults actually flatten into nearly horizontal planes as they are traced to deeper levels in the crust. At the surface, such normal faults may dip downward at angles of about 40–70°, but seismic studies of their deeper orientation indicate that they may become nearly horizontal 6–9 miles (10–15 km) down. Normal faults that exhibit this concave up-curvature are known as **listric** faults, and there are many of them in the Great Basin. Because the plane of a listric fault is curved, blocks of rock must rotate as they are displaced along the fault plane. If, for example,

Figure 8.12 Geometry of classic normal faults (right), compared to listric faults (left). Note that listric faults pass into nearly horizontal detachment faults at depth and generally create half-grabens at the surface.

Figure 8.13 East-tilting layers of Cambrian limestone in the Wah Wah Mountians of western Utah cut by a listric fault. Note the down-to-the-left (west) displacement of the light gray limestone strata along the fault that runs through the shadowed gully.

Figure 8.14 Principal normal faults in the Great Basin. Note the strong north-south orientation of most of these late Cenozoic extensional faults. The arrows indicate the direction of tilting of the mountain blocks bounded by the faults.

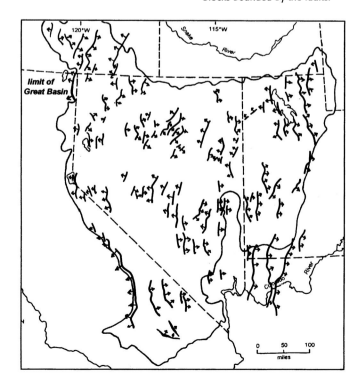

the layers of rock in a hanging wall block were initially horizontal, they would gradually become tilted to higher angles as the block moved down along a curved listric fault. This type of rotational displacement is very common in the Great Basin, and it has created a pervasive tilt, or "lean," in many of the mountain ranges of the region (Figure 8.13). Such mountain ranges bounded by listric faults generally lack the symmetry of simple horsts and will exhibit a steep face on one side and a gently sloping flank on the other. There is an interesting, but complex, pattern to the tilt of Great Basin mountains. The mountain blocks in the eastern part of the region tend to lean to the east, while those in the west commonly tilt westward. However, there are almost as many exceptions to this general rule as there are conformities (Figure 8.14). Instead of a perfectly symmetrical arrangement of tilted mountain blocks, the pattern in the Great Basin appears to consist of several localized zones of uniform orientation arranged with only a very crude directional symmetry. Within each of these zones, called "regional tilt-domains" by geologists, the listric and normal faults dip in a consistent direction, west or east, and the fault-bounded mountain blocks lean in the opposite direction (as depicted in Figure 8.13). So far, dozens of regional tilt domains have been identified in the Great Basin, and, collectively, they segment the landscape of the Great Basin into a crude sort of geomorphic mosaic. We will return to the significance of the tilt domains when we examine the ongoing geological evolution of the Great Basin in Chapter 10.

Another consequence of the listric faults in the Great Basin is that most of the major basins in the region are not true grabens, bounded by equally prominent, high-angle normal faults. Because the hanging walls moving along a listric fault must rotate as they descend the curved fault plane, one side the graben is almost always deeper than the other side. The side of the basin opposite the most prominent listric fault is marked by either a subordinate normal fault or by no fault at all. Thus, geologists commonly describe the basins separating mountain ranges as **half-grabens** (see Figure 8.12). It can be difficult to observe the tilted floor of half-grabens

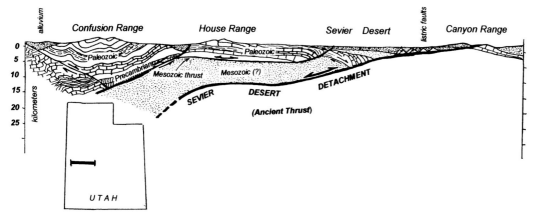

Figure 8.15 The Sevier detachment fault in the eastern Great Basin. Named for the Sevier Desert between the Canyon and House Ranges of western Utah, the Sevier detachment is a low-angle extensional fault that formed when an ancient (Cretaceous) thrust fault was reactivated in mid-Tertiary time.

once they have filled with sediment, but many of the flat-floored lowlands in the Great Basin have such an asymmetric subsurface structure (as sketched in Figure 1.5). The sediment that accumulates in such a tilted half-graben always tends to be much thicker on one side of the valley than on the opposite side, no matter how perfectly level the valley floor may look. The deep side of Great Basin half-grabens may contain as much as 15,000 feet of sediment! We will return to the basin-filling sediments a little later in this chapter.

All normal faults accommodate extension by allowing the crust to break apart in blocks that spread out a little as they shift up or down relative to each other. Listric faults can accomplish even more crustal stretching than high-angle normal faults, because the flattening of the fault plane allows the hanging wall to move directly away from the foot wall. With a predominance of listric faults, the Great Basin has become highly extended since normal faulting began 30 million years ago. The actual amount of extension varies considerably from place to place in the Great Basin, but most geologists agree that the entire region is now about 250 miles (400 km) wider than it was at the onset of normal faulting. This value is about half of the current width of the Great Basin, which means the province is now about twice as wide as it was in mid-Tertiary time. Imagine that if Salt Lake City and Reno had existed at the beginning of the Miocene Epoch, they would have been separated by only 250 miles. This extension continues even to the present time. Incremental movement along the many active normal faults continues to widen the Great Basin at an average

rate of around 10 mm (about one-half inch) per year. Coincidentally, this widening produces several acres of new real estate in the Great Basin each year, but I haven't yet devised a legal way to claim it! In any event, with 100 percent extension over the past 30 million years, the Great Basin ranks as one of the most highly extending regions in the world.

There are some interesting ironies in the late Cenozoic extension of the Great Basin. Remember that for more than 100 million years prior to the normal faulting the region had been primarily under the influence of compression from the convergence of plates along the western margin of North America. As we have already seen, low-angle thrust faults commonly resulted from these stresses. When the normal faults began to develop, they originated at the surface and propagated downward, intersecting the nearly horizontal Mesozoic thrusts. In some cases, these inactive thrust faults were reactivated by the Cenozoic extension, but with the opposite displacement. A partial explanation for the dominance of listric fault geometry in the Great Basin is that the normal faults, as they were developing, commonly merged with the older, low-angle Mesozoic thrusts in the subsurface. The powerful crustal extension reversed the displacement on the ancient thrusts, and blocks driven east in the Mesozoic Era were pulled, or perhaps slid, back to the west along the gently inclined thrust ramps. In recent years, geologists have identified several subsurface extensional faults in the Great Basin that appear to be reactivated Mesozoic thrusts, and have called them **detachment faults**, or just "detachments" (see Figure 8.12). One excellent example of a

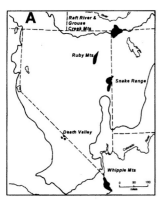
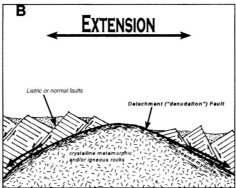

Figure 8.16 Metamorphic core complexes of the Great Basin. A, a map of the location (in black) of the major metamorphic core complexes in the Great Basin and adjacent areas. B, the generalized geologic structure of a metamorphic core complex.

detachment fault that developed from a Mesozoic thrust in Cenozoic time is the Sevier Desert detachment in western Utah (Figure 8.15). This low-angle fault originated in the late Cretaceous as a thrust related to the Sevier Orogeny, but in late Cenozoic time the upper slab of deformed rock slid back to the west when extensional forces began to affect the eastern Great Basin. The block above the Sevier Desert detachment appears to have moved about 35 miles (60 km) to the west during the extensional episode. Similar structures developed in numerous places in the eastern Great Basin, where the Cretaceous thrusts were most abundant. Elsewhere in the region, some of the older Mesozoic thrusts in the central Great Basin behaved in a similar manner when extension began in mid-Tertiary time.

Unlike the wave of mid-Tertiary volcanic activity that swept through the region, no uniform timing pattern seems to accompany the Cenozoic extensional faulting in the Great Basin. Recent studies suggest that normal faulting began as early as late Eocene time, 45 million years ago, in a few localities such as the Eureka area, Ruby Range, and Wood Hills of north-central Nevada. By about 30 million years ago, in the late Oligocene, extensional deformation had spread across the central Great Basin, but it was still widely dispersed and localized. Then, in early Miocene (about 20 million years ago), the rate of extension increased dramatically and seems to have spread rapidly across the Great Basin. About 10 million years ago, the intensity of extension diminished markedly in most areas, but extension remains active to the present time across the region. Thus, the association between the mid-Tertiary volcanic activity and extensional faulting is only a very general one,

and there appears to be little basis for any cause-and-effect relationship between these two mid-Tertiary events. For example, in the northern Great Basin, some faulting preceded the volcanic activity, and some followed it. In the central part of the province, the two phenomena were more or less contemporaneous, while in the south, faulting commonly occurred both before and after the explosive eruptions began.

In addition to the variable timing of normal faulting in the Great Basin, the amount of crustal stretching that occurred during late Cenozoic time was far from uniform, with some areas experiencing as much as four to five times the extension of other tracts. Among the more highly extended domains in the Great Basin are areas known as **metamorphic core complexes**, which appear to have developed primarily during mid-Tertiary time (Figure 8.16). Metamorphic core complexes are characterized by a dome-like core of metamorphosed crystalline rock, above which blocks of unaltered rock, bounded by high-angle normal faults, have moved laterally along a gently dipping detachment fault (Figure 8.16B). The "detachments" in metamorphic core complexes, originally called "denudation" faults in 1972 by geologist Richard Armstong, are not reactivated Mesozoic structures, as are other detachments in the Great Basin. Instead, they appear to be slip surfaces that developed as the core rock beneath them was arched upward while the overlying rocks were stretched into blocks and transported to either side of the rising dome. Just beneath the detachments, streaky or banded metamorphic rocks called **mylonites** are almost always present, attesting to the high pressures and temperatures that must have been generated when the overlying blocks slid along the detachment. The mylonites and other rocks near the detachments commonly appear to have been broken, shattered, and partly liquified during the movement of the overlying unaltered rocks. All of this evidence suggests that at least some of the metamorphism of the core complexes occurred at high temperatures and while the rocks above the detachment were being sheared off the top of the dome by extensional stresses. Metamorphic core complexes appear to have developed in areas where the crust was unusually thick, where mid-Tertiary extension at the surface was accompanied by the growth of a blister-like dome from deeper levels. As the dome rose, the younger rocks above it broke into fault-bounded blocks and slid (or were pulled) a considerable distance over the underlying detachment. The metamorphic rocks that comprise the crystalline core may be altered Mesozoic plutonic masses or they may be older rocks that were metamorphosed during both the Mesozoic and Cenozoic Eras. Therefore, the origin of the metamorphic complexes in the Great Basin appears to be primarily related to the mid-Tertiary history of the region, but it probably was influenced by earlier events, such as Cretaceous magma intrusion and Mesozoic compression, as well. There is no doubt that the metamorphic core complexes have complex, multistage histories; but, during mid-Tertiary time, these areas experienced much more extension than occurred elsewhere in the Great Basin.

The combination of normal faults, listric faults, and detachment faults played a key role in the stretching of the Great Basin in Cenozoic time, but there was another mechanism of extension that may have been equally important. All of the faults described thus far developed in the upper or middle crust, where rocks tend to behave as brittle materials, breaking when stresses are applied to them. Deep underground, where temperatures and pressures are much higher, rocks are less brittle and can actually bend and flow in response to directed stress. Such behavior, bending or flowing to accommodate stress rather than breaking under it, is known as **ductile** deformation. This may be a difficult concept to grasp because we often see rocks break and we never see them bend or flow. The idea of ductile deformation in rocks is contrary to our everyday experience because the low pressure and temperature at the earth's surface, along with the instantaneous manner in which we commonly stress rocks, favors brittle deformation, or breaking. Deep underground, where the conditions are much different from those at the surface, ductile deformation is not only possible, but it is the dominant mode of accommodating stresses in the deep earth. Brit-

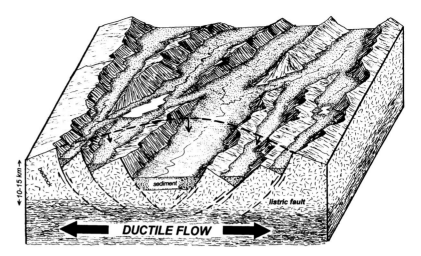

tle fracturing is as rare in the lower crust as ductile deformation is at the surface. Thus, all of the extensional faults—normal, listric, and detachment—should be limited to the upper part of the crust, because they all represent brittle fracturing. At deeper levels, we might expect the extensional stresses to result in ductile flow.

In recent years, scientists have employed seismic profiling techniques to develop fairly good models of the crustal structure in the western United States. These models are not always as precise and detailed as we might wish, but they do indicate that the top of the ductile zone is currently located about 15–20 km (10–13 miles) beneath the Great Basin. The brittle-to-ductile transition may actually occur over several miles, and structure of this transformation zone is probably quite complicated. Nonetheless, most geologists believe that the brittle-ductile boundary was probably at least 10 km (6 miles) deeper prior to Cenozoic extension than it is now. It is assumed that the heating of the crust related to the mid-Tertiary volcanic outburst softened some of the rocks closer to the surface, lifting the brittle-ductile boundary to its present position. These considerations suggest that at crustal depths exceeding about 30 km, Cenozoic extension was accomplished primarily through the ductile flow of soft rocks (Figure 8.17). Some geologists have even speculated that a fluid layer developed within the crust in late Tertiary time. As the soft (or fluid) layer began to flow under extensional stress, the rigid blocks above partially sank into this zone as they were broken apart by the extensional faults. As they

Figure 8.17 Ductile flow and collapse of the Great Basin. Beginning in mid-Tertiary time, the Great Basin region was stretched through the combination of ductile flow at depth and the development of listric faults nearer the surface. The extension of the region was accompanied by the nearly simultaneous collapse of the older highland, producing the modern basin-and-range terrain.

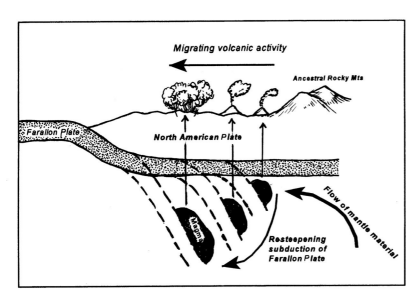

Figure 8.18 The resteepening of the Farallon Plate in mid-Tertiary time may be an explanation for the surge of volcanic activity in the Great Basin. Note that in addition to the location of the volcanic activity, the style and intensity of the eruptions shifted as well.

sank into the lower ductile layer, the blocks tipped to one side because their thickness exceeded their width, a geometry which was (and is) gravitationally unstable. This leaning of the sinking blocks may be an explanation for the tilting that we observe so commonly in Great Basin mountains and half-grabens. Thus, the ductile lower crust acted like a rug that was pulled out from under the brittle rocks of the upper crust during late Cenozoic time. As the extensional faults developed and the soft layer beneath them spread laterally, the whole Great Basin began to collapse, block by block (Figure 8.17). This collapse was not instantaneous, but million of years later it would result in the broad area of internal drainage that we know today. We will return to the more recent history of the Great Basin in later chapters, but what we have explored thus far prompts an interesting question: Could there have been any fundamental cause for *all* of these mid-Tertiary events? Was there a connection between the return of igneous activity, the advent of extensional stress, the "shallowing" of the ductile-brittle transition, and the collapse of the surface? Once again, to find the answers, we must shift our attention to lithospheric plates.

PLATE TECTONICS OF THE MID-TERTIARY

The nearly simultaneous outburst of volcanism and extension in the Great Basin during mid-Tertiary time presents a dramatic contrast to the tranquil early Cenozoic interval. This profound shift suggests that some major change in the interaction of lithospheric plates under western North America must have taken place. The silicic magma that was erupted across the region in the Oligocene and Miocene Epochs probably signifies the renewed subduction of the Farallon Plate that was previously scraping against the lower surface of the North American Plate. Subduction could have been reinitiated by the steepening of the angle at which the Farallon Plate slipped under the continent. This resteepening of the subduction angle in the mid-Tertiary Era represents a reversal of the events that ended the magmatic activity of the late Meso-zoic Era. If such a plate tectonic event occurred gradually, beginning with the part of the plate that had skidded farthest under the continent, then we might envision the Farallon Plate falling away from the bottom of the overriding continent like a trap door that slowly opens (Figure 8.18). As the subduction angle steadily in-creased, magma would be generated in different places at different times under the continent. Because the zone of subterranean melting would migrate beneath the continent, a moving ripple of volcanic activity would sweep across the surface of the overriding plate. In a general way, this conforms to the pattern that we have described for the mid-Tertiary volcanic flare-up in the Great Basin. In addition, this model also explains the increase in the intensity of volcanic activity as the Oligocene and Miocene Epochs progress. We would expect the amount of silicic, gas-charged magma produced under the continent to increase steadily and reach a maximum when the entire Farallon Plate was once again descending at a uniform high angle. If the postulated resteepening occurred in such a gradual and progressive manner, then we must assume that the Farallon Plate was sliding under North America from the southwest, and the resteepening began under (or beyond) the northeast edge of the Great Basin. Assuming that subduction of the Farallon Plate was reestablished in the manner described above, we might wonder why such a shift in the plate interactions occurred in the first place. No one knows precisely what happened about 40 million years ago to cause the plate tectonic setting to change, but the apparent approach of the Farallon Plate from the southwest may be a clue. Though the evidence is scanty, it seems very likely that the Farallon Plate was sliding under the continent from nearly a due west direction throughout the late Mesozoic–early Cenozoic interval. The apparent change in the direction of convergence between the North American and Farallon Plates in mid-Tertiary time is probably related to a major reorganization that affected the trajectories of the oceanic plates on the floor of the ancient Pacific Ocean. As we will soon learn, there is indisputable evidence for a momentous rearrangement of plates along the West Coast about 30 million years ago, and the restoration of high-angle subduction under North America may be an early indicator of a major change in the direction and speed of plate motion. Though a complete and detailed understanding of the mid-Tertiary flare-up in the Great Basin still eludes geologists, it seems likely that the resteepening of the Farallon Plate played a (if not *the*) principal role in igniting the great conflagration. Undoubtedly, there is more to the story of the mid-Tertiary volcanic awakening in the Great Basin, and just now geologists are beginning to glimpse some of those details by examining later Cenozoic events. When we explore the more recent history of the Great Basin in Chapter 10, we will return to the Farallon Plate, with emphasis on the ultimate fate of this slab of oceanic rock.

Once it was reestablished, the subduction of the Farallon Plate did more than rekindle igneous activity in the Great Basin. It also may have set the stage for the extension that accompanied the great volcanic frenzy. As the Farallon Plate fell away from the North American Plate, the compression generated by the interaction of the two plates vanished. These squeezing forces had previously uplifted the entire Rocky Mountain region, and must have been transmitted into the eastern Great Basin as well, though no major mountain systems rose in the latter area. When this immense compression was relaxed, the entire western edge of North America may have rebounded a little, stretching out after some 30 million years of being gripped in a giant geological vise. Some of the extension in the Great

Basin may have begun as a response to the end of compression along the eastern margin of the province as the Laramide Orogeny diminished. However, after the extension started, other factors accelerated it in late Cenozoic time, as we will explore later in this chapter.

Geologists commonly use the phrase "slab roll-back" to describe the way the Farallon Plate fell back into the upper mantle as subduction was reestablished. This "roll-back" would probably have generated a surge of hot, semisolid rock drawn upward from the mantle into the space created between the sinking oceanic plate and the continent above (Figure 8.18). This material was not initially molten, but it may have melted as it migrated upward into zones of lower pressure. The magma that arose from such melting would have represented molten mantle material, which is relatively deficient in silica, compared to the crustal rocks that melt in the subduction process. The mantle-derived magma would have been more mafic (or less silicic) and more fluid than the pasty lavas that were exploding from the calderas on the surface. Eventually, when this mafic magma reached the surface, it erupted as very hot and fluid lava, flowing from elongated fissures that opened in response to the extensional stress that was simultaneously sweeping across the Great Basin. Once on the surface, the mantle-derived magma was so fluid that it formed **flood basalts**, vast sheets of black hardened lava, rather than the calderas and tuffs that are associated with more silicic magma. Millions of years after the resteepening of the Farallon Plate, mafic basalt flows became a dominant product of the bimodal volcanic activity in the Great Basin. However, this notable shift in the composition of the magma probably began as soon as the Farallon Plate descended deep enough to displace mantle materials upward. While there was considerable overlap in time between the two styles of volcanic activity, the lavas derived from mantle sources become increasingly common in later epochs of the Cenozoic Era. Once again, there is more to the story of the late Cenozoic surge of bimodal volcanic activity in the Great Basin, but the *beginning* of the phenomena might be linked to mid-Tertiary plate tectonic events.

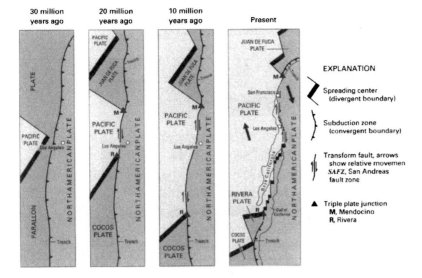

EXPLANATION

Spreading center
(divergent boundary)

Subduction zone
(convergent boundary)

Transform fault, arrows
show relative movement
SAFZ, San Andreas
fault zone

▲ Triple plate junction
M, Mendocino
R, Rivera

Figure 8.19 Development of the San Andreas Fault system. This diagram, adopted from the U.S. Geological Survey, illustrates the overrunning of the oceanic ridge system by the North American continent beginning about 30 million years ago. Note that between the migrating Mendocino and Rivera triple junctions, transform plate interactions replace the earlier convergent boundaries (compare with Figure 8.20). The Great Basin developed where extensional stress was generated inland from the growing San Andreas system.

Figure 8.20 The relationship between extension in the Great Basin and the origin of the San Andreas Fault system. The extension of the Great Basin began when North America overran the ridge system in the Pacific Basin about 30 million years ago. Since that time, the Pacific plate has been moving to the northwest, against the western edge of North America. In the bottom panel, this lateral motion is represented by the symbols ⊗ (away from viewer) and ⊙ (toward viewer).

The renewed subduction of the Farallon Plate was relatively short-lived. About 30 million years ago, in mid-Cenozoic time, a plate tectonic event along the west coast of North America not only ended subduction but also accelerated the extension of the region, kindled new surges of bimodal volcanic activity, and set the stage for a radical transformation in the geologic setting of the entire western edge of the continent. North America, unimpeded in its westward march since the early Mesozoic Era, finally met an obstacle when the leading edge of the continent began to collide with an oceanic ridge system (Figure 8.19). Through this colli-

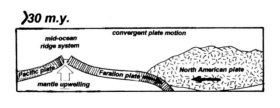

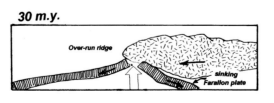

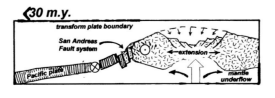

sion, a small portion of the North American Plate made contact with the Pacific Plate, while other parts of the continental margin were still engaged with the Farallon Plate. For the first time, three different plates—the Farallon, Pacific, and North American—all met at a single point along the west coast (Figure 8.19). Such a point is known as a **triple junction**, and the development of one 30 million years ago along the west coast was a pivotal event in the geological history of the Great Basin and adjacent regions. The subsequent history of the mid-Cenozoic triple junction involves some fairly complex geometric interactions among the three involved plates, as presented in the simplified diagrams in Figure 8.19. As North America continued to overrun the spreading ridge, the continent passed over more and more of the Pacific Plate, causing the single triple junction to separate into two, each of which began to migrate in opposite directions along the leading edge of the continent. The northern triple junction is known as the **Mendocino triple junction**, because it is currently located near the town of Mendocino, on the north coast of California. To the south, the **Rivera triple junction** is the counterpart of the Mendocino triple junction. When the Mendocino triple junction first developed in late Oligocene–early Miocene time, it was located much farther south, near the present-day site of Los Angeles. As North America overran more of the oceanic ridge in later Cenozoic time, the Mendocino triple junction migrated northward, while the Rivera triple junction drifted south, as illustrated in Figure 8.19. Between the two migrating triple junctions, North America was sliding to the west against (and perhaps a bit over) the Pacific Plate, which was moving to the northwest. Thus, because the North American and Pacific Plates were moving in the same *general* direction, there was no convergence and no subduction in the expanding area between the Rivera and Mendocino triple junctions (Figure 8.20). Inasmuch as the North American Plate was moving very slowly to the west while the Pacific Plate was moving somewhat faster to the northwest, the boundary between the two plates involved mainly side-by-side motion, as the oceanic plate scraped northward along the leading edge of the continent.

Such plate boundaries that involve primarily lateral relative motion are known as **transform plate boundaries**. Beginning about 30 million years ago, the convergent boundary between the Farallon and the North American Plates was gradually replaced by a transform boundary between the Pacific and North American Plates. Throughout late Cenozoic time, the transform plate boundary grew more extensive, as larger areas of the Pacific Plate came into contact with the western edge of North America. Today, this transform plate boundary is more than 1,300 km (780 miles) long, and it is represented by the famous **San Andreas Fault system** in California. In the vicinity of the Great Basin, the old Farallon Plate has now disappeared under western North America, most of it probably consumed in the now-defunct subduction zone. However, north and south of California and the Great Basin, small remnants of the Farallon Plate still exist. The Rivera and Cocos Plates offshore of Central America, and the Juan de Fuca and Gorda Plates adjacent to the Pacific Northwest, represent parts of the oceanic Farallon Plate that have not yet been overrun by the continent.

The development of the transform plate boundary along western North America roughly coincided with the onset of extension in the Great Basin. As we have already learned, some of this extension can be related to the relaxation of compression that accompanied the end of plate convergence. In addition, it is important to remember that the Pacific Plate is, and was, moving rapidly to the northwest along the San Andreas Fault system. This motion produces powerful shear stresses (leading to devastating earthquakes) along the plate boundary and, at the same time, tends to "drag" the area inland from the plate boundary to the northwest. The extension of the Great Basin accelerated into later Cenozoic time as the evolving transform plate boundary pulled the crust of the continent to the northwest. Of course, the evolution of the transform boundary (San Andreas Fault system) was a gradual event, and is still occurring today. Thus, the extension of the Great Basin began on a small and relatively localized scale about 30 million years ago, accelerated dramatically in Miocene time, and continues today. We will return to the modern geologic activity in the Great Basin in Chapter 10, but first let us turn our attention to the landscape that resulted from the effects of the crustal stretching that accompanied the birth of the San Andreas transform plate boundary.

THE MID-TERTIARY LANDSCAPE

As the Great Basin began to collapse into a region of internal drainage, many small and isolated basins developed where blocks of rock fell along normal faults or rotated downward along listric faults. In response to the extensional forces that were beginning to sweep through the region, the earliest grabens and half-grabens formed in the late Oligocene Epoch. The sediments that accumulated in these early basins tends to be relatively fine-grained silt and mud that was deposited under the influence of sluggish streams or in lakes. With the intensification of extension in the Miocene, the basins became larger, deeper, and more numerous. As the land fell, the rivers that drained the increasingly rugged mountains were directed into numerous rapidly subsiding basins. By mid-Miocene time, these swift streams were washing enormous amounts of coarse sand and gravel into expanding basins. Therefore, in contrast to the early Cenozoic sediments, mid-Miocene and younger strata in the Great Basin are generally dominated by coarse-grained sandstone and conglomerate. Recall, also, that the explosive volcanoes were still active throughout the Great Basin, even as the early basins were beginning to develop. Consequently, Oligocene and Miocene sedimentary rocks are also commonly associated with layers of volcanic ash and/or lava flows, as they are at The Sump (Plate 12) and in Rainbow Basin (Figure 8.21). Because extensional faulting continued throughout the Cenozoic Era, the basins fell steadily for million of years, and thousands of feet of mud, sand, gravel, and volcanic ash accumulated in many of them. In Esmeralda County of western Nevada, for example, Cenozoic strata older than 6.9 million years (pre–late Miocene) comprise a sequence in

Figure 8.21 Sedimentary and volcanic rocks of the Miocene Barstow Formation at Rainbow in the Mojave Desert.

Figure 8.22 Middle Tertiary mammals from the Great Basin. A, the skull (left) and restoration (right) of an Oligocene titanothere similar to *Protitanops* from the Death Valley region. The skull is approximately two feet long. B, the skull (left), about eight inches long, and restoration (right) of *Merycoidodon*, a common Miocene oreodont. C, the skull of *Gomphotherium*, a distant relative of the modern elephants. Some of the larger Miocene gomphotheres had skulls in excess of three feet long.

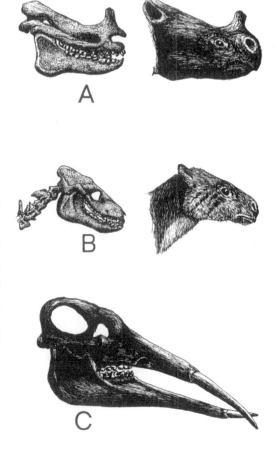

excess of 9,000 feet thick! Even more impressive are the 11,000 feet of sediment and volcanic material that accumulated during the Oligocene and Miocene Epochs in the vicinity of the Tintic Mountains in western Utah. Even in the Mojave Desert, the remote southern pocket of the Great Basin, Miocene strata alone comprise a sequence more than a mile thick!

Oligocene and Miocene sedimentary rocks of the Great Basin have produced a wealth of fossil material demonstrating than a rich fauna and flora thrived amidst the volcanic rampage and seismic havoc of the mid-Tertiary Period. These fossils have been found primarily in the lake, stream, and swamp deposits that accumulated in the localized basins. More rarely, fossil scraps are sometimes found in the tuffs, indicating that the volcanic blasts did occasionally claim some prehistoric casualties. In the House Range area of western Utah, late Oligocene lake beds, approximately 28–30 million years old, have produced well-preserved fossil leaves of trees very similar to modern cottonwoods and poplars, willows, maples, sumac, and holly. These plants suggest that a lush temperate forest cloaked the Great Basin highlands just as the great extensional collapse was beginning. Even more intriguing is the elevation suggested by the ancient flora: paleobotanists have estimated that the trees in the Oligocene forests of the eastern Great Basin grew at an elevation of around 9,800 feet above sea level. Today, this same region has an average elevation of about 5,600 feet, suggesting more than 4,000 feet of collapse over the past 30 million years or so. The same rock succession that produces the Oligocene plant fossils in western Utah also contains layers of minerals such as anhydrite, halite, and dolomite that form when large bodies of freshwater evaporate. Sediments that contain these minerals are therefore commonly referred to as **evaporites.** The Oligocene evaporites indicate that the climate of the mid-Cenozoic was, at least at times, warm and relatively dry. Evidently, the Oligocene lakes and ponds of the Great Basin experienced several cycles of growth and decline linked to climatic variations between warm-dry and cool-moist conditions.

In addition to the plant fossils, some spectacular fossils of large animals have surfaced from the Oligocene strata in the Great Basin. The most sensational of such fossils have been recovered from the Titus Canyon Formation in the Death Valley region. The Titus Canyon Formation consists of lake bed and stream deposits that were washed into an ancient basin near Death Valley in early Oligocene time, about 36 million years ago. At that time, a verdant wooded savannah covered the region, in strong contrast to the austere desert that exists in the same region today. The lush flora and varied fauna that thrived in the ancient woodland are documented by hundreds of fossils collected from the Titus Canyon Formation over the past seventy years. The larger animals in the Titus Canyon fauna include primitive tapirs, rhinos, dogs, rodents, camels, and deerlike creatures. Some of the animals in this ancient assemblage are so bizarre that they cannot easily be compared to anything that lives today. For example, in 1936, paleontologist Chester Stock discovered the remains of an odd beast he called *Protitanops curryi*. This large creature belongs to an extinct group of mammals known as **titanotheres,** or "brontotheres" (Figure 8.22A). In their general appearance, titanotheres were similar to the living species of rhinoceros but had a pronounced hump over the shoulders and a very bizarre set of horns on the face. The titanotheres generally had four hoof-bearing toes on their forefeet, but only three toes on their hind feet. Though they evolved from small ancestors, most Oligocene titanotheres were very large animals, standing as much as 6 or 7 feet tall at the shoulder. Judging from their large mashing molar teeth, these gigantic beasts probably browsed the softer, leafy vegetation that grew near the lakes and streams in the Great Basin during Oligocene time. The titanotheres probably lived in herds and constantly moved along the edges of Oligocene lakes or roamed the floodplains adjacent to the rivers of the region. Collectively, the primitive mammals, fossil leaves, petrified wood, and invertebrate fossils from the Titus Canyon Formation suggest that a rich and well-watered terrestrial ecosystem flourished at the beginning of the turbulent mid-Tertiary interval.

≈

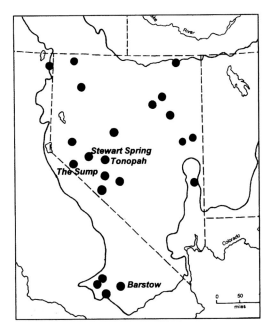

Figure 8.23 Major Miocene fossil localities of the Great Basin. The circles represent only the largest and most important concentrations of Miocene fossils among the hundreds of localities documented by paleontologists. Each of these sites has been given a formal name, but only those mentioned in the text are labeled.

While the Oligocene fossils in the Great Basin are intriguing, they are only abundant in a few locales, occurring here and there in relatively restricted sites such as Titus Canyon. In Miocene rocks, however, the abundance of fossils literally explodes across the entire region. Scores of Great Basin Miocene fossil localities have been excavated by paleontologists over the past century, some of them ranking among the richest fossil-bearing sites in the world (Figure 8.23). This sudden profusion of fossils probably reflects the continuing development of the basins that began to subside in the Oligocene Epoch. By the middle part of the Miocene Epoch, the extensional basins of the Great Basin were large and extensive, and sediment was accumulating in them at a rapid pace. Many primitive mammals appear to have preferred the fertile, well-watered lowlands to the more rugged highlands, and during the Miocene there was much more of this prime habitat available than during earlier epochs. More habitat means larger populations of plants and animals, all at a time when sediment was accumulating in the basins at a furious rate. The dramatic increase in the abundance of fossils in the Miocene is, in essence, a consequence of the continuing evolution of the Great Basin—bigger basins, more animals, and more sediment almost guarantees more fossils. So spectacular are some of the Miocene fossil localities in the Great Basin that they have become the standard of comparison for the entire continent. For example, paleontologists have divided the Miocene Epoch of North America into five formal ages, based on the changing character of land mammals that lived during the time. One of the five subdivisions is the Barstovian Land Mammal Age (about 16.5 to 11.5 million years ago), which is based on the fossils from the Barstow Formation of the Mojave Desert area. The Barstow Formation consists of more than 3,000 feet of sandstone, conglomerate, mudstone, and volcanic ash that has produced the remains of an amazingly varied assortment of primitive mammals that includes ancient horses, camels, dogs, rhinos, rodents, bears, pigs, antelope, and even elephants. Elsewhere in the Great Basin, Miocene fossils have been recovered from dozens of sites that are nearly as rich as the fossil beds in the Mojave region (Figure 8.23). As we might expect, the thousands of Miocene fossils from the Great Basin reveal some rather strange creatures in the prehistoric menagerie of mammals that evolved amidst the exploding volcanoes and trembling mountains.

Prominent among the odd-ball Miocene mammals from the Great Basin are the **oreodonts,** of which several different species have been documented. Because no living mammal is much like an oreodont, these primitive beasts are somewhat difficult to describe and envision. In their general appearance and size, the oreodonts looked a bit like a cross between a sheep, a pig, and a dog (see Figure 8.22B). Oreodonts were hoof-bearing herbivorous mammals, and the four toes on each foot suggests a relationship with modern pigs, even though their teeth, limbs, and skull were not very swinelike. Oreodont fossils are so common in the Great Basin that sometimes their remains help to characterize a specific horizon within a thick sedimentary sequence. Such is the case with the Oreodont Tuff, a layer of volcanic ash about 16 million years old, which has been designated as one of the formal subdivisions of the Barstow Formation. Other strange Miocene beasts that have been identified in several places in the Great Basin are the distant relatives of elephants known as **gomphotheres,** named for the best-

known genus of this extinct group, *Gomphotherium.* The mere presence of elephant relatives in North America may seem surprising, but the gomphotheres were extraordinary creatures, wherever they lived. Gomphotheres were nearly as large as modern elephants and belonged to the same group of mammals, the Order Proboscidea. In fact, *Gomphoterium* was similar to modern elephants in its general features, though its barrel chest gave the body a somewhat bulkier appearance. However, compared to the skulls of modern proboscideans, those of the gomphotheres were highly elongated and bore long tusks that extended from both the upper and lower jaws (Figure 8.22C). Unlike the ridged teeth of modern elephants, the molars of *Gomphotherium* had high, blunt cones that fit into pockets on the opposing teeth, indicating that these elephants probably fed on leafy vegetation. Gomphotheres also appear to have had a somewhat shorter and thicker trunk than do the modern elephants. Thus, while we would easily recognize gomphotheres as elephant relatives if we were to see them today, their barrel bodies, stubby trunk, and four prominent tusks would seem strangely comical compared to their living relatives. By the end of the Miocene Epoch, there were evidently several different types of gomphotheres present in the Great Basin along with a horde of other primitive mammals. Accompanying the curious oreodonts and gomphotheres in Miocene time were several kinds of camels, rhinos (*Aphelops*), bears (*Hemicyon*), pronghorns (*Merriamoceros*, et al.), peccaries and pigs (e.g., *Hesperhys* and *Dyseohyus*, among others), many kinds of dogs (*Cynarctus, Aelurodon, Euoplocyon*, et. al), deerlike ungulates, and a multitude of rodents (*Copemys, Peridiomys*, et al.).

Most of the famous Miocene fossil localities also yield plant fossils that provide a glimpse of the floristic habitats that supported the rich mammal fauna. In particular, the plant fossils from the Stewart Spring and Cedar Mountain areas of western Nevada have revealed an incredibly diverse plant community. At least forty-two different kinds of plants have been identified from Miocene strata in the Stewart Valley area, including several types of oaks, hickory, elm, poplars, maple, spruce, alder, and willows.

The 11–16-million-year-old lake beds that produce these plants fossils also yield exquisite fossils of fish and aquatic invertebrates such as clams and snails. In addition, delicate fossils representing the remains of bees, wasps, flies, moths, cicadas, and beetles also occur in these same lake beds. Similar plant, insect, and fish fossils have been found in dozens of other localities in the western Great Basin, suggesting that the numerous Miocene lakes might have been linked with a system of rivers and streams. From this evidence, it appears that the Great Basin was anything but a desert in the mid-to-late Miocene Epoch. Instead, it was a land of lakes that supported a dense forest of mixed hardwoods with a few conifers. Beyond the lakes, rivers meandered across shrub-covered plains that extended for miles. The air was alive with the humming of insects, while great herds of mammals roamed the forested valleys and lush slopes. The Miocene panorama in the Great Basin might have reminded us of the great savannahs of East Africa that, until recently, supported magnificent herds of many different kinds of animals. Coincidentally, the same geological processes of rifting, extensional faulting, and volcanism that dominated the mid-Tertiary history of the Great Basin are active in East Africa today. Imagine the African savannah with a little more water and somewhat larger forested tracts, with gomphotheres replacing elephants, with oreodonts instead of antelope, with titanotheres in place of rhinos. Add a few lakes, mix in the insects, sprinkle in some smaller primitive mammals, and you've re-created the essence of the Miocene landscape in the Great Basin.

By the time the Miocene Epoch ended about 5 million years ago, the rudimentary features of the modern Great Basin landscape were beginning to emerge. Two major differences distinguished the Miocene countryside from the modern panorama: 1) the Miocene landscape was less rugged, with somewhat lower mountains and higher valley floors; and 2) the entire Great Basin, with its forested plains and abundant lakes and rivers, was much less arid and desertlike in the Miocene than it is now. Thus, the evolution of the Great Basin in its present form hinged on two late Cenozoic phenomena: continuing uplift and faulting to create the more robust topography, and a profound change in climate to more desertlike conditions. Both of these events must have occurred in the relatively brief span of the 5 or 6 million years that separates the late Miocene Epoch from the twenty-first century. The geological and climatic changes that brought the Great Basin to its present state are the final two chapters in a 3-billion-year legacy of changing land and life. In a general sense, the late Cenozoic geological history of the Great Basin is dominated by the themes of bimodal volcanism and normal faulting that were established during the mid-Tertiary interval. Over the past 5 million years, these two complementary geological phenomena have continued in most areas of the Great Basin, albeit with some interesting new wrinkles that we will explore in the final chapter. In contrast, the climatic history of the Great Basin over the past few million years includes some dramatic events that were unprecedented in any previous epoch, and had striking effects on the flora and fauna of the region. It was the tumultuous climatic schizophrenia of the late Cenozoic Era that eventually transformed the Great Basin into the stark desert we know today. But the environmental metamorphosis from a lush woodland to a dry wilderness was not the result of a simple climatic shift; it was a fitful trend toward aridity so complicated that it is not completely understood, even after more than a century of study. And we probably have not yet seen the end result: the great climatic transformation is still underway!

The frigid, gray April dawn came slowly over Mono Lake. Too slowly, at least for the shivering students in the small tent city scattered among the ponderosa pines overlooking the shallow lake basin from the south. Three inches of snow had fallen overnight, and the temperature had dipped to single digits, while the brisk wind created a mournful sigh from the tree tops. Though they were curled into tight balls within their downy cocoons, and bundled together inside protective shells of rip-stop nylon, none of the students had slept more than a hour or two since retiring from the campfire the previous night. Even those who had adequate sleeping bags were kept awake most of the night by the shivering and restlessness of their more poorly equipped classmates. As soon as the frosty campsite became visible in the pallid light that filtered through the heavy overcast sky, students began to emerge from the tents, hoping to resuscitate the extinguished fire. Piling anything dry and combustible on top of the wet, smoldering ashes, they worked in unison to ignite some desperately needed flames. Eventually, their efforts were rewarded by a small flicker that rose from the pile of crumpled paper, torn cardboard, and damp pine cones. Crouched shoulder to shoulder around the pit, with hands held outward over the tiny flame, the students attempted to capture every morsel of heat before it dissolved into the morning chill. Judging from the chattering teeth, bleary eyes, and shivering moans that greeted me as I walked toward the pit, they were only marginally successful.

To the north, Mono Lake and a portion of the surrounding basin were visible through a gap in the snow-clad pines. A sage-covered plain of volcanic ash sloped gently toward the distant lake, which lay beneath the overcast sky with a somber gray color. The icy austerity of the scene was a stark contrast to the Sierra Club poster of a sunny Mono Lake under crystalline blue skies that hangs in my office. One of the students who had seen the poster commented aloud on its cruel deception. West of our campsite, I could glimpse the yawning glacial canyons that opened toward us from the ragged Sierra Nevada crest, fully bound in deep snow, just as the Spanish name implies. The frigid vista seemed to fortify the chill of the morning, and though the panorama was breathtaking, it brought little to comfort to any of us. It did, however, provide the perfect stimulus for pondering such phenomena as glaciers, ice ages, and climate change. As the students gradually recovered from, or just became accustomed to, the cold, our conversation drifted toward things glacial. After offering a few points of ice age trivia to the mostly unreceptive assembly, I decided to reserve further discussion of cold things for warmer times. Except for an occasional sniffle or cough, silence settled over the students as they continued to nurse the trifling fire. Still shivering against the cold bite of the morning air, I turned to look out over the morose Mono basin. I began to think about Israel C. Russell.

Israel Russell was one of many illustrious geologists to visit the Great Basin during the grand era of geological and geographic exploration of the region in the late nineteenth century. Russell was born in New York in 1852 and educated at New York University. After gaining practical experience in geology and natural history at Ward's Natural Science Establishment, Russell embarked on a series of government scientific and mapping expeditions to the West. In 1878, he accompanied the famous wheeler Survey to

CHAPTER NINE

FIRE AND ICE: THE MODERN GREAT BASIN EMERGES

western Utah and Nevada as an assistant to geologist J.J. Stevenson. During his time with the Wheeler Survey, Russell became familiar with the work of Grove Karl Gilbert, who had served as geologist on the expedition for three years, from 1871–74, before joining the Powell Survey of the Colorado Plateau from 1874–79. Gilbert was a remarkable geologist, with keen powers of observation and interests as broad as the Great Basin. In western Utah, Gilbert recognized the evidence for an enormous prehistoric body of water that he named Lake Bonneville in 1875. Russell, also a very perceptive observer of natural phenomena, learned a great deal about the reconstruction of ancient landscapes from his association with Gilbert and Stevens. It is no surprise that when Russell was dispatched to the western Great Basin with the fledgling U.S. Geological Survey in 1881 he became captivated by the unequivocal evidence of great prehistoric lakes and glaciers in the presently hot and dry desert region. Almost everywhere in the western Great Basin, Russell observed many of the same features that Gilbert had used to reconstruct ancient Lake Bonneville in Utah. For three years (1881–84), Russell surveyed northwestern Nevada and eastern California under the direction of geologist Clarence King, who had also recognized the evidence for ancient lakes in the region during his famous survey of the 40th parallel, from 1867 to 1872. In fact, King had named the ancient body of water Lake Lahontan, in honor of Baron de LaHontan, a French explorer and adventurer. Russell accurately mapped the ancient shorelines of Lake Lahontan and gathered a wealth of information on the glacial features preserved in mountains that rimmed the prehistoric lake. Russell visited the Mono Lake region during his first year in Nevada and was particularly struck with the unusual chemistry, landforms, and biota of this enclosed lake. He was equally impressed with the abundant evidence for a much larger prehistoric lake in the Mono Basin, which he recognized must have been a companion of Lake Lahontan. By carefully mapping features related to the ancient lake and adjacent glaciers, and by studying the deposits left behind by both, Russell was able to unravel the complicated pattern of changing ice age landscapes in the region. Russell published his impressive work on Lake Lahontan in 1885 as U.S. Geological Survey Monograph 11. Gilbert, delayed by administrative duties with the USGS, did not complete his study of Lake Bonneville until 1890, when it was published as Monograph 1. These two beautiful volumes still stand as classics of North American geology. Though many details and refinements have since been added to both studies, the basic conclusions reached by both Gilbert and Russell about the evolution of Quaternary landscapes in the Great Basin remain valid.

We were preparing to leave camp on that snowy morning when a few streaks of sunlight managed to penetrate the gloomy clouds overhead. The overnight snowfall, coupled with the angled illumination, highlighted the subtle concentric ridges that swept along the eastern shore of Mono Lake far to the northeast. Those hummocks, like so many bathtub rings inscribed on the sage plain, were among the features mapped by Israel Russell that documented the large primeval ancestor of the modern lake. As I threw a water jug, completely frozen, into the back of the pickup truck, I wondered what hardships Russell endured during his three years of fieldwork in the Mono Lake and western Nevada area. Surely, he too must have shivered in the cold around a restive fire after an overnight snowfall. Just as surely, he must have been burnt by sun, stung by wind, and parched by the wilting dryness. Russell probably endured other discomforts, such as saddle sores and rawhide chafing, that most of us today can scarcely imagine. But anyone who explores the incredible Quaternary history of the Great Basin can understand the fascination that kept Russell at his task. He was undoubtedly so mesmerized by the beauty of the land, and so lost in its deep history, that no hardship could diminish his zeal for uncovering it. It is only fitting, in consideration of the difficulties of his task and the quality of his work, that the ice age forerunner of Mono Lake is known today as Lake Russell. It is also fitting that two of the greatest geological monographs written in the late nineteenth century feature the Quaternary landscapes of the Great Basin.

This is not just a tribute to the personal qualities of Russell and Gilbert; it also reflects the extraordinary changes that swept across the Great Basin during the past 2 million years. In painting the landscape of the Great Basin on the canvas of North America, Mother Nature saved the boldest strokes of her brush for the very last interval of geologic time: the Quaternary Period.

THE QUATERNARY PERIOD

The Quaternary Period is the final major subdivision of the geological time scale, encompassing the past 2 million years or so. It is a period of remarkable changes in the earth's climate, geography, and life. It is also the time when modern humans emerged, an evolutionary milestone unlike any that came before. Perhaps because our own history as a species is so inextricably entwined with Quaternary events, this period has become one of the most heavily debated increments of geological time. Among scientists, there are continuing controversies, sometimes very heated, on such matters as the definition, the limits, and the proper subdivisions of the Quaternary Period. When it was originally established in 1829 by the French geologist Paul Desnoyers, the Quaternary "System" consisted of rocks, mostly loosely consolidated sediments, that rested on Tertiary strata in the Seine River basin of France. The Quaternary was so named by Desnoyers because it was envisioned as the fourth increment of the European geological record, joining the Primitive, Secondary, and Tertiary that had been established in the 1760s. Thus, the original definition of the Quaternary was based on position of the strata (overlying the Tertiary beds) and their weakly lithified character, both of which suggest geological youth. In the 1830s, British geologist Charles Lyell developed a system of subdividing the Cenozoic strata of Europe based on the percentage of fossils in the various layers that were still present in the modern world. Lyell's percentages served as the original basis for the epochs of the Cenozoic periods, such as the Paleocene, Eocene, and Miocene. For the Quaternary, Lyell established two subdivisions: the **Pleistocene Epoch** ("mostly new," in Greek), for rocks from

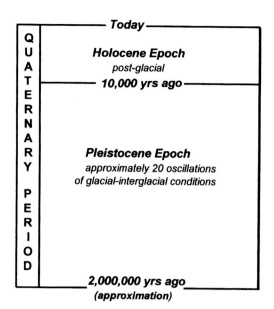

Figure 9.1 The Quaternary Period. The last period of the Geologic Time Scale consists of two unequal epochs, the Pleistocene "Ice Age" that began about 2 million years ago, and the nonglacial Holocene, or modern epoch, that commenced around 10,000 years ago.

which 90 percent of the fossils are still living and 10 percent are extinct, and the **Holocene Epoch** ("entirely new," in Greek), for sediments that produce only fossils of organisms that still exist, without any extinct forms at all (Figure 9.1). In effect, Lyell had shifted the definition of the Quaternary from one based on stratigraphic position and sediment character to one based on the fossil fauna. By 1846, geologists in Europe had demonstrated conclusively that much of the Pleistocene sediment accumulated from melting glaciers, and the concept of an "Ice Age" in the recent geologic past became firmly established. Gradually, as a detailed glacial history of Europe was delineated by geologists, the Pleistocene Epoch became synonymous with the Ice Age. By 1873, even Lyell embraced the concept that the Pleistocene Epoch *was* the Ice Age. Thus, the definition of the Pleistocene Epoch, and consequently the beginning of the Quaternary Period, shifted again, this time from a faunal basis to climatic criteria. The climatic definition of the Pleistocene seemed reasonable in the late nineteenth century, but more recently scientists have learned that the "Ice Age" actually includes many cycles of climatic oscillation that began and ended at different times in different places. However, even before scientists were aware of the complexity of climatic cycles of the past 2 million years, there were three different ways of defining the Quaternary Period and its two epochs. Now that geologists have recognized as many as fifty separate glacial episodes since Miocene time, the beginning of the Quaternary Period has been even more controversial. Hoping to establish an unambiguous definition for the beginning of the Pleistocene and Quaternary intervals, scientists have attempted to use new criteria based on radiometric dating, the chemical record of climate change in deep-sea sediments, and the pattern of reversals in the polarity of the earth's magnetic field. None of these approaches has won universal acceptance, and the issue is still hotly debated among geologists. Estimates for the beginning of the Quaternary Period range from 1.65 million years (end of the last period of reversed magnetic polarity), to 1.88 million years (beginning of the last interval of reversed magnetic polarity), to 2.4 million years (beginning of extreme climate instability in the Northern Hemisphere), to 2.48 million years (another reversal in magnetic polarity). The end of the Pleistocene Epoch, which corresponds to the beginning of the Holocene Epoch, is somewhat less controversial. The vast majority of scientists accept the climatic definition of the Pleistocene-Holocene boundary as the end of the last major glacial episode in the Northern Hemisphere. Of course, this event was not perfectly synchronous everywhere north of the equator, ranging from about 13,000 to about 9,000 years ago. Realizing that the climatic phenomena that provide the modern basis for delineating the Quaternary Period are not instantaneous events, geologists have learned to live with the "soft" boundaries of this increment of time and its subdivisions. In practice, we can approximate the beginning of the Quaternary Period at about 2 million years, and the boundary between the Pleistocene and Holocene Epochs at around 10,000 years ago (Figure 9.1). Though these are workable estimates of the time frame of the Quaternary Period, be forewarned: the actual values used by the hundreds of geologists that have studied Quaternary events in the Great Basin vary considerably. This discomforting confusion simply reflects the historical ambiguity of the Quaternary Period and the erratic complexity of the

"Ice Age." In confronting the uncertainty of the Quaternary time scale, we should probably follow the lead of geologists: just get used to it!

PATTERN AND CAUSES OF QUATERNARY CLIMATE CHANGE

When contemplating the Ice Age, many people envision an ice-bound landscape locked in the frosty depths of Mother Nature's deep freeze. Survival in such a brutally cold world seems difficult, at best, to most of us accustomed to the warmth of the modern world. Even though this popular image of the Ice Age has become consistently reinforced in popular culture and myth, it suffers from several inaccurate misconceptions. For one, there was no such thing as *the* Ice Age. By studying the geological records of climate, such as the temperature-dependent ratio of oxygen isotopes in marine sediment, we are now able to recognize about twenty significant climatic oscillations over the past 800,000 years, and nearly fifty fluctuations over the past 2.5 million years (Figure 9.2). Thus, there were many "ice ages," not a single one, and geologists consistently use the plural form of the term in alluding to Pleistocene history. We have even learned that the earth has experienced numerous severe glacial episodes millions (or even hundreds of millions) of years before the Pleistocene Epoch began. Such ancient ice ages occurred in the late Precambrian (about 700–800 million years ago), in the late Ordovician Period (about 450 million years ago), and in the Permian Period (about 250 million years ago). At least some of these older ice ages were much more severe than the climatic deterioration of the Pleistocene Epoch. The numerous Pleistocene cold intervals each lasted around 50,000 years, and were separated by brief warm intervals, known as **interglacials**, each encompassing from 10,000 to about 25,000 years. Thus, the climate of the Pleistocene ice ages was not uniformly cold, because many warm interglacials regularly punctuated the longer cold snaps. It is much more accurate to envision the Pleistocene Epoch as a time of great instability in climate, rather than thinking of it as an era of persistent and unyielding cold. There were many times during the ice ages when conditions were much warmer than they are today; but such episodes were brief, and most of them were terminated by the abrupt return of glacial circumstances. In this regard, it is interesting to ponder the fact that the earth has only enjoyed the current warm climate for about 10,000 years. Will we, in the next ten millennia or so, slip back into an ice age? Is the earth's Holocene climate still subject to the long-term instability that affected it for more than a million years during the Pleistocene? No one knows the answers to these questions, but there is at least a good possibility that we are still living in the ice ages!

In addition to overlooking the multiple and episodic nature of the ice ages, most people also tend to overestimate the amount of cooling that accompanied the glacial oscillations. During the peak of the last glacial cycle about 18,000 years ago, the surface of the sea was only about 8°F (4–5°C) cooler than it is today, though the interiors of the continents may have been as much as 15°F colder. In the valleys of the northern Great Basin, the average ice age temperature on a July day was about 75–80°F, far from the biting cold most people associate with ice ages. In this same area today, the average July high is about 90°F in Elko, Nevada, which is not as great a post-glacial difference as one might imagine. Some of the earlier Pleistocene glacial intervals might have been much colder than the most recent one, but none of them was characterized by persistent sub-freezing average annual temperatures, at least at the latitude of the Great Basin. Though the ice ages were not typified by extreme cold, the lower temperatures did result in significant increases in rain and snow. The increased precipitation of the ice ages was a prime factor in the numerous cycles of environmental change that characterized the Pleistocene Epoch. All of the available information suggests that the glacial-interglacial cycles of the ice ages were thus triggered by relatively subtle changes in temperatures and moisture instead of by any drastic shifts in climate. The world was not catastrophically plunged into the deep-freeze during the glacial advances, nor was it suddenly grilled into interglacial warmth. Chances are, if we could have lived through one of the Pleistocene

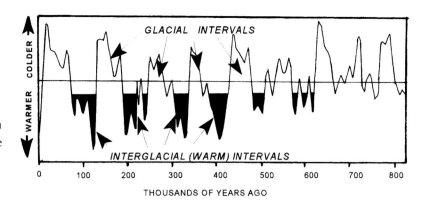

Figure 9.2 Climatic oscillations of the late Pleistocene Epoch. There were actually numerous "ice ages" during the Pleistocene Epoch, when glaciers advanced during repeated cold intervals that lasted around 50,000 years. The glacial episodes were separated by brief warm periods, known as "interglacials," that lasted from 10,000 to about 25,000 years. Note that many small-scale climate changes are superimposed on longer-term glacial and interglacial cycles.

climatic swings, we might not even have noticed it happening from year to year. After all, we currently may be living during just such a time of global climate change.

Even though the onset of the Pleistocene glacial intervals was gradual, they each resulted in profound environmental and geographic changes throughout most of North America. In the Midwest, great masses of glacial ice expanded southward from Canada, eventually reaching the latitude of Kansas. At their maximum stage of development, these ice sheets were more than a mile thick and covered 30 percent of the continent, including the present sites of Minneapolis, Chicago, and Detroit. When the vast ice sheets receded about 11,000 years ago, they exposed a flat, scoured landscape that had been bulldozed by billions of tons of ice and covered with an undulating veneer of glacial rubble. The Midwest ice sheets did not penetrate the mountainous regions of the West, but long tongues of ice known as **alpine glaciers** formed along the mountain crests and slid slowly down canyons under the influence of gravity. As the great alpine glaciers filled mountain canyons, the valleys were scoured into deep U-shaped troughs, while ragged crags of rock were sculpted in higher elevations by the moving ice. The rugged glaciated terrain in the Rockies, the Sierra Nevada, the Cascade Range, and the highest Great Basin peaks are among the most scenic landscapes in the world, and present a sharp contrast to the subdued glacial landscape of the Midwest. By 18,000 years ago, so much water had accumulated as glaciers in North America and Eurasia that the sea level had fallen by about 350 feet. As sea level fell, the oceans drained away from coastal regions, exposing wide tracts of land behind the retreating surf. The site of present-day New York City was nearly sixty miles inland, and San Francisco Bay was a broad river valley through which water released from the Sierra glaciers rolled toward the Pacific shore, miles to the west. Alaska was, oddly enough, mostly unglaciated in the late Pleistocene, but was connected to the tip of Siberia by a narrow strip of land which is today submerged beneath the Bering Strait. This vanished land connection between North America and Asia is thought to have played a critical role in the migration of humans to the New World sometime in the late Pleistocene. Thus, the geography and physiography of North America was much different in the ice ages from the landscapes we are accustomed to in our own time. As we will see later in this chapter, there were also comparable contrasts between the creatures that inhabited the continent during the general "Ice Age" and the flora and fauna of our own post-glacial time.

Before narrowing our focus to the Ice Age Great Basin, let us briefly consider the ongoing mystery concerning the cause of the Pleistocene ice ages. Over the past thirty years or so, geologists have made steady progress in charting the wild climatic oscillations of the Pleistocene. As it is now known, the pattern of climate change over the past two million years was so erratic and capricious that no single mechanism can be linked to it as a causal event. Analysis of ice cores taken from the Antarctic and Greenland ice sheets, as well a variety of geological evidence, suggests that the initial cooling of the global climate began as early as 20–25 million years ago in the early Miocene. Some scientists, impressed with the fierce intensity of Miocene volcanic activity in the Great Basin and other regions, have speculated that ash and dust from the volcanic explosions may have blocked enough energy from the sun to produce a significant and persistent cooling of the atmosphere. However, this volcanic activity was probably too early, and too localized, to have resulted in enough cooling to trigger the global Pleistocene ices ages millions of years later. Nonetheless, the cooling effect of volcanic eruptions is well documented, and the Miocene activity may have destabilized the climate just enough to make it susceptible to different sorts of disruptions that followed in Pliocene and Pleistocene time. Over the past fifty years, geologists have ruled out a number of other possible causes for the Pleistocene ice ages, including plate tectonic events, disruptions of the deep ocean currents, and variations in the energy output from the sun. In the 1920s, the Yugoslavian scientist Milutin Milankovitch first suggested that small periodic changes in several characteristics of Earth's orbit could have caused enough variation in the amount of solar energy reaching the surface to initiate the glacial advances. Milankovitch calculated (by hand!) the combined effects of several cyclic orbital perturbations and discovered that the resulting variations in solar energy reaching the earth matched the timing of the Pleistocene glacial advances, insofar as geologists had worked them out. This work gave rise to the well-known theory that the climatic cycles of the Pleistocene were the result of such orbital "forcing" mechanisms. In honor of their discoverer, the cyclic variations in solar heating that stem from orbital causes are now known as Milankovitch cycles. The theory that such cycles caused the Pleistocene ice ages has been vigorously debated ever since the idea was first advanced more than seventy years ago. However, with the advent of modern computing power, unimaginable in Milankovitch's time, the accuracy of his original calculations has been verified and many details have been added to his concepts that linked glacial history to orbital cycles. The connection between the ice ages and Milankovitch cycles still suffers from some inconsistencies and uncertainties, such as the precise timing, the relative severity, and extent of the orbital cycles and glaciations. We know, for example, that the climate did not deteriorate from the orbital cycles prior to the great volcanic eruptions of the Miocene. It appears that the Milankovitch cycles became effective only after the climatic was already in a downward spiral with respect to temperature. Nonetheless, most geologists endorse the concept of orbital forcing mechanisms, though they are certainly not the entire story behind the Pleistocene ice ages. While it is beyond the scope of our discussion to examine all the other geological, oceanic, and atmospheric factors that might have contributed to the ice ages, it is clear that the Pleistocene climatic events were not the result of anything simple, singular, or sudden.

Each time the climate dipped into a cool interval, glaciers began to expand incrementally as more snow fell in the winter than was melted in the summer. Each year, as long as the cooling trend persisted, residual snow was buried under an increasingly thick blanket. Eventually, the snow was pressed into glacial ice by the weight of the overlying material, and the glaciers began to advance. Glacial ice formed most readily at high elevations and in high latitudes, because these areas received the greatest amount of snow, the raw material of all glaciers. As temperatures continued to decline, the vast ice sheets in Canada slowly expanded and the alpine glaciers in the West grew longer and thicker. At lower elevations, and at more southerly latitudes, the cooler ice age temperatures brought increasing amounts of rain rather than snow. In such areas, rainfall began to exceed evaporation and the excess water accumulated in lakes. Such lakes, primarily the result of the increased rainfall related to the ice age cold cycles, are known as **pluvial lakes**, from the Latin word for rain. Huge pluvial lakes developed many times across the lowlands of North America during the Pleistocene Epoch. Such lakes expanded and contracted in rhythm with the advancing and retreating glaciers as the climate oscillated between glacial cold and interglacial warmth. Thus, it was not just ice that expanded during the ice ages; liquid water, in the form of huge pluvial lakes, was an equally prominent feature of the Pleistocene landscapes of North America.

THE ICE AGES IN THE GREAT BASIN

As we have already learned, the early beginnings of the ice ages can now be traced back to the late Miocene Epoch. However, these early cooling episodes were relatively localized, brief, and mild. It was not until about 2.4 million years ago, in the late Pliocene Epoch (or early Pleistocene, according to some geologists), that the climatic deterioration appears to have become a significant global phenomenon. In the Northern Hemisphere, there was a noticeable increase in the instability of the climate beginning about this time, and the earliest glacial advances began. By the time the ice ages were well underway, the general topography of the Great Basin had evolved to essentially its modern form. The broad upland had collapsed under the continued influence of extensional faulting, and the overall pattern of alternating north-south mountains and valleys had been established. The highest peaks in the modern Great Basin were also the highest peaks in the Pleistocene, though they may have been a few hundred feet lower than they are now. Likewise, the valley floors were close to their modern elevations, and most of the rivers in the region had already lost their connection to the sea. Within the Great Basin, the Pleistocene drainage patterns were somewhat different than those of today (as we shall soon see), but the overall configuration of the landscape was very similar to the modern pattern.

It is very difficult to reconstruct the earliest Ice Age events in the Great Basin. This is because both glaciers and pluvial lakes have a tendency to obliterate or obscure the evidence of early activity each time they redevelop in subsequent glacial cycles. The history of an alpine glacier, for example, is recorded by the abrasions it leaves on bedrock surfaces, by the depth and shape of the canyons it flows through, and by the sediment deposited from the melting ice as it recedes at the end of a glacial cycle. These effects of glaciation may be conspicuous during an ensuing interglacial period, but when the glacier returns down the mountain valley it tends to re-abrade the bedrock, destroy or modify the canyon profile, and plow up the old sediment and/or bury it under a blanket of new rubble. In a similar manner, pluvial lakes leave both erosional and depositional records of their existence on the landscape, but each successive stage of lake development makes it more difficult to recognize the effects of earlier stages. Consequently, very little is known about the glaciers and pluvial lakes that formed in the Great Basin 2 million years ago. However, much more data is available for the subsequent episodes of glaciation and lake development, especially for the most recent cold cycle that peaked around 15,000 years ago (see Figure 9.2). Geologists are currently working to reconstruct the details of the early Pleistocene history in the Great Basin, but currently only those glacial and pluvial events that occurred during the past several hundred thousand years have been deciphered in detail. In spite of this limited knowledge, the evidence suggests that the late Ice Age environments in the Great Basin were strikingly different from those of today. Moreover, the environmental changes of the late Pleistocene Epoch were as rapid as they were profound, taking place over just a few thousand years, in some cases. No period in the multibillion-year history of the Great Basin brought greater, or quicker, changes to the land and life of the region than did the Quaternary.

Glaciers of the Great Basin

Alpine glaciers developed in the highest mountains of the Great Basin several times during the Pleistocene Epoch, but they were not particularly abundant in the region. Evidence of alpine glaciation, such as U-shaped canyons, abraded rock surfaces, and glacially deposited sediment, has been found in only about 30 of the more than 400 mountain ranges in the Great Basin. Not surprisingly, most of the evidence for Pleistocene glaciation comes from such places as the Ruby Mountains, the Snake Range, the Toiyabe Range, the Deep Creek Range, and the White Mountains, all of which have large tracts of land higher than 10,000 feet above sea level. In addition, the 9,733-foot-tall Steens Mountain in southeastern Oregon was heavily glaciated, probably due to its northerly location rather than its elevation. From such lofty breeding grounds, the alpine glaciers, moving like great tongues of ice, descended mountain canyons repeatedly during the late Pleistocene. In the Ruby Mountains, for example, glaciers moved down Lamoille Canyon at least twice, carving the most impressive glacial canyon in the region (Plate 13). During the most recent glaciation, about 15,000 years ago, the Lamoille Canyon glacier

was nearly fifteen miles long and several hundred feet thick. The earlier period of glacial activity in the Ruby Mountains appears to have occurred about 150,000 years ago. Thus, the two episodes of glaciation in the Ruby Mountains correspond well to the last two cold intervals of the Pleistocene Epoch (Figure 9.2). Almost certainly, there were earlier glacial events in the Ruby Mountains, but the evidence for them is meager to nonexistent. Glacial features similar to those of the Ruby Mountains can also been seen in the Snake Range to the south, where elevations reach as high as 13,063 feet at Mt. Wheeler. Tucked away in a protected recess of Mt. Wheeler, at an elevation of about 11,200 feet, is the sole surviving remnant of the Pleistocene glaciers. The little Mt. Wheeler glacier, sheltered from the warmth and aridity of the Great Basin by the shadow of the majestic peak above it, covers only about fifty acres, and is most often mantled by a blanket of rock and sediment. But the widespread glacial features in the Snake Range ramparts suggest that alpine glaciers were much more numerous and extensive in that location during the Ice Age. In most of the other glaciated ranges of the central Great Basin, the data are either too scant, or have yet to be compiled in sufficient detail, to allow geologists to work out the exact pattern of Pleistocene glacial activity.

Along the periphery of the Great Basin, in the Sierra Nevada and the Wasatch Range, evidence of Pleistocene glaciation is much more extensive than it is in the interior of the province. This is probably because these bordering ranges received much more snow during the Pleistocene than did any of the interior ranges of the Great Basin, a situation that still exists today. In particular, the Sierra Nevada, the highest mountain system in the region, experienced a long history of recurring glacial advances separated by brief recessions. Geologists recognize evidence for at least five glacial cycles in the Sierra Nevada over the past 800,000 years or so: the Tioga glaciation (10,000–25,000 years ago), the Tenaya glaciation (about 37,000 years ago), the Tahoe glaciation (56,000–118,000 years ago), the Mono Basin glaciation (131,000 years ago), the

Sherwin glaciation (more than 760,000 years ago), and possibly others. On the opposite side of the Great Basin, the high Wasatch Mountains of Utah experienced a comparable history of repetitive glaciations, but there appears to have been fewer glacial episodes and they cannot be easily correlated with the complex glacial history of the Sierra Nevada. In both of these peripheral ranges, alpine glaciers formed along the high crest of the mountains and plunged down steep canyons, scouring spectacular U-shaped canyons in the process (Figure 9.3). Some of the most stunning glacial scenery in the world can be seen the Wasatch Mountains southeast of Salt Lake City and in the High Sierra west of Owens Valley. In both of these locations, the Pleistocene glaciers sometimes extended beyond the mountain front out across a portion of the adjacent valley floor. When such far-reaching glaciers receded back up the canyons during the warm interglacials, the sediment released from the melting ice accumulated as ridges and looping mounds of coarse rubble called **moraines** (Figure 9.4). The moraine deposits indicate the former position and extent of the glacial ice, and are key sources of data on the glacial history of the Great Basin. Unless the moraine deposits were washed away by glacial meltwater, dark-colored soils commonly developed on the piles of rubble during interglacial periods. Such soil horizons sometimes became buried under younger moraine deposits when the glacial cycle resumed again during subsequent cold intervals.

The separation of multiple moraine deposits by nonglacial soils is a common feature of the sedimentary record of the Pleistocene and reflects the repetitive nature of the ice ages. Occasionally, volcanic eruptions (more on these later) resulted in the accumulation of layers of ash on the moraine and soil deposits. Radiometric dating of the ash layers, coupled with radiocarbon dating of some of the nonvolcanic sediment, provides valuable time references for the various Pleistocene glaciations. This is especially significant on the west side of the Great Basin, where the numerous glacial advances from the Sierra Nevada overlapped with areas of extremely vigorous volcanic activity. The temporal connection between glaciers and volcanoes is an

intriguing juxtaposition of opposites. Fire and ice, just as the title of this chapter suggests, were two dominant themes in the Quaternary history of the Great Basin. But, there was a third element in the Ice Age panorama of the Great Basin as well, one that presents a striking contrast to the modern desert landscape. Water, a scarce commodity in the Great Basin today, was everywhere during the ice ages.

Figure 9.3 Aerial view up Little Cottonwood Canyon in the Wasatch Mountains, carved by glaciers that descended several times during the past 70,000 years, and perhaps earlier.

Pluvial Lakes of the Great Basin

The development and advance of alpine glaciers in the mountainous areas can be thought of as the high-elevation response to the cold climatic conditions that periodically befell the Great Basin during the Pleistocene Epoch. At lower

Figure 9.4 Ridges of Pleistocene moraine extending into the Mono Basin from the Sierra Nevada in the distance. Note the U-shaped canyon descending toward the valley in the foreground.

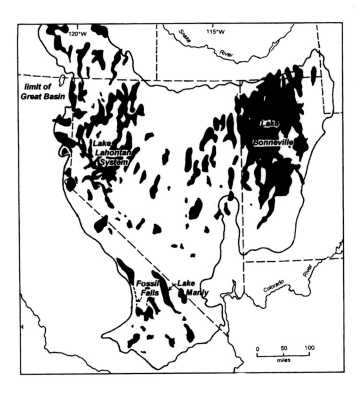

Figure 95 Pluvial lakes of the Great Basin. More than 200 lakes developed at various times during the Ice Age. The largest of these, Lake Bonneville and the Lake Lahontan system, formed in the low-elevation tracts near the periphery of the region.

elevations, where temperatures were slightly higher, the cold intervals brought increased rainfall rather than deepened snow. Today, the average annual rainfall in the valleys of the Great Basin varies from as much as 16 inches in the northwest to as little as 2 inches in the Death Valley area to the south. During the Pleistocene cold intervals, when the average temperatures were at least 10°F cooler, annual rainfall may have increased to more than 60 or 70 inches, at least in some locations. At the same time, the cooler temperatures and cloudier skies would have reduced the amount of water lost through evaporation. It has been estimated that the average rate of evaporation in the Great Basin during the last Pleistocene cold interval was only about 25 inches/year, compared to the modern rates that vary from about 40 inches/year to more than 70 inches/year. Thus, each time the alpine glaciers began their incremental advances from the high mountain crests, the cooler conditions also resulted in surplus water in the lower valleys and basins. Because the Great Basin, as a whole, had already collapsed into a region of internal drainage, most of the surplus water accumulated in enclosed pluvial lakes that grew steadily larger as long as the rainfall exceeded evaporation. At the peak of the

last cold interval, about 15,000 years ago, no fewer than 120 pluvial lakes existed in the Great Basin (Figure 9.5), collectively covering at least 28 million acres. The largest of the pluvial lakes were comparable in size to the modern Great Lakes, each storing an enormous amount of water.

Like the glaciers that accompanied them, the pluvial lakes of the Great Basin record a very complex pattern of Pleistocene climatic fluctuations. It is important to remember that the development of pluvial lakes was a gradual and repetitive process. Pluvial lakes filled most of the enclosed basins more than once during the Pleistocene Epoch, and successive deep-lake cycles (sometimes called "pluvials") were commonly separated from each other by times when the lake basins were completely dry ("interpluvials"). The dry episodes generally correspond to the warm interglacials, while the deep-lake pluvials roughly, but not always precisely, correlate with the episodic glacial advances. The near synchroneity between pluvial lake and glacial cycles dispels a popular misconception about the Ice Age lakes of the Great Basin: they did not form from glacial meltwater. For the most part, the glaciers were advancing, not melting, during the times that the pluvial lakes were growing. In addition, many of the pluvial lakes of the Great Basin formed within watersheds that did not drain any areas where glaciers existed, further weakening the perceived causal link between melting ice and growing lakes. Instead of cause-and-effect, the relationship between the glaciers and pluvial lakes was a climatic and temporal one: the pluvial lakes represent the lowland response to the Pleistocene cold cycles that simultaneously caused the glacial advances in loftier elevations. More snow at higher elevations resulted in the advance of glaciers, while increased rainfall on lower terrain led to expanding lakes. Both phenomena were triggered by the periodic cold intervals that came and went throughout Pleistocene time.

At the beginning of each pluvial lake cycle, small bodies of water collected in the numerous enclosed valleys of the Great Basin. Each of the pluvial lakes would grow larger, as long as rainfall exceeded evaporation. Two general types of

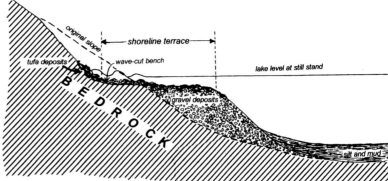

geological evidence record the history of each pluvial lake in the Great Basin: 1) erosional features on the mountain slopes adjacent to the basins, and 2) sediments that accumulated in and around the lakes (Figure 9.6). As the lakes grew larger, water began to rise against the mountain slopes that descended into the basins. Waves crashing against the slope, currents seeping along the shore, or both, eroded a flat notch in the slope commonly called a **bench**. Some of the rock and soil eroded from the wave-cut bench was commonly deposited just offshore as a sand or gravel bar or spit, broadening the bench into a **terrace** by extending the flat notch toward the lake basin (Figure 9.6). Farther offshore, in the middle of the lake basin, finer-grained sediment such as muddy silt accumulated under deep, calm water. Sometimes, minerals such as calcium carbonate precipitated from the lake water to form a crusty mantle of **tufa** along the ancient shoreline of the lake. There are many types of Ice Age tufa in the Great Basin, but they all consist primarily of carbonate minerals formed along the edge, or sometimes on the floor, of the pluvial lakes. In many places, tufa deposited along ancient shorelines commonly bound the terrace gravel into a solid mass of material that looks much like concrete (Figure 9.7), enhancing the topographic expression of the old shoreline. Thus, the terraces, which have often been described as "bathtub rings," are partly erosional and partly depositional features related to the ancient pluvial lake shorelines. Some of the pluvial lake terraces in the Great Basin are so prominent that they are impossible to miss, while others are very subtle features that are discernable only through detailed mapping and careful study of the mountain slopes.

Because all terrace features require time to form, the prominence of a lake terrace reflects the stability of the ancient lake: the longer the shoreline stays at the same elevation, the more conspicuous the shoreline terrace will be. If the lake expands or contracts steadily, such that the shoreline continually migrates up or down the slope without pause, no terrace features are likely to develop. In such cases, the topographic evidence of the ancient lake will be very subtle,

at best. Only when the lake remains at the same level for several centuries or millennia will a noticeable terrace result from the slow processes of shoreline erosion and deposition. Such periods of lake stability are sometimes called **still stands**, and reflect brief interludes of stability superimposed on the longer-term climatic dissonance of the Pleistocene Epoch. Terraces can be developed while pluvial lakes are expanding, or they can represent still stands that occur when the lakes are in decline. For this reason, there is no straightforward relationship between the actual elevation of a shoreline terrace and its age. When several terraces exist around a single basin, the highest terrace above a valley floor is not necessarily the oldest one, and, in fact, rarely is. Most of the pluvial lake basins in the Great Basin are surrounded by multiple ancient shorelines, an indisputable record of erratic oscillations in lake levels driven by the undulating climate of the ice ages. As is also true of glaciers, the cyclic nature of pluvial lakes makes it very difficult to unravel their histories. Each cycle of lake decline exposes older shorelines and sediments to erosion. When a pluvial lake rises again, new shoreline features and sediments are superimposed on whatever remains of the older

Figure 9.6 Evidence of pluvial lakes in the Great Basin. A typical pluvial lake terrace consists of both erosional elements, such as a wave-cut bench, and depositional features including gravel bars and tufa drapes. The development of a shoreline terrace results in a modification of the original slope that becomes more conspicuous the longer the lake level remains stable. Offshore deposits of fine-grained silt and mud provide additional information on lake history.

Figure 9.7 Pleistocene shoreline gravel deposits of the Great Basin, such as this material from the slopes of the Wasatch Mountains, are commonly cemented by tufa and other forms of calcium carbonate.

evidence. Thus, each cycle of lake growth and decay obscures the features that record earlier episodes. Though the capricious behavior of pluvial lakes was apparent even to Russell and Gilbert in the late nineteenth century, the individual lake histories are so complex that modern geologists have thus far deciphered only a few of them in any detail. We know that extensive pluvial lakes appeared in the Great Basin as long ago as about 660,000 years, perhaps earlier. In general, however, much more is known about the more recent (generally within the last 25,000 years) lake cycles than of earlier Pleistocene events.

Another problem that vexes geologists attempting to reconstruct the history of Pleistocene lakes in the Great Basin is that the pattern of lakes was constantly shifting. To envision the problem, imagine several enclosed basins separated by divides of different elevations. At the beginning of a cold cycle, several small lakes might develop within their individual basins. At this early stage of a pluvial episode, the geological record of each lake will be unique because no two basins are ever *exactly* the same in terms of the size of the watershed, the terrain, the amount of runoff, the soils, the rock types, the vegetation, and other factors that influence the lake over time. As the cold conditions persist, the lakes expand and the water levels rise, each lake still recording its own history. Eventually, however, one of the lakes will rise above the lowest divide surrounding its basin and will overflow to join with an adjacent lake. Merged into one lake by rising water levels, the two nearby basins will subsequently record uniform histories because they are inundated by the same body of water. If the merged lake continues to rise, it may capture another adjacent lake by breaching the divide that formerly separated them. Thus, as the pluvial lakes in the Great Basin expanded, they continuously merged with each other, creating an ever-changing pattern of land and water across the region. In the ensuing decline, the mergers were reversed as the lakes separated from one another and fell back into their original basins. Sometimes, however, the original basins had been filled with sediment during the deep lake stage, or shifting rivers had modified the regional drainage, so the pattern of lakes that would develop during the next climatic downturn was much different from that of earlier lake cycles. This constantly shifting pattern of lakes and rivers left an extremely complex sedimentary and erosional record of pluvial lake systems. So tangled are the histories of the various pluvial lakes in the Great Basin that most of them have still not been completely decoded from the geological evidence. Nonetheless, decades of research on some of the larger systems has allowed geologists to reconstruct the major events of the more recent pluvial cycles. Even though the behavior of these lake systems is only known for a small fraction of Pleistocene time, some very interesting events appear to have occurred in the Great Basin during the last few pluvial cycles. Let's explore those events by examining a few lake systems in a little more detail.

Lake Bonneville

Lake Bonneville, named by Gilbert in honor of Captain Benjamin L. E. de Bonneville, was the Ice Age predecessor of the modern Great Salt Lake and the largest of the pluvial lakes in the Great Basin (Figure 9.8). At its maximum stage of development, about 15,000 years ago, this immense lake, often called an "inland sea," covered nearly 20,000 square miles and was comparable in size to modern Lake Michigan. In the deepest part of the Bonneville basin, about where the Great Salt Lake is now located, the prehistoric lake was more than 1,000 feet deep. Unlike its small saline remnant, Lake Bonneville was a freshwater lake, receiving runoff principally from the Wasatch and Uinta Mountains to the east and the Sevier River system to the southeast. Several of the most prominent mountain ranges of northwest Utah, such as the Oquirrh, Stansbury, and Newfoundland Ranges, stood as islands in Lake Bonneville. The lake extended from its eastern shoreline along the foothills of the Wasatch Mountains westward to slightly beyond the present Utah-Nevada border. To the south, the lake stretched all the way into the Escalante Desert region, northwest of Cedar City, Utah. All along the extensive and meandering shoreline, there were numerous pockets, bays, and coves nestled between projecting peninsulas and spits.

When Gilbert first documented the evidence for Lake Bonneville in 1890, he recognized two cycles of lake development, based primarily on his mapping of shoreline features around the old basin in northern Utah. Since that time, scientists have mapped the shoreline features with greater precision, discovered additional shorelines, conducted detailed studies of exposed lake sediments (Figure 9.9), and acquired thick core samples of subsurface lake deposits. It is fair to say that Lake Bonneville is probably the best-studied pluvial lake system in North America. Ironically, in spite of the abundance of geological information pertaining to Lake Bonneville, there is still considerable disagreement among scientists over the details of its history. It is clear, though, that several lakes have come and gone within the Bonneville basin over the past million years or so. As is usually the case with the pluvial lakes of the Great Basin, the earlier stages of Lake Bonneville are the most obscure. Nonetheless, there is excellent evidence for the events that occurred in the past 30,000 years, as the major stages of this relatively recent phase of the lake's history are relatively well documented. For this reason, many scientists favor restricting the term "Lake Bonneville" to just the most recent pluvial cycle in the Bonneville basin. On the other hand, some researchers have used the same term to describe *all* of the Quaternary lakes that developed in the basin, of which there were many. Though few of the details about the early Quaternary lakes in the Bonneville basin are known, there is substantial and indisputable evidence that even the most recent phase of the lake's history included some fascinating events linked to the climatic schizophrenia of the Pleistocene ice ages.

Around 30,000 years ago, following a long history of earlier pluvial lake cycles, the Bonneville basin appears to have been completely dry. The vacant plain, situated about 4,200 feet above sea level, covered thousands of square miles of the eastern Great Basin. About 28,000 years ago, probably in response to a cooling trend in the climate, water began to accumulate in a small lake in the Bonneville basin (Figure

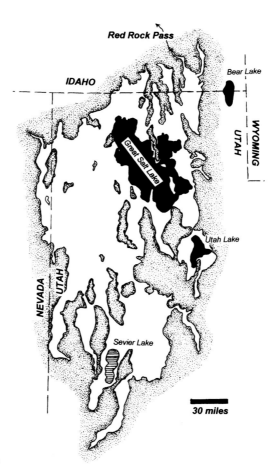

Figure 9.8 Lake Bonneville, the largest pluvial lake in the Great Basin, at its peak stage in the late Pleistocene. Remnants of this immense lake include the modern Great Salt Lake, Utah Lake, and the Sevier Lake basin.

Figure 9.9 Lake Bonneville sediments in the remote Tule Valley of western Utah.

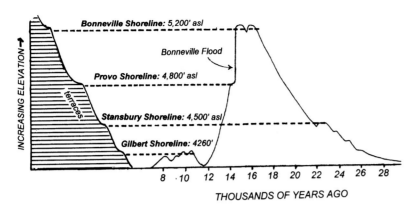

Figure 9.10 Late Pleistocene–early Holocene history of Lake Bonneville, and the development of shoreline terraces surrounding the Bonneville basin. Elevations of the major shorelines are approximate and vary from place to place due to post-Bonneville isostatic adjustments.

Figure 9.11 The Provo and Bonneville shoreline complexes on a mountain slope near Wendover, Utah.

9.10). The lake rose steadily, but erratically, for the next 15,000 years or so. There were several brief still stands, and even some minor recessions, during this time of lake expansion. About 22,000 years ago, when the surface of the lake had risen to an elevation of about 4,500 feet, there was a prolonged still stand that resulted in what geologists call the Stansbury Shoreline. The Stansbury Shoreline can be difficult to observe around the Great Salt Lake because it is not very prominent, and occurs so low on the adjacent mountain slopes that it is commonly masked by post–Lake Bonneville deposits, vegetation, roads, and buildings. After the Stansbury still stand, Lake Bonneville resumed its deepening trend until it reached an elevation of about 5,200 feet, around 16,000 years ago. Lake Bonneville might have grown even larger, except that it began to spill through Red Rock Pass (see Figure 9.8) in southeastern Idaho, the overflow cascading north to join the Snake River–Columbia River drainage system. This

means that for a brief period in the late Pleistocene, this northeastern corner of the Great Basin was not part of the region of internal drainage. Water flowing from Lake Bonneville through the Red Rock Pass outlet reached the Pacific Ocean via the Snake and Columbia Rivers. There were evidently some minor fluctuations in the level of Lake Bonneville at this peak stage, but the water remained at nearly the same elevation for at least several centuries. During this high stand of Lake Bonneville, the Bonneville shoreline complex was developed. This shoreline complex is conspicuous as the highest prominent lake terrace on the slopes of the mountains surrounding the Bonneville basin (Figure 9.10, Figure 9.11).

The most dramatic event in the recent history of Lake Bonneville occurred about 14,500 years ago. At this time, water rushing through Red Rock Pass washed away enough soil and rock to unleash one of the most cataclysmic torrents in earth history, an event generally known as the great Bonneville Flood. Evidently, the rock and soil that confined Lake Bonneville at Red Rock Pass suddenly failed, and at least *1,100 cubic miles* of water was released into the Snake River drainage. As the Bonneville Flood raced across southern Idaho, river channels were filled and breached, and walls of water surged toward the Snake River Canyon. Only the highest hills and mountains of southern Idaho escaped inundation under the Bonneville Flood. Even after more than 14,000 years, the scoured bedrock and piles of flood rubble can still be seen across thousands of square miles of the Snake River plain in southern Idaho. At its peak flow, the Bonneville Flood discharged more than *33 million* cubic feet of water each second (cfs), according to estimates by geologists. For comparison, in the flood year of 1983, when the Colorado River inundated parts of the Grand Canyon, its discharge peaked at slightly more than 110,000 cfs, a trivial fraction of the flow from Lake Bonneville. Consider also that the discharge of the Bonneville Flood was *five times* the average flow of the Amazon, the world's largest river! The Bonneville Flood was equivalent to more than three-fourths of the total discharge of *all* the world's rivers combined! It is

impossible to imagine the fury of the Bonneville Flood for a good reason: in spite of the hundreds of devastating floods that humans have endured in the post-Pleistocene world, no one has ever seen anything even remotely comparable to the great Bonneville Flood. No one knows for sure how long the Bonneville Flood raged, but it probably did not last more than a few weeks. The flood led to a rapid decline in the level of Lake Bonneville from the prehistoric high stand to what is called the Provo level, at an elevation of about 4,800 feet (see Figure 9.10). At this elevation, some 350 feet lower than its highest stage, the level of Lake Bonneville was evidently stabilized by the hard, resistant rock that had been exposed at Red Rock Pass during the flood. This still stand, which resulted in the broad Provo Shoreline terrace, was relatively brief, lasting only a few hundred years. Around 14,000 years ago, the beginning of the end commenced for Lake Bonneville, as the climate shifted toward more arid conditions and the lake began a steady decline. During this terminal ebb, Lake Bonneville began to separate into several interconnected lakes. Sevier Lake, a large southern arm of Lake Bonneville, was connected to the dwindling lake at this time by a north-flowing river, which no longer exists in western Utah. However, the old valley of this river is still evident as the Old River Bed in the desert of western Utah (Figure 9.12).

As Lake Bonneville withered under the increasingly dry climate, the chemicals dissolved in the freshwater remained in the lake as water continued to evaporate. The evaporative concentration of minerals eventually resulted in the famous salinity of the Great Salt Lake, a body of water approximately eight times saltier than the ocean. By 11,600 years ago, Lake Bonneville had dwindled to a salty remnant no larger than the modern Great Salt Lake. The once colossal lake may have disappeared completely at this time. Thus, the demise of Lake Bonneville resulted from two unrelated events: 1) a catastrophic flood, and 2) a climatic shift from pluvial (cool and wet) to interpluvial (warm and dry) conditions. About 10,500 to 11,000 years ago, however, Lake Bonneville rallied briefly, as a cool climatic interlude lifted the lake to an elevation of

at least 4,260 feet, slightly above the present level of the Great Salt Lake. Some scientists have suggested that the lake may have briefly risen nearly to the Provo level at this time, but the evidence for such a high level is very controversial. This temporary resuscitation resulted in the cryptic Gilbert Shoreline, the youngest, lowest, and most obscure of the principal shoreline complexes in the Bonneville Basin. Almost as soon as the lake rebounded to the Gilbert level, it began its final decline. By 8,000 years ago, Lake Bonneville had once again fallen to the historic level of the Great Salt Lake, where it has since remained, except for some minor weather-induced fluctuations. One such fluctuation, the 1986–87 rise of the Great Salt Lake to the historic high elevation of 4,212 feet, resulted in nearly a half-billion dollars of damage to lake-side industries and facilities. Given the nearly million-year history of erratic lake behavior, synchronized with the climatic instability of the Quaternary Period, no one should trust a lake in the modern Bonneville basin to behave according to human design. Lake Bonneville is not dead and gone; it waits in salty diminution to return to its Pleistocene glory. Judging from the Ice Age behavior of lake, we can safely assume that someday in the future it will do just that.

The Lake Lahontan System

Lake Bonneville was not the only large pluvial system in the Great Basin. On the opposite side of the province, the Lake Lahontan system

Figure 9.12 The Old River Bed near the Simpson Mountains of western Utah. As Lake Bonneville declined in the late Pleistocene, water from the Sevier arm drained northward through this channel into the central basin.

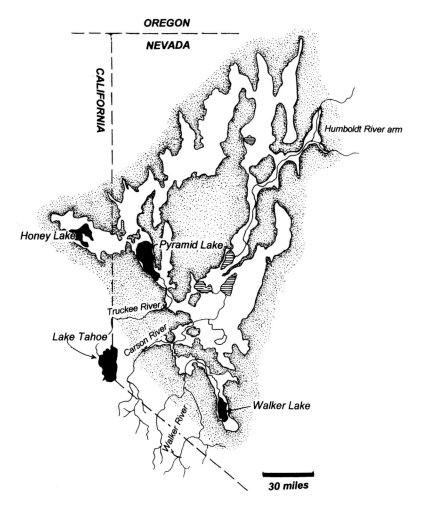

OREGON

NEVADA

CALIFORNIA

Humboldt River arm

Honey Lake

Pyramid Lake

Truckee River

Lake Tahoe

Carson River

Walker Lake

Walker River

30 miles

Figure 9.13 Lake Lahontan at its maximum stage, about 13,000 years ago. This large and complex pluvial lake consisted of numerous interconnected basins that joined and separated as the water level fluctuated in rhythm with Ice Age climatic cycles. Remnants of Lake Lahontan include modern Pyramid, Walker, and Honey Lakes.

developed in the low valleys of western Nevada during the pluvial cycles of the Quaternary Period. Compared to Lake Bonneville, Lahontan was a smaller lake, but it still covered more than 8,600 square miles at its maximum stage, about 13,000 years ago. The deepest point in Lake Lahontan was near modern Pyramid Lake in northwest Nevada, where the prehistoric water was 900 feet deep. The location of the two largest pluvial lakes of the Great Basin on the western and eastern periphery of the region is not a coincidence. When the dome-like highland of the Great Basin collapsed under the extensional forces generated in mid-Cenozoic time, the lowest tracts of the fallen landscape were situated where the lowest flanks of the dome had been. On a regional scale, Pleistocene runoff tended to drain away from the high axis of Great Basin to the lower areas to the east and west. No large pluvial lake systems formed in the central Great Basin because the basins in

that region were, and still are, much higher than the valleys far from the crest of the fallen dome. Pluvial lakes existed in the central Great Basin during the ice ages, but they were small and isolated compared to the great Bonneville and Lahontan systems.

Lake Lahontan was surrounded by a highly irregular shoreline that meandered across the uneven landscape of the western Great Basin (Figure 9.13). There were even more islands, peninsulas, coves, bays, and narrow straights in Lake Lahontan than there were in Lake Bonneville. This great pluvial lake can be pictured as a series of flooded valleys, interconnected by narrow passages and serpentine inlets. Runoff into Lake Lahontan came from six different rivers: the Truckee, Carson, and Walker, flowing from the Sierra Nevada; the Susan River flowing from the southern Cascade Range; and the Quinn and Humboldt Rivers that originate within the Great Basin. The Pleistocene ancestors of these modern rivers brought water into Lake Lahontan from the same watersheds that they drain today, but they did not necessarily follow the same courses in the ice ages that they now do. In addition to Pyramid Lake, other remnants of Lake Lahontan include Walker Lake and Honey Lake, neither of which are strongly saline, along with the Carson and adjoining Humboldt sinks, which receive water in only the wettest years. The area that drained into Lake Lahontan was actually very large, but there was evidently less rainfall in this part of the Great Basin than elsewhere in the region during late Pleistocene time. The western side of the Great Basin, directly in the rain shadow of the Sierra Nevada, is the driest part of the Great Basin today, and appears to have been so in the ice ages as well. Still, there was enough water to form an impressive pluvial lake.

The history of Lake Lahontan is recorded by geological evidence similar to that documenting the rise and fall of Lake Bonneville. Throughout western Nevada, multiple lake terraces are incised on hillsides (Figure 9.14) and accumulations of lake sediments are exposed in river banks and gullies. However, the geological record of Lake Lahontan is spiced with another important ingredient generally missing from the sedimentary archives of Lake Bonneville: volcanic ash. Late Pleistocene volcanoes were evidently more active and more widespread in the western Great Basin than they were near the Bonneville basin. Periodically, volcanic ash fell into Lake Lahontan, or settled onto the gentle slopes of the bordering hills. More than fifty layers of hardened volcanic ash have been found preserved between strata of lake, stream, or wind-blown sediment that accumulated in or around Lake Lahontan. Amenable to relatively precise radiometric dating techniques, these ash layers provide valuable reference points in time for geologists attempting to decode the history of Lake Lahontan from the sedimentary record. Consequently, even though less scientific scrutiny has been directed toward Lake Lahontan than toward Lake Bonneville, its history has been deciphered in comparable detail.

Like other pluvial lakes, the earliest stages in the history of Lake Lahontan are cloaked in the uncertainty fostered by scant geological evidence and the complexity of the basin. Part of the mystery stems from the multiple drainage systems that fed the lake, but some of it is attributable to the many sub-basins that existed in the Lahontan system. There were seven major sub-basins in the Lahontan system, each separated by thresholds of different elevations. Remember that as adjacent sub-basins connect when the lake is rising, or disconnect as it falls, the geological record in each basin is affected. Correlating the sedimentary record of one sub-basin with another can sometimes be extremely difficult. The numerous ash deposits are of great value in this effort, but it is still a very difficult task that commonly produces ambiguous results. Nevertheless, there is good evidence for at least three major cycles of inundation over the past 650,000 years. Late Pliocene–early Pleistocene lake beds along the course of the Humboldt River near Elko, Nevada, indicate that the river did not reach the western Great Basin until sometime early in the Pleistocene. Since the Humboldt is one of the main conduits of water into the Lahontan basin, it is doubtful that a lake, at least a large one, existed in western Nevada in the earliest Pleistocene. However, lake sediments ranging in age from about

Figure 9.14 Multiple shoreline terraces of Lake Lahontan on the slopes of the Wassuk Range near Hawthorne, Nevada.

635,000 years to 610,000 years do occur in the vicinity of the Rye Patch Reservoir, a part of the Lahanton basin. These sediments clearly indicate an early cycle of pluvial lake development in the Lahontan basin. This lake appears to have fluctuated up and down slightly for about 25,000 years during the early ice ages, before it eventually vanished. After it disappeared, the Lahontan basin was dry for nearly 300,000 years before it was inundated again about 350,000 years ago. This time, the pluvial lake system lasted more than 200,000 years, during which time there were at least three prominent fluctuations. From about 130,000 years ago to 35,000 years ago, there appears to have been only small temporary lakes in the Lahontan basin, separated from each other by dry valleys. The last major pluvial cycle in the basin began about 35,000 years ago, roughly the same time that Lake Bonneville began to develop. As is also true of its companion to the east, this final episode in the history of Lake Lahontan is known with considerably more detail than the earlier cycles of lake growth and decline.

The final filling of Lake Lahontan began slowly about 35,000 years ago. By about 22,000 years ago, following a series of minor ups and downs, the lake had risen to an elevation of more than 4,100 feet above sea level. At this time, Lake Lahontan completely covered the basins now occupied by Pyramid Lake, Winnemucca (dry) Lake, and the Black Rock Desert, submerging these low areas under a single pool

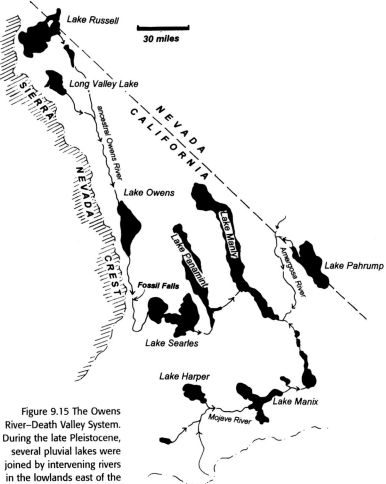

Figure 9.15 The Owens River–Death Valley System. During the late Pleistocene, several pluvial lakes were joined by intervening rivers in the lowlands east of the Sierra Nevada crest. Lake Manly, situated at the end of the system in Death Valley, also may have received some inflow from the ancestral Mojave and Amargosa Rivers.

about 12,000 years ago, Lake Lahontan had fallen so low that all of the sub-basins were detached from each other, and the residual lakes within them were rapidly withering. Though there have been many minor climate-induced fluctuations in the remnant lakes over the past ten millennia, for all practical purposes, Lake Lahontan was long gone when the Holocene Epoch began, 10,000 years ago. The modern remnants of Lake Lahontan, leftovers of the pluvial age in the western Great Basin, may one day rise again should our presently warm interpluvial climate give way to a prolonged chill.

The Owens River–Death Valley System

We learned earlier (in Chapter 1) that the climate of the southern part of the Great Basin is noticeably warmer and drier than anywhere else in the province. Though the overall climate of the Pleistocene pluvial intervals was relatively cool and wet, the southern Great Basin was still the "banana belt" of the ice ages. With somewhat less rainfall and warmer temperatures, no Ice Age lakes comparable in size to Lahontan or Bonneville developed in the southern Great Basin, even during the peak of the pluvial cycles. However, in the late Pleistocene, and perhaps earlier in the ice ages as well, a very interesting system of interconnected lakes and rivers developed along the eastern foot of the Sierra Nevada and extended more than 350 miles to the southeast, into the most arid parts of the Great Basin. We will refer to this ancient drainage as the Owens River–Death Valley system (Figure 9.15). Vestiges of this drainage still exist today; but, under the influence of the modern post-pluvial warmth, the lakes have dried, the rivers have largely disappeared, and the various components of the system have become disconnected from each other. However, the region between the Sierra Nevada escarpment and Death Valley is adorned with abandoned river channels, heaps of river-washed sand and gravel, and old lake beds that provide evidence of the former existence of the Owens Valley–Death Valley system. Such clues have allowed geologists to reconstruct the overall system; but there are still many questions concerning such details as when, how often, and for what lengths of time it

of water. For several millennia the level of Lake Lahontan remained nearly constant, until a brief drought-induced recession occurred around 15,000 years ago, when the water level temporarily dropped by several hundred feet. By 14,000 years ago, Lake Lahontan was on the rise again, and reached its highest level about 13,000 years ago. At this time, the lake level was about 4,360 feet above sea level and all the sub-basins were connected to each other. The timing of this pluvial maximum is close to, but not exactly synchronous with, the late Pleistocene high stand of Lake Bonneville in the eastern Great Basin. A climatic shift toward warmer and drier conditions began almost as soon as Lake Lahontan reached its culminating peak, and the lake remained at its highest level for only a few centuries, at most. No dramatic flood events were associated with the decline of Lake Lahontan, but the climatic changes were such that the lake fell by about 350 feet in less than 1,000 years. By

operated. As we have seen with pluvial lakes elsewhere in the Great Basin, the earliest Pleistocene history of the Owens River–Death Valley system is virtually unknown, but considerable data is available about the later history of at least some of the elements in this fascinating Ice Age drainage.

When it was fully connected and operating at the highest level, the Owens River–Death Valley System began at Lake Russell, the Ice Age ancestor of modern Mono Lake. Though it came within fifteen miles of the Walker Lake arm of Lake Lahontan, Lake Russell was never a part of that large pluvial system. When its depth exceeded about 750 feet, Lake Russell spilled into Adobe Valley, which separates the hills rimming the Mono Basin from the White Mountains to the east. Lake Russell appears to have been disconnected from the Owens River–Death Valley system several times during later Pleistocene time, and thus was not a permanent element in the system. The outflow from Lake Russell soon filled the shallow Adobe Valley, overflowing southward into the upper part of the ancestral Owens River drainage. At this point, the ancient river gathered additional water from the outflow of pluvial Long Valley Lake, from Sierra Nevada runoff, and probably from many springs in the region. Even when Lake Russell was disconnected from the system, the ancestral Owens River was large enough to carve an impressive gorge through an elevated volcanic tableland as it continued south from the Long Valley region (Plate 14). After it left the deep canyon at a point near modern Bishop, California, the ancestral Owens River began to spread over the broad valley floor between the Sierra Nevada escarpment on the west and the White-Inyo Mountains to the east. The river gained additional water from numerous streams emerging from the mountains on either side, and continued south into pluvial Lake Owens, which was centered slightly north of the modern town of Olancha, California.

At its largest stage, Lake Owens was more than 35 miles long, 200 feet deep, and covered some 270 square miles. During its pluvial peaks, Owens Lake rose over a low divide to the south, where the ancestral Owens River was reborn

Figure 9.16 Fossils Falls, site of a spectacular Ice Age cascade along the ancestral Owens River in eastern California.

and continued its southward flow. South from Lake Owens, the Owens River flowed through a maze of volcanic cones, many of which were active during the Pleistocene Epoch. In the area around the modern California hamlet of Little Lake, the Coso Range to the east converges toward the escarpment of the Sierra Nevada to squeeze the Owens Valley into a narrow gap between the adjacent mountain slopes. Many times, lava from the active volcanoes, as well as water from the Owens River system, flushed through this gap. Occasionally, basaltic lava that erupted from the small volcanoes filled the channel of the Owens River, blocking the flow of water and creating temporary lakes behind natural lava dams. When the ancient river eventually rose above the lava dams, it cascaded over the basalt flows in what must have been impressive waterfalls, graced by clouds of mist rising from rocky pools and potholes created as millions of tons of water pounded the black rock. The best-known site of such an ice age waterfall on the ancestral Owens River is known today as Fossil Falls (Figure 9.16), a popular stop along Highway 395 just north of Little Lake. The torrent that plunged more than 100 feet over the

Figure 9.17 The Pinnacles, composed of Ice Age tufa that formed on the floor of Pleistocene Lake Searles in the Mojave region.

two lips of basalt at Fossil Falls must have created a spectacular scene. The abundant artifacts and stone flakes recovered by archaeologists at this site suggest that early humans in the Great Basin lingered here for extended periods of time to fashion tools, catch fish, hunt game, prepare skins, trade goods, and conduct other activities. Given the likely beauty of the falls around which they congregated, it seems reasonable to assume that they also might have simply stood around awe-struck, doing nothing at all but admiring the scenery.

About fifteen miles downstream from Fossil Falls, the ancestral Owens River reached the Indian Wells Valley, where it fed pluvial China Lake. The basin here was only about forty feet deep, however, and it filled quickly to spill over into the larger Lake Searles basin to the east. During the low ebbs of interpluvial times, Lake Searles was disconnected from China Lake, while at pluvial peaks, they coalesced into one large lake. Perhaps because more than $2 billion worth of chemicals (soda ash, potash, borates, etc.) have been extracted from the Pleistocene lake beds in Searles Valley, geologists have conducted more detailed studies of the pluvial history in this basin than anywhere else in the Owens River–Death Valley System. Several times over the past 130,000 years, Lake Searles expanded and contracted, alternating between deep-lake cycles and dry-basin interludes. Undoubtedly, this extreme cyclic behavior is closely related to the great concentration of evaporite

minerals in the Pleistocene sediments that filled the Searles basin. Each time the pluvial lake diminished, the water in it became more saline. On those occasions when the lake dried up completely, a thick crust of evaporite minerals accumulated from the residual water. Even when the lake was fully engorged, some interesting things happened on its floor. Calcium-bearing water evidently gushed from numerous springs on the bottom of the lake, mixing with the alkaline water fed in by the Owens River. The combination of spring and lake water, assisted by algae that grew around the mouths of the submerged springs, resulted in the precipitation of calcite in the form of tufa. The crusty tufa formed most readily around the orifice of the springs and, over time, mounds and spires of tufa rose as much as 150 feet above the lake bottom. More than 500 of these odd towers and knobs can still be seen along the southwest edge of the Lake Searles basin, where they are known as The Pinnacles (Figure 9.17). During the past 40,000 years, the lake reached its greatest depth of about 660 feet twice: an early peak 35,000 to 40,000 years ago, and a later climax between 15,000 and 20,000 years ago. Between these two high stages, around 28,000 years ago, Lake Searles appears to have completely dried up. The most recent peak of Lake Searles corresponds roughly to the last high stands of both Lake Bonneville and Lake Lahontan.

Though it did not do so continuously throughout the Pleistocene Epoch, Lake Searles did occasionally rise high enough to spill over into Panamint Valley, situated between the Panamint Mountains and the Argus Range. During times of abundant water, the Owens River–Death Valley system reached this remote valley and formed a pluvial lake more than 60 miles long and 900 feet deep. No saline minerals are found in the Pleistocene lake beds of Panamint Valley, so it does not appear that the overflow from Lake Searles occurred continuously, or even very often, during late Pleistocene time. However, the high mountains adjacent to the valley would have provided excellent local watersheds in the Pleistocene Epoch. Thus, there was probably significant local inflow to help maintain Panamint Lake at times when there was little or no input from the Owens River system. The history of pluvial Lake Panamint is still very poorly known, but it does appear to have spilled over a low threshold near Wingate Pass at the southern end of the valley several times. Water flowing over this low rim would have drained into Death Valley to the east, but there is no water-carved canyon or great accumulations of Pleistocene river sediment to document such events.

The terminus of the Owens River–Death Valley system was Lake Manly, a large pool that occupied the floor of Death Valley during pluvial epochs. Named for one of the rescuers of the 1849 immigrants stranded in Death Valley, Lake Manly is probably the most enigmatic of all the pluvial lakes in the Owens River–Death Valley system. Evidence of Ice Age inundation of Death Valley is widespread throughout this harsh desert region, and includes such features as multiple shoreline terraces, gravel spits and beach ridges (Figure 9.18), and isolated patches of tufa and other lake sediments. This evidence suggests that Lake Manly was at times as deep as about 600 feet, but the determination of the water depth for the ancient lake(s) is not a simple task. This is because Death Valley is so tectonically active that the valley floor is significantly lower today than it was in the Pleistocene Epoch. Also, the mountains adjacent to the valley, particularly the Black Mountains on the east side, are rising so fast that some of the shoreline features have now been carried to elevations higher than the lakes probably ever reached. Finally, the rapid uplift has produced rugged mountains that have shed great amounts of sediment into Death Valley, where accumulations of such alluvial material raise the floor and partly compensate for the ongoing down-faulting. All of this recent geological ruckus in Death Valley makes the simple determination of pluvial lake depth a difficult problem. It results in the simultaneous lifting the ancient shoreline features, lowering the floor of the Pleistocene lakes, and invigoration of erosion, which erases some of the shoreline evidence and washes sediment over features that document the earlier Ice Age terrain. Six hundred feet seems to be a good estimate of the maximum depth of Lake Manly,

Figure 9.18 This gravel bar in Death Valley accumulated near the edge of Lake Manly between 10,000 and 30,000 years ago, when the lake was about 300 feet deep.

but it many have been even deeper, and was certainly shallower, at various times in its mysterious history.

Though the evidence for Ice Age lakes in Death Valley is undeniable, geologists still have not developed a precise chronology of pluvial events in the basin. Similar to the names attached to pluvial lakes elsewhere in the Great Basin, the term Lake Manly refers to a series of lakes that gathered at the end of the Owens River system at different times over the past several hundred thousand years. One of the problems in reconstructing the history of Lake Manly is that the ancestral Owens River system was not the only source of water into Death Valley during Pleistocene time. The Amargosa River, which today follows a looping course to enter the southern end of Death Valley, began to flow into the basin in the late Pleistocene after an older intervening pluvial lake near Tecopa breached its northerly divide. The Mojave River, flowing from the southwest through its own series of pluvial lakes, is also thought to have reached Lake Manly at various times during the Pleistocene Epoch. There might even have been a fourth input to Lake Manly from the Saline Valley area to the northwest. With so many possible connections, all constantly shifting during the Pleistocene Epoch, it is easy to understand the current confusion about the history of Lake Manly. Recently, geologists have examined several deep subsurface cores from the Badwater Basin in central Death Valley for clues about the

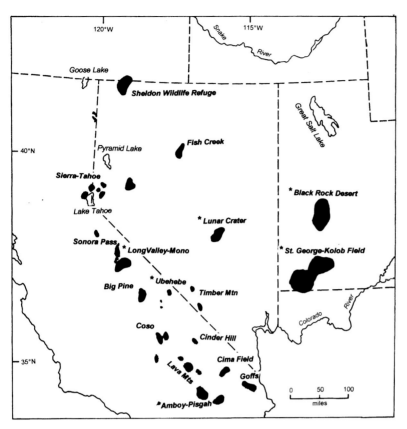

Figure 9.19 Principal areas of Quaternary volcanic activity in the Great Basin. Asterisks (*) indicate sites that have experienced eruptions during the past 10,000 years (Holocene Epoch).

ancient pluvial lakes. Though these studies are still in progress, at least two phases of lake development, from about 130,000 to 180,000 years ago and between 10,000 and 35,000 years ago, are indicated by the sedimentary record. The earliest lake appears to have been connected to the Owens River system, but during the latest pluvial cycle, drainage appears to have been from the Amargosa and Mojave River systems alone. Detailed studies of the younger lake deposits suggests that the water depth and salinity of Lake Manly fluctuated many times, rising to high stands on three different occasions over the past 26,000 years. As geologists continue to unravel the complex history of Lake Manly, it appears that this lake was even more erratic than most of the other pluvial systems in the Great Basin. There is still much to learn about the Owens River–Death Valley system.

Elsewhere in the Great Basin, Pleistocene pluvial lakes existed almost everywhere that modern lakes, either permanent or ephemeral, occur. In southeast Oregon, for example, modern Harney and Malheur Lakes were merged into one much larger pluvial system that covered almost 1,000 square miles. To the west of pluvial Harney Lake, another Ice Age lake encompassed modern Christmas, Silver, and Fort Rock Lakes, uniting them under a body of water that covered 600 square miles to a depth of 100 feet. In the central Great Basin, between the Lahontan and Bonneville systems, several isolated pluvial lakes formed in the valleys of Nevada. The vacuous Railroad Valley held a pluvial lake more than 300 feet deep, while east of the Ruby Mountains, pluvial Franklin Lake covered nearly 2,000 square miles of valley terrain under at least 200 feet of water. Along the ancestral Mojave River in the southwestern Great Basin, a series of pluvial lakes developed in the Manix, Cronise, Soda, and Silver Lake basins. The pluvial lakes along the Mojave drainage were sometimes joined, sometimes separated, at various times in the Pleistocene Epoch. Not all of the Ice Age lakes of the Great Basin were as impressive as these examples, but one may safely assume that wherever water accumulates in our modern post-pluvial world, there were beautiful semipermanent lakes shining during the cold cycles of the Pleistocene Epoch. If we could return to the last pluvial peak, about 15,000 years ago, we would experience a Great Basin far different from the austere deserts and steppes that prevail today. We would have been surrounded by a verdant land of lakes, laced with rivers and thundering cascades. The sight and sound of water would have been inescapable in the pluvial panorama. Depending on where in the region we visited, we probably would have heard another kind of thunder, as well—the roar of volcanic detonations.

QUATERNARY VOLCANIC ACTIVITY

The bimodal volcanic activity that began in late Tertiary time continued into the Pleistocene Epoch, and, as we will soon see, can even be considered a threat in the Great Basin today. In the mid-to-late Pleistocene Epoch, as the pluvial lakes and glaciers were undergoing their rhythmic expansions and contractions, eruptions of basaltic lava were widespread across the Great Basin, though most of the activity was concentrated along the east and west peripheries of the region and in the Mojave Desert area (Figure

9.19). The Ice Age eruptions were dominated by the effusion of high-temperature, low-silica basaltic magma that appears to have risen from the lower crust or upper mantle. Magma from such deep sources is generally rich in iron, magnesium, and other metals compared to fluids generated nearer the surface, and the Quaternary volcanic rocks certainly reflect this distinction. The basaltic rocks of the Quaternary volcanic fields not only are rich in iron and magnesium-bearing minerals but also commonly contain embedded fragments of upper mantle rock as well. The basaltic volcanism commonly resulted in extensive fields of black lava flows dotted with isolated cinder cones (see Figures 9.20 and 9.21). However, in some places, partial crystallization of the magma, and/or assimilation of granitic material into it during ascent, shifted the composition of the magma to a more silicic composition. When this happened, the sticky magma either squeezed out onto the surface to form stubby rhyolite flows or domes, or was blasted into ash during a violent volcanic explosion. The explosive eruption of silicic magma was not as common in the Great Basin during the Quaternary Period as it was in mid-Tertiary time, but when it did happen, the results were devastating.

Many of the centers of Quaternary volcanic activity are superimposed on areas of active extensional faulting. Evidently, the stretching and breaking of the crust localized volcanic activity by providing fractures that served as conduits for the rising magma. In addition, the extension and thinning of the crust may have helped to generate the magma by reducing the pressure on the rocks below. To visualize the connection between extension and melting, recall that the melting point of rocks drops whenever pressures are reduced. Thus, it is possible for rocks to melt solely in response to a reduction in pressure, even when temperature remains constant. As the crust in the Great Basin was further thinned and stretched in Quaternary time, the mantle below would have been decompressed, and the subterranean rocks could have liquefied to form bodies of iron-rich magma. Remember, also, that upward currents in the deeper mantle, related to the over-run oceanic ridge, were prob-

Figure 9.20 Pahvant Butte, a small volcano in the Black Rock Desert volcanic field, was erupted into the waters of Lake Bonneville.

ably situated beneath the Great Basin region in Quaternary time. These currents would have carried heat into the zone of decompression, enhancing the melting of upper mantle rocks. These mechanisms generated voluminous pools of fluid rock in the subsurface at the same time that normal faults were opening in the crust above. The result was a widespread flare-up of bimodal volcanic activity in Quaternary time, some of it continuing to late in the Holocene Epoch. It is well beyond our scope to examine each of the hundreds of volcanoes that erupted in the Great Basin over the past 2 million years; to do so would require another book. However, inasmuch as some of them may still be potentially active, it is definitely worth the effort to review the history of some of the more notable Quaternary volcanic centers.

Figure 9.21 Pisgah Crater, surrounded by extensive lava flows, may be the youngest volcano in the Great Basin.

Along the eastern edge of the Great Basin, lava that erupted from a cluster of at least two dozen different Quaternary volcanoes resulted in the extensive basalt flows and cinder cones of the Black Rock Desert, west of Fillmore, Utah. Not to be confused with the flat desert basin of the same name in northwest Nevada, Utah's Black Rock Desert is blanketed by extensive lava flows that range in age from early Pleistocene (1.5 to 2 million years old) to as young as just a few thousand years. As part of the Sevier arm of Lake Bonneville, the Black Rock Desert region was inundated by the great pluvial lake while the most recent eruptions were in progress. In particular, the cinder cone known as Pahvant Butte, between Delta and Fillmore, was clearly built during a series of eruptions that began under the waters of Lake Bonneville, and continued until the cone rose more than 800 feet above the lake bottom and about 300 feet above the water. The main phase of the eruptions at Pahvant Butte occurred about 16,000 years ago, just before Lake Bonneville reached its most recent high stand. The shoreline terraces along the flanks of Pahvant Butte are the result of about 1,500 years of shoreline erosion that occurred before the lake receded to leave the small volcano high and dry. There may be several other places in the Black Rock Desert volcanic field where lava was erupted into Lake Bonneville, but Pahvant Butte is the most obvious result of such an event. Imagine the ash-laden clouds of hissing steam that would have churned skyward as this volcano emerged from beneath the cold waves of Lake Bonneville!

Though basaltic flows and cones are the most common products of Quaternary volcanism in the Black Rock Desert region, numerous shallow dome-like intrusions of rhyolite were also emplaced in the area between 1 million and 500,000 years ago. Such rhyolite domes and flows are particularly prominent in the Mineral Mountains near Milford, Utah. Residual heat from these mid-Pleistocene igneous bodies is the basis for the significant geothermal energy resources that occur in this part of the Great Basin. South of the Black Rock Desert, the St. George–Kolob volcanic field is entirely basaltic in composition and includes many volcanoes

that were active in the late Pleistocene. This field straddles the Great Basin–Colorado Plateau boundary and includes the well-preserved cinder cones near Snow Canyon as well as the black basalt flows along the base of the Hurricane Cliffs northwest of St. George, Utah. The St. George–Kolob volcanic field can be traced to the south, beyond the Great Basin, where it merges with the Uinkaret volcanic field of the western Grand Canyon region. Few Quaternary volcanoes occur along the lower Colorado River corridor west of the St. George–Kolob field, but in the Mojave Desert–Death Valley portion of the southwestern Great Basin, hundreds of square miles are covered by basalt flows erupted from such youthful vents.

Quaternary volcanic activity in the Mojave Desert was most intense in the Lava Mountains, the Cima area, and in the Amboy-Pisgah field (Figure 9.19). In the Amboy-Pisgah field, Quaternary lava flows cover a total of about 80 square miles, and numerous small cinder cones rise above this black expanse. Pisgah Crater (Figure 9.21), actually a small cinder cone in the Amboy-Pisgah field, probably erupted last about 400 years ago, making it one of the youngest volcanoes in the Great Basin. The fascinating Cima volcanic field, covering about 55 square miles of the East Mojave Preserve, consists of at least 65 different flows of basaltic lava that issued from more than 50 vents. This field is dotted with cinder cones ranging in height from about 100 feet to more than 500 feet, and with basal diameters as great as a half-mile. Though the earliest volcanic activity in the Cima field began 10 million years ago (in the late Miocene Epoch), the majority of the vents were active in Quaternary time as well, some as recently as about 1,000 years ago. The basalt flows in the Cima field commonly contain fragments of iron-rich mantle rocks, confirming the deep source of lava in this region. The Lava Mountains in the northwest Mojave Desert are composed primarily of domes and thick flows that are more silicic in composition than the basaltic fluids erupted in adjacent regions. The silicic masses in the Lava Mountains, along with smaller exposures of similar rocks exposed nearby, illustrate the bimodal nature of Quaternary volcanism that is typical of the Great Basin as a whole.

The volcanic activity at Ubehebe Crater, at the north end of Death Valley, and in the Lunar Crater volcanic field in central Nevada resulted in some interesting variations in the eruptive style of Quaternary volcanoes in the Great Basin. In both of these areas, basaltic lava reached the surface in late Quaternary time, before the Great Basin climate had acquired its modern desert character. Water was much more plentiful in the Great Basin at that time, including great amounts of groundwater beneath the surface. Occasionally, the hot basaltic magma that rose toward the surface came into contact with the groundwater and instantly converted much of it into steam. As the groundwater flashed into vapor, great steam explosions were triggered just below the surface. Such explosions left deep craters, known as **maars**, where pulverized rock, soil, and lava droplets were blown sky-high. Both Ubehebe Crater and Lunar Crater are maars, the products of volcanically induced steam explosions. Ubehebe Crater (Figure 9.22), one-half mile wide and about 600 feet deep, is the largest of a dozen maars that exploded in a area of less than a square mile. Though the timing of the steam explosions in the Ubehebe area is difficult to establish with precision, charcoal in the soil blanketed by material blown from the crater is only between about 140 and 300 years old. Thus, the steam explosions must have occurred after that time and, if so, then Ubehebe Crater is the youngest-known volcanic feature in the Great Basin. Lunar Crater, in Nye County, Nevada, is a maar very similar to Ubehebe Crater but is part of a much larger volcanic field in the central Great Basin. Volcanoes in the Lunar Crater field began to erupt more than 4 million years ago, and continued to as recently as about 15,000 years ago. During this lengthy period of eruption, numerous volcanic cones were constructed along a northeast-trending zone in the central Great Basin. Elsewhere in the Great Basin, Quaternary maars also occur in the Carson Desert volcanic field near Fallon, Nevada, and in the Harney Lake area of southeast Oregon. Though the Quaternary basaltic magmas of the Great Basin generally erupted

Figure 9.22 Ubehebe Crater resulted from a recent steam explosion triggered by the upward migration of basaltic magma into the zone of groundwater. Note the explosion rubble that partially covers the late Miocene sedimentary strata exposed in the crater walls.

with little violence, some impressive volcanic explosions occurred wherever magma met groundwater.

Perhaps the most intriguing of all the centers of Quaternary volcanism in the Great Basin is the Long Valley–Mono Craters area (Figure 9.23) along the eastern escarpment of the Sierra Nevada. This area, which includes the Inyo Craters near the ski resorts of Mammoth, California, is unusual in that the effusive eruption of basaltic magma, so common elsewhere in the Great Basin, was not the dominant style of activity. Instead, most of the volcanic activity involved the eruption of highly viscous silicic magma that produced rhyolite domes and knobs throughout the region. Basaltic eruptions occurred in the Long Valley area during Quaternary time, but such events were relatively small and infrequent in this region. The thick and sticky rhyolitic magma that oozed toward the surface in this area encountered groundwater seeping from the High Sierra watersheds to the west, and produced numerous maars. Such explosion craters are particularly well developed in the Inyo Craters chain north of Mammoth Mountain. The strong north-south alignment of the Inyo Craters suggests that the rhyolitic magma ascended along a linear fault that parallels the Sierra-Mono normal fault along the base of the mountains to the west. Radiocarbon dates on wood embedded in the debris blasted from the Inyo Craters suggest that explosions in this

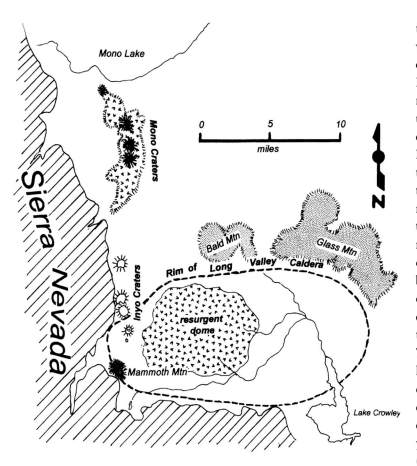

Figure 9.23 Sketch map of the Long Valley–Mono Craters region. Quaternary eruptions have occurred in dozens of places in this volcanically active part of the western Great Basin. There is a strong possibility of future volcanic events in this area.

chain occurred as recently as about 500 years ago. The north-south alignment of the Inyo Craters continues to the Mono Craters, a chain of about thirty explosion craters filled with, or sometimes perched atop, large rhyolite domes. The geological evidence suggests that each of the craters in this eleven-mile-long chain began with an explosion pit that was later filled with, and sometimes raised by, effusions of viscous rhyolite magma. The most recent eruptions in the Mono Craters chain appear to have occurred about 600 years ago.

By far the most impressive Quaternary volcanic structure along the Sierra escarpment is the Long Valley caldera (Figure 9.23). This immense oval depression, approximately 10 by 20 miles in size and more than a mile deep, formed 730,000 years ago when a detonation of unimaginable ferocity blasted ash as far away as Nebraska. Geologists estimate that more than 140 *cubic miles* of ash was discharged during the great explosion, and blown as high as ten miles into the atmosphere. Much of the ash settled back to the ground in the immediate vicinity of

the caldera, burying glacial deposits and older volcanic rocks under more than 1,000 feet of incandescent embers. The hardened ash from the Long Valley caldera comprises the Bishop Tuff, a rock unit so distinctive and widespread that it is used to correlate Quaternary events across the entire western portion of North America. The Long Valley caldera is so large that it is difficult to see in its entirety from any vantage point on the ground. Besides its gargantuan size, another reason why viewing the caldera is difficult is that since the great eruption occurred more magma has welled up in the center of the depression, raising the floor of the old caldera by several thousand feet. Such a mound of post-explosion volcanic rock is called a **resurgent dome**, and represents the rebuilding process that sometimes follows volcanic catastrophes. The bulbous high ground within the Long Valley caldera is a classic example of a resurgent dome, and it has partially refilled the colossal cavity created by the initial blast. The extrusion of viscous lava between 150,000 and 50,000 years ago resulted in the Mammoth Mountain (Figure 9.23) volcano, suggesting that the caldera-filling activity is still in progress. Further concealing the great depression were the several episodes of glaciation that have occurred in the Long Valley area since the caldera-forming explosion occurred. With an elevation close to 12,000 feet above sea level, portions of the original rim of the caldera were repeatedly eroded by glaciers and/or buried under glacial sediment. Given the extreme youth and violent nature of the volcanic activity in the Long Valley–Mono Craters region, it is natural to wonder about the likelihood of catastrophic eruptions in the future. Could slumbering Mammoth Mountain, the Inyo Craters, or the Mono Craters awaken soon? If so, what kind of activity could we expect? We will return to these fascinating questions in our final chapter.

THE QUATERNARY ENVIRONMENT OF THE GREAT BASIN

Our brief review of the glacial, pluvial, and volcanic phenomena in the Great Basin suggests that the landscape was anything but static during Quaternary time. Repetitive changes in climate over the past 2 million years fostered continual shifts in drainage patterns, the distribution and size of lakes, and the extent of glaciers throughout the region. Factor in the effects of the Quaternary volcanic events on the landscape, superimposed on those changes driven by climate, and the transience of the Ice Age environment becomes almost unfathomable. Nonetheless, it was not just the weather and the land that was subject to the wild cycles of change during Quaternary time. Life not only existed but also flourished in great profusion throughout the Great Basin amidst the ecological chaos of the ice ages. The climatic fluctuations had profound effects on the vegetation that grew in the Great Basin, and throughout Quaternary time the ranges of various plants shifted up and down, or north and south, in response to the changing conditions. Ice Age animals responded to these floral changes by migrating across, and sometimes beyond, the region as they adjusted their ranges to match the distribution of favored habitats and food resources. Consequently, life during the Ice Age in the Great Basin was as variable in space and time as the land that sustained it. If we could return to the Great Basin during Quaternary time, the kinds of plants and animals we would witness would depend on exactly when and where our time travels led us.

During the interglacial (or interpluvial) episodes, the ancient landscape would have looked very much like the warm and dry deserts and steppes we know today. There were several times during the Quaternary Period when the Great Basin was probably even warmer and more arid that it is today. As such periods of warmth developed, the great pluvial lakes would shrink or even disappear, leaving expansive lake beds encrusted with saline minerals. The pluvial rivers would diminish into trickles, and some of them would vanish completely, leaving only their dry stream beds as monuments to a past age of abundance. Scrubby desert plants such as the modern sagebrush, shadscale, and blackbrush would have decorated the desolate landscapes during the warm intervals. Coniferous forests generally would be restricted to the cool mountain peaks and higher ramparts. When the cold

glacial (or pluvial) cycles commenced, all of these austere conditions would be gradually reversed. Year by year, the pluvial lakes expanded over their dry basins while the rivers grew steadily larger in response to cooler and wetter conditions. The dry-adapted shrubs of the preceding interglacial either migrated south, where warmer conditions persisted, or were simply submerged by the rising pluvial lakes. At higher elevations, the alpine glaciers began to advance down mountain canyons, the largest of them sometimes reaching the shores of the pluvial lakes. Icebergs floated in Lake Bonneville! Simultaneously, trees adapted to the cooler conditions of the high mountains inched their way downward during the cold intervals, colonizing some of the areas abandoned by the vanished desert shrubs. Conifers (various types of pine, spruce, and fir) that grow today in the higher mountains of the Great Basin lined the shores of the great pluvial lakes, three thousand feet below their current ranges. The famous bristlecone pines, Ice Age relics that are restricted to the elevations above 9,000 feet in the modern Great Basin, grew at elevations as low as 6,000 feet during glacial times. In the warmer Mojave Desert region, a relatively lush pinyon pine and juniper woodland covered areas that now support only a sparse cover of creosote, yuccas, and cacti.

The Ice Age fauna of the Great Basin consisted of a mix of familiar modern forms and some rather bizarre and exotic extinct creatures. The large pluvial lakes, when fully expanded, would have been filled with freshwater and supported thriving aquatic ecosystems. Sediments deposited in Lake Bonneville, for example, have produced abundant remains of Ice Age insects, snails, and clams. In addition, the Pleistocene fossils of the Bonneville basin indicate that at least eight different kinds of fish swam in the great lake, including sculpin, suckers, whitefish, trout, chubs, and cisco. Fossils of terrestrial Pleistocene vertebrates are fairly abundant in the Great Basin, occurring most commonly in lake-edge sediments, river-deposited sand and gravel, and in cave sites scattered across the region (Figure 9.24). From these fossils, it is clear there were many rabbits, rodents, cats, dogs,

Figure 9.24 Fossil forelimb bones from an Ice Age musk ox (genus *Symbos*) found near Taylorsville, Utah, in Lake Bonneville gravel deposits.

bears, bison, antelope, deer, peccaries, snakes, lizards, and birds living throughout the Great Basin in pluvial times. Most of these were similar to creatures that live in equally chilly climates today. However, there were also some very strange elements in the Ice Age fauna of the Great Basin, particularly among the larger mammals. For example, musk oxen wandered the shores of the pluvial lakes, several types of large ground sloths ambled along the hillsides, and groups of camels, llamas, and horses loped across the wooded lowlands. Among the exotic beasts, some achieved gigantic proportions: the largest of several types of Pleistocene mammoths in the Great Basin stood almost 13 feet tall at the shoulder, and an extinct bison was about twice as large as the modern buffalo and sported horns that measured more than 10 feet from tip to tip! The Ice Age mammal predators included the familiar weasels, skunks, coyotes, wolves, bears, bobcats, and mountain lions, accompanied by some extraordinary extinct kin. The stocky saber-tooth cats, with 8-inch-long canine teeth and weighing up to 600 pounds, must have been the most fearsome predators of the Ice Ages. In addition, another Pleistocene cat

that lived in the Great Basin was 25 percent larger than the modern African lions. *Arctodus simus*, the extinct short-faced bear, was about 5 feet tall at the back, comparable in stature to a modern horse! So many Ice Age fossils have been found in the Great Basin that it is utterly impossible to provide a complete review in the space available here. Nonetheless, it is important to understand that the flora and fauna of the Great Basin not only survived the profound environmental oscillations of the Pleistocene ice ages but flourished throughout them. And what an incredible menagerie of beasts roamed the Great Basin during the Pleistocene (Plate 15).

Even our abbreviated review of the Ice Age fauna suggests that something significant happened to the animals in the Great Basin since the last glacial/pluvial peak. By the time the Holocene Epoch began, about 10,000 years ago, many of the most striking Pleistocene mammals had forever disappeared. For the past ten millennia, there have been no saber-tooth cats, or mammoths, or camels, or ground sloths, or giant bison in the Great Basin, or anywhere else in North America, for that matter. The list of extinct Pleistocene species is a very long one, and the modern mammal fauna is but an impoverished remnant of the rich array of animals that graced the Pleistocene woodlands and forests. What could have created, in so short a span of time, such a biotic calamity within a virtual garden of Pleistocene life? It is tempting to attribute these rapid extinctions to the profound climate change that accompanied the transition from Pleistocene to Holocene time, an event that ultimately transformed the Great Basin from woodlands and forests to deserts and steppes. However, such environmental fluctuations had occurred many times over the past 2 million years without causing any significant extinction. In fact, a few of the Quaternary climatic cycles probably involved even more extreme changes than those that occurred at the beginning of Holocene time. No, we cannot look to recent climatic shifts, and their environmental consequences, as the sole explanation for the demise of so many Ice Age creatures. Likewise, the geologic events of the Pleistocene, dominated by extensional faulting and volcanism, did not change significantly at the Pleistocene-Holocene boundary. While scientists continue to debate the causes of the great Pleistocene extinction, most agree that it was a complex event and that no simple answers satisfy all the data that pertain to it. However, there is a growing suspicion that, in the continuing search for extinction mechanisms, we will find the most important part of the answer in one place: the mirror.

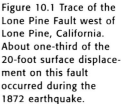
CLOSING CIRCLE, UNFINISHED STORY

A bright moon hung over the small town of Lone Pine, California, in the wee hours of March 26, 1872. A few miles west of town, the jagged crest of the Sierra Nevada was still mantled with snow, and glistened a ghostly white above the Great Basin town. The low, rounded knobs of granite in the Alabama Hills, between the town and the Sierra escarpment, were laced with eerie dark shadows cast by the moonlight into the deep clefts in the massive rock. It was quiet in Lone Pine that night, as all three hundred or so of the residents had been deep in slumber for hours. Nothing stirred in the crisp predawn darkness, except in the bedroom of Antonia Montoya. The young Mexican woman had been visited by a male admirer that evening, and her pleasure with his call must have been powerful, for it remained undiminished, even at that late hour. Lying wrapped in each other's embrace, the lovers, like everybody else in Lone Pine, must have been completely oblivious to the Owens Valley fault zone, a system of deep rifts in the earth's crust that lay under the adobe buildings comprising the small town. Antonia's disregard of the Owens Valley fault zone ended at 2:30 that morning. Sadly, so did her life.

At that moment, one of the faults in the Owens Valley zone suddenly jerked to life. The Alabama Hills were instantly thrust up more than seven feet along the Lone Pine Fault, and simultaneously wrenched northward by a few feet. The vibrations created by the sudden twitch of rock against rock began to shake all fifty-nine of the buildings that existed in Lone Pine. Fifty-two of them began to crumble, including the home of Antonia Montoya. As the adobe bricks collapsed upon her, her lover made a hasty escape from the disintegrating structure through a bedroom window. According to a re-

port published a month later in the *San Francisco Chronicle*, he made no attempt to rescue his sweetheart from the rubble, and was never found after the earthquake. At the very moment of Antonia's agony, the illustrious naturalist John Muir was awakened by the rumbling from the earthquake at his cabin in Yosemite Valley, more than 100 miles to the northwest. In the predawn moonlight, Muir heard rock falls rumbling down the walls of the canyon and saw clouds of dust rising from the places where the rocks smashed onto the valley floor. The earthquake was also felt in Elko, Nevada, and its vibrations were powerful enough to stop mechanical clocks in San Diego at 2:30. In addition to Senorita Montoya,

Figure 10.1 Trace of the Lone Pine Fault west of Lone Pine, California. About one-third of the 20-foot surface displacement on this fault occurred during the 1872 earthquake.

DISASTER IN 1872

ON THE DATE OF MARCH 26, 1872 AN EARTHQUAKE OF MAJOR PROPORTIONS SHOOK OWENS VALLEY AND NEARLY DESTROYED THE TOWN OF LONE PINE. TWENTY SEVEN PERSONS WERE KILLED. IN ADDITION TO SINGLE BURIALS, 16 OF THE VICTIMS WERE INTERRED IN A COMMON GRAVE ENCLOSED BY THIS FENCE.

twenty-six other residents of Lone Pine never saw the sunrise of March 26, and it took days for the survivors to excavate their bodies from the wreckage of the fallen buildings. Sixteen of the victims were buried in a mass gravesite that still exists north of town (Figure 10.2), while the others were laid to rest in individual plots. Among those who survived the earthquake, fifty-six people suffered serious injuries, and $250,000 worth of property was destroyed. In 1872, that value was close to the total worth of the entire town.

Because it occurred before modern seismographs existed, there are no instrumental records of the Lone Pine earthquake, and no one knows exactly how powerful the temblor was. Based on the documented effects of the earthquake, and the 100-mile-long wound that it ripped in the earth, modern geologists estimate that its magnitude was between 7.8 and 8.5 on the Richter scale. It was, in all likelihood, the most powerful earthquake to strike California in historic times. Not that the actual magnitude of the Lone Pine earthquake mattered much to the victims of the event, but it does suggest that the geological evolution of the Great Basin is far from finished, a conclusion substantiated by the many strong earthquakes that have occurred

since 1872. On October 2, 1915, for example, a magnitude 7.6 earthquake struck Pleasant Valley, Nevada (about fifty miles south of Winnemucca), hoisted the Tobin Range by as much as 20 feet, and left a prominent escarpment along the base of the mountains (Figure 10.3). On December 16, 1954, two earthquakes of approximate magnitudes 7.2 and 7.0 occurred within four minutes of each other in the remote Dixie Valley area of west-central Nevada. These earthquakes were felt over an area of 200,000 square miles, including such distant sites as Sacramento, California. The Mojave Desert region has a long history of strong earthquakes, including the Hector Mine event (magnitude 7.1) fifty miles southeast of Barstow on October 16, 1999. Even more powerful was the 1992 Landers earthquake (magnitude 7.3) along the southern margin of the Mojave region. The Landers event left a rupture more than fifty miles long on the desert floor, along which the maximum surface offset was almost 20 feet. The 1934 Hansel Valley, Utah (magnitude 6.6), and 1993 Klamath Falls, Oregon (magnitude 5.9), earthquakes demonstrate that powerful seismic events can occur even in the remote corners of the Great Basin. These earthquakes, along with the thousands of lesser temblors that occur every year, signify that the crust of the Great Basin is still shattering under the tectonic stresses related to modern plate interactions. But, as the Great Basin continues to evolve, there will be one very important distinction between its geologic past and its protracted future: humans will be firsthand witnesses of the coming epochs of geologic activity. We have, in fact, been observers, participants, and victims of geologic events in the Great Basin for thousands of years.

Though the archaeological evidence is meager and controversial, it appears that humans first inhabited the Great Basin sometime between 11,000 and 13,000 years ago. If so, then the earliest human residents in the region probably witnessed some of the rapid changes in climate and environment that occurred as the Pleistocene woodlands gave way to the modern deserts. The first human immigrants to reach the Great Basin probably saw large lakes with forested shores, cowered at the sight of saber-tooth cats, hunted mammoths and giant bison, and sought shelter in caves and rock overhangs. Prehistoric humans were also likely to have been awestruck, perhaps terrified, by the eruptions of Holocene volcanoes in the Long Valley–Mono Craters chain, in the Black Rock Desert of Utah, in the Mojave region, or in the Death Valley area. While there is no evidence whatsoever of human tragedy resulting from any prehistoric volcanic calamity, pluvial flood, or earthquake, such events can't be ruled out. Long before Europeans arrived in the West, and even millennia before the modern Native American tribes became established in the region, humans must have glimpsed the geologic vigor and wonder of the Great Basin. The Lone Pine disaster of 1872 is the most tragic meeting between humanity and geology ever *known* to have occurred in the Great Basin, but it was certainly not the only time that the continuing geological evolution of the Great Basin collided head-on with desert folk. More importantly, future engagements between people and planet are almost guaranteed in the Great Basin, given the rapid increase in the human population of the region and its dynamic geologic setting. This final chapter concerns the present, both human and geological, as the terminus of the past and the threshold of the future. Now that our journey has taken us through more than 3 billion years of natural history, it's time to examine what is happening now in the Great Basin, what may occur in the future, and how we (or our descendants) might be affected.

At the dawn of the Holocene, about 10,000 years ago, the Great Basin was virtually indistinguishable from the province we know today (see Plate 16). The geological foundation had been laid over the preceding eons, the pluvial lakes and glaciers were gone, all but a few of the victims of the Ice Age extinctions had vanished, and the dry desert climate prevailed. Through the remainder of Holocene time, sporadic volcanic eruptions occurred in several places, most commonly along the western side of the region from eastern California to west-central Nevada. Many of the normal faults in the region exhibit evidence of Holocene displacement, demon-

Figure 10.3 The dislocation, or scarp, created by the 1915 earthquake in Pleasant Valley, Nevada, is still evident along the base of the Tobin Range.

strating that extensional faulting continues to the present day. Finally, as we have already seen, some the faults in the region were subject to instantaneous slip during severe seismic events. Thus, there are four interrelated phenomena—environmental (climatic) change, volcanism, faulting, and earthquakes—that constitute the prevailing themes of the modern geological era in the Great Basin. Not only will these themes herald the geological future, each of them will also greatly affect the lives of people in the Great Basin.

MODERN FAULTING AND EXTENSION IN THE GREAT BASIN

To many people living in our modern age, technology is both a blessing and a curse. In the sciences, the efficiency and accuracy of electronic number-crunching, coupled with the error-free execution of repetitive tasks by computers, has fostered sweeping technological revolutions in virtually all fields during the past several decades. Though modern high-tech approaches have undeniably expanded our awareness of the natural world, they can also separate us from it, if we are not careful to retain a solid connection to real phenomena. Geology, one of the most traditional sciences, has become increasingly reliant on modern technology in recent years, but still clings steadfastly to an insistence on first-hand, verifiable field observations and data as the *sine qua non* of all proper investigations.

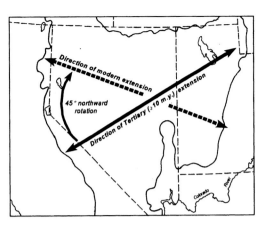

Figure 10.4 Rotation of extensional stress in the Great Basin. Initial extension occurred in a northeast-southwest direction but rotated clockwise by about 45° after late Miocene time. This rotation of extensional stress probably resulted from the northward movement of the Pacific Plate against the western edge of North America.

And, well it should; no technological innovations can turn flawed data into sound insights about any aspect of nature. Nonetheless, modern technology has provided marvelous new ways to observe the results of slow geologic processes that would otherwise have been undetectable for millions of years. In the Great Basin, the blending of modern technology with classical fieldwork has revealed much about the ongoing geological evolution of the region. There is no better example of how high-tech methods can complement traditional geological approaches than the study of recent faulting and extension in the region, and the means by which the recent history is extrapolated to the present.

As we have already learned, traditional geological field studies suggest that most of the extension of the Great Basin has occurred after the beginning of the Miocene Epoch. Since the extension began more that 25 million years ago, several different mechanisms have collectively stretched the crust several hundred miles. The actual amount of ancient extension varies considerably across the region, as do the styles (normal faults, listric faults, core complexes, or ductile flow) of deformation. The orientation of normal and listric faults of varying ages suggests that the earliest extension was aligned in a southwest-northeast direction, and the younger (less than 10 million years old) prehistoric faults developed from extension that was oriented in a northwest-southeast direction. Thus, the direction of extension appears to have rotated clockwise by about 45° during late Tertiary time (Figure 10.4). The net result of the rotating extension was the generation of hundreds of normal faults that today are oriented in a domi-

nantly north-south trend (see Figure 10.5). During this period of late Tertiary extension, the Great Basin was stretched at an average overall rate of about 20 mm (slightly less than an inch) per year.

The rotation of the stretching forces in the Great Basin appears to be related to the development of the San Andreas Fault system along the western edge of the North American Plate during the time that extension was in progress. As the Pacific Plate slid north along the San Andreas fault system, it pulled or "wrenched" southwestern North America with it, causing the extensional stress field to rotate under the influence of the northbound slab. In addition, geologists have identified several zones of dextral-slip faults, similar to those in the San Andreas system, along the western margin of the Great Basin. For example, the **Walker Lane belt** of western Nevada and eastern California is a narrow zone of dextral-slip faults that disrupt the normal north-south grain of the Great Basin landscape. The seventy-mile-wide Walker Lane belt (Figure 10.5) is aligned roughly parallel to the San Andreas system, and includes several major lateral faults such as the Owens Valley, Death Valley, and Furnace Creek fault zones, along with many normal faults. The Walker Lane belt appears to have originated about 25 million years ago, just after the transform boundary between the North American and Pacific Plates developed. The total offset of all the faults in the Walker Lane belt is approximately 45 miles, with the blocks west of the individual faults consistently shifted to the north. This is the same sense of motion, right-lateral or "dextral," that characterizes the San Andreas Fault system. The **Eastern California Shear Zone** (ECSZ) is another prominent zone of dextral faults that trend to the northwest across the Mojave Desert region (Figure 10.5). Both the Walker Lane belt and the Eastern California Shear Zone are probably inland expressions of the dextral shear stress that was generated between the North American and Pacific Plates after the transform boundary (the San Andreas system) developed between the two plates. These two zones of relatively young dextral-slip faulting indicate that the southwest margin of the Great Basin, that portion nearest to the transform boundary, was sliced into slivers of rock, even as it was being stretched in Tertiary and Quaternary time. The slivers were shuffled to the northwest as the Pacific Plate scraped against the edge of the continent. This scenario explains what has been called the San Andreas Discrepancy. Relative to the stable interior of North America, the Pacific Plate is moving to the northwest at a rate of about 50 mm/yr (roughly 2 inches/yr). However, recent offset along the San Andreas Fault system, the principal boundary between the two plates, varies from 15 mm/yr to 35 mm/yr, substantially less than the overall displacement between the two plates it separates. So, where does the rest of the dextral-slip motion occur? Almost certainly, the sliding slivers of rock in the Walker Lane belt and the East California Shear Zone accommodate some of the "missing" displacement between the North American and the Pacific Plates. Both of these belts of dextral-slip faults, along with the San Andreas system itself, can thus be considered as components of a broad zone of deformation that arises from the shearing of rock in western North America by the northwest-moving Pacific Plate. As we will soon learn, the zones of dextral-slip faulting in the western Great Basin may have originated millions of years ago, but they are still very active, and dangerously so.

There is another intriguing element in the pattern of faults in the Great Basin that discloses additional insights about the stretching forces that affect the region today. Despite the great number of normal faults in the Great Basin, there are several linear zones crossing the region in which the normal faults terminate and no major uplifted blocks stand (see Figure 10.5). On either side of these narrow northwest-trending "fault-free" zones, the orientation of extensional faults and tilted mountain ranges commonly reverses from a west-tilt to an east-tilt, or vice versa. Thus the transverse zones are boundaries between the several "regional tilt-domains" we described earlier. Geologists attribute the tilt-domain boundaries to the variable amount of extension in the Great Basin: when one area is stretched more than an adjacent zone, the normal faults in each will termi-

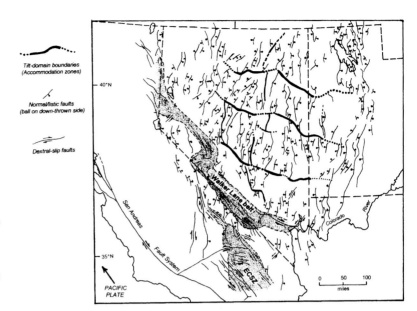

Figure 10.5 Fault map of the Great Basin region. Northwest- trending accommodation zones in the central Great Basin reflect the variation in extension across the region. Note the predominance of dextral-slip faults in the western Great Basin. ECSZ = Eastern California Shear Zone.

nate in a boundary zone where the differential stress between the two area is "accommodated." In other words, the boundaries between tilt-domains act somewhat like the great fracture zones on the sea floor that chop the mid-oceanic-ridge system into short segments that are slightly offset from each other. Both the ridge-segmenting oceanic faults and the tilt-domain boundaries exist because extension varies from place to place, whether it occurs on the sea floor or in the Great Basin. The tilt-domain boundaries in the Great Basin are now referred to as **accommodation zones**, because they compensate for the variable amount of extension in adjacent parts of the region. Thus, geologists concluded years ago that the extensional deformation in the Great Basin was not uniform. The accommodation zones clearly indicate that some areas have been stretched more than others. But, how much stretching is occurring today? Is the re-gion still being extended? If so, has the rate of extension changed since the normal faults originated in mid-Tertiary time? Do the dextral-slip faults to the west still affect the deformation of the Great Basin today? Such questions can be difficult to resolve with traditional geologic methods alone. However, modern high technology tools have revealed some surprising new perspectives on the continuing evolution of the Great Basin.

It is impossible for humans to directly detect movement that occurs at rates of a few milli-

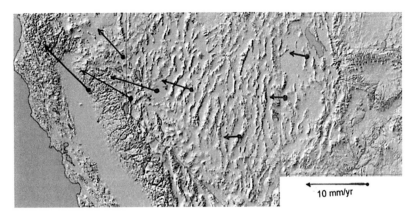

10 mm/yr

Figure 10.6 Velocity vectors for selected locations in the Great Basin. Arrows indicate the direction of movement of some of the GPS stations used in the BARGEN project. The length of the arrows is proportional to the actual ground motion. Note that the extensional velocity increases from east to west in the Great Basin, and that this motion is deflected to the northwest by the influence of the transform boundary between the Pacific and North American Plates.

meters per year, unless we observe it over long periods of time. That is why geological mapping, which portrays the cumulative effect of slow movements over million of years, so effectively documents long-term changes in landscapes. However, recent technological advances have now made it possible to detect the short-term effects of slow, long-term processes without waiting million of years. Using very sophisticated satellite-based global positioning systems, geologists can now locate points on the surface with millimeter-scale accuracy. If such accurate measurements are repeated over several years, even very slow geological movements become apparent. During the 1990s, geodetic surveying projects using such precise satellite-based GPS (Global Positioning System) technology focused on the Great Basin in an effort to learn more about the geological forces currently affecting the region. The project known as BAR-GEN (Basin and Range Geodetic Network), for example, employed an array of more than fifty ground stations scattered across the central Great Basin, mainly along the Highway 50 corridor. The precise locations of these sites were measured by satellite and monitored over a period of several years. Coupled with other information from satellite radar interferometry (another high-tech method of surface mapping), very tiny movements of the ground sites were detected. When the small-scale (millimeters/year) motion of the sites was mapped on a regional scale (Figure 10.6), the pattern of current deformation in the Great Basin was revealed in a way that no traditional geologic technique could rival. The long-held perception of variable extension in the Great Basin is clearly substanti-

ated by these measurements. The actual amount of extension in the Great Basin currently varies from 2 or 3 mm/year along the eastern edge to more than 12 mm/year near the western margin. These values suggest that the magnitude of extension is generally less now than it was in mid-Tertiary time, when the geological data suggests extension rates as high as 20 mm/year. Even though the Great Basin appears to be stretching less now than it was millions of years ago, the majority of the normal faults in the region are still very active. Moreover, the influence of the plate boundary to the west appears to be even stronger now than it was earlier in the Cenozoic era. Note on Figure 10.6, that areas in the western Great Basin not only are moving rapidly but also that the motion is deflected to the northwest, parallel to the displacement on the San Andreas system and Walker Lane belt. In the eastern portion of the Great Basin, the slower motion is directed due west, evidently unaffected by the strain of the Pacific Plate.

These remarkable observations clearly demonstrate that the extension of the Great Basin that began some 25 million years ago continues today. Oddly, the new measurements of actual motion suggest that there is really no single axis of extension in the Great Basin, no central zone of lateral spreading. Instead, it appears that the entire crust of the region is being stretched westward as if it was a block of taffy anchored to the more stable Colorado Plateau–Rocky Mountain area to the east. The accommodation zones and the regional tilt-domains signify the segmentation of the crust as the stretching forces pull it from the undeformed interior of the continent. These observations leave us with one inescapable conclusion: the Broken Land, from the Rockies to the Sierra Nevada, is breaking apart!

EARTHQUAKES IN THE GREAT BASIN

With so many active faults in the modern Great Basin, it is not surprising that the entire region is prone to earthquakes. Thousands of earthquakes occur every year in Nevada, western Utah, and eastern California, and some of them have been very powerful (Table 10.1). Seismic

Table 10.1. The Ten Greatest Historic Earthquakes in the Great Basin

	year	location	magnitude
1.	1872	Lone Pine, Calif.	7.8–8.5 (est.)
2.	1915	Pleasant Valley, Nev.	7.6
3.	1992	Landers. Calif.	7.3
4.	1932	Gabbs (Cedar Mts.), Nev.	7.2
5.	1954	Fairview Pk./Dixie Valley, Nev.	7.1,6.9
6.	1999	Hector Mine, Calif.	7.1
7.	1860	Carson City, Nev.	6.8 (est.)
8.	1934	Hansel Valley, Ut.	6.6
9.	1947	Manix, Calif.	6.5
10.	1993	Klamath Falls, Ore.	5.9

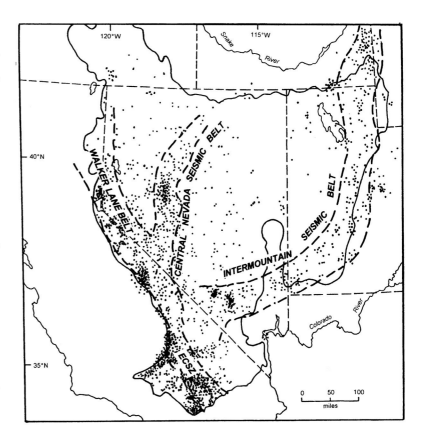

Figure 10.7 Seismicity of the Great Basin. Each point represents the epicenter of an earthquake that occurred over a thirty-year period, from 1950–1980. Note that the earthquakes are concentrated along the western and eastern borders of the Great Basin.

tragedies like that of Antonia Montoya are rare only because the desert terrain is currently so sparsely populated. However, as cities such as Reno, Las Vegas, and Salt Lake City continue to grow, the potential for earthquake disasters in the region likewise increases. Therefore, understanding the pattern and nature of seismic activity in the Great Basin is important for more than purely academic reasons. Given the modern plate tectonic setting of the region, we can be assured that humans in the area will continue to feel the ground shake beneath them. The many powerful temblors of the past century suggest that it is only a matter of time until those vibrations result in catastrophe.

The seismicity in the Great Basin, though it is widespread, is not randomly dispersed. The most frequent and powerful earthquakes are concentrated in three prominent belts of seismic activity (Figure 10.7). Along the eastern margin of the Great Basin, the Intermountain Seismic Belt stretches from Las Vegas northward through Salt Lake City, and continues beyond the Great Basin to the Yellowstone area of northwest Wyoming. Straddling the western border of the area, a northwest-aligned zone of intense seismic activity is superimposed on the Walker Lane belt and includes the East California Shear Zone in the Mojave region. A few miles south of Hawthorne, Nevada, near the California border, this western belt of earthquakes splits into a Y-shaped pattern. The western branch continues to the northwest through the Lake Tahoe region, and eventually passes west of Pyramid Lake into the northwestern

corner of the Great Basin. The eastern branch of the "Y," known as the Central Nevada Seismic Belt, trends almost due north through west-central Nevada, before it dies out north of Winnemucca. Though these zones of seismicity are all related to the ongoing extensional deformation of the Great Basin, each of them is unique with respect to seismic history, earthquake mechanics, and potential for future activity. Let's examine each zone in a little more detail.

The Intermountain Seismic Belt (ISB) is a zone of earthquake activity nearly 1,000 miles long, stretching from the southern Great Basin into southwestern Montana. Only the southern portion of this belt lies within the Great Basin, where it generally parallels the eastern boundary of the province. Because this belt passes through regions of different tectonic settings, there are probably several different mechanisms that produce the earthquakes along the ISB. The largest historic earthquakes along this belt have generally occurred north of the Great Basin segment. The Hebgen Lake, Montana, earthquake of 1959 (magnitude 7.5) and the 1983 Borah Peak event in Idaho (magnitude 7.3) demonstrate the capacity of the ISB to produce devastating earthquakes. The most powerful earthquake to occur on the Great Basin portion of the ISB was the 1934 Hansel Valley event (magnitude 6.6), which released only a fraction of the energy liberated by the Montana and Idaho quakes. Nonetheless, the Hansel Valley event was powerful enough to have resulted in widespread destruction had it not occurred in such a remote and sparsely populated area. The large and rapidly growing population centers of the eastern Great Basin—the Wasatch Front communities of Utah—all lie directly astride the ISB. Seventy-five percent of the residents of Utah live under the constant threat of a seismic disaster within the Intermountain Seismic Belt.

The serious seismic risks associated with the ISB, along with its importance in understanding the contemporary tectonic setting of the Great Basin, have compelled intensive study of this zone over the past several decades. The main part of the ISB in the eastern Great Basin coincides with the Wasatch Fault zone, a complex system of extensional faults that runs along the base of the Wasatch Range for more than 200 miles in northern and central Utah. More than 15,000 feet of cumulative vertical displacement has occurred along the Wasatch Fault zone since its inception in mid-Miocene time. The many earthquakes along the fault zone indicate that it is still very active. In fact, geologists have determined that the current rate of uplift of the Wasatch Range block is about 1.5 inches per century. As blocks on either side of the Wasatch Fault shift vertically, east-west horizontal displacement occurs simultaneously at about the same rate. While no catastrophic earthquakes have emanated from the Wasatch Fault zone in historic time, earthquakes in the range of magnitude 5–6 occur commonly enough to raise some concern about the seismic safety of Utah's largest cities. Moreover, there is ample geological evidence of powerful prehistoric temblors on the Wasatch system that would have devastating impacts were they to occur today. Decades of seismological monitoring along the Wasatch Fault zone has demonstrated that the earthquakes in this area tend to occur at shallow depths, generally less than about 15 km (10

miles) below the surface. This means that there is relatively little subsurface material to absorb the energy released by an earthquake and to dampen the waves it generates underground. Also, the communities at greatest risk of seismic impacts in central Utah are mostly situated on loose sediment and unconsolidated soil that tends to soften and lose rigidity as seismic vibrations pass through it.

The Intermountain Seismic Belt and the Wasatch Fault zone are the principal reasons that central Utah is earthquake prone. However, the risk of earthquakes is not uniform everywhere in the ISB, or in the Wasatch Fault zone, because the earthquakes tend to occur in small, localized areas separated by "quiet" zones. The discontinuities in earthquake activity, along with the pattern of actual surface ruptures, have allowed geologists to subdivide the Wasatch Fault zone into several segments. Each segment of the fault zone has its own unique seismic history, and the risk of future earthquakes also varies from segment to segment. Overall, seismologists have estimated that the probability of a major (magnitude 7+) earthquake along the Wasatch Fault portion of the ISB in the next one hundred years is about 30 percent. In some segments of the fault zone, the probability may be as high as 60 percent; in others it may be only 20 percent. In any case, the issue of seismic safety in Utah is certainly much more than the media-generated disaster hysteria that is so often used by news outlets for self-promotion. There are legitimate scientific reasons to be concerned about the Wasatch Fault, and someday residents of the Wasatch Front *will* encounter firsthand the dynamic geologic turbulence of the Great Basin.

South from central Utah, the Intermountain Seismic Belt curves toward the west, and sweeps into southern Nevada before it gradually merges with zones of seismic activity in the southwest portion of the Great Basin. Near the southwestern terminus of the ISB, there is a noticeable concentration of seismicity in southern Nevada just north of Las Vegas (Figure 10.7). This is not a natural feature of the tectonic framework of the Great Basin, because that cluster of earthquakes emanates from the Nevada Test Site, where underground detonations of nuclear devices have resulted in many artificial earthquakes over the past several decades. Though the tests have now been curtailed (or so we are informed) by treaty, the vibrations they formerly produced were recorded by seismographic monitoring stations and plotted on maps depicting the overall seismicity of the region. While the concentration of seismic activity in south-central Nevada may be artificial, elsewhere in the western Great Basin the levels of natural seismicity are, in some areas, even more intense!

The Walker Lane–East California Shear Zone

The most seismically active part of the Great Basin is the western limb of the province, where earthquake activity is superimposed on the zones of dextral-slip faulting we have previously identified as the Walker Lane belt and the East California Shear Zone (ECSZ). It is in this area that the largest earthquake in the region struck Lone Pine in 1872. In addition, recent events such the 1992 Landers and 1999 Hector Mine earthquakes, both in the ECSZ, clearly indicate the extreme potential for damaging earthquakes to occur in the western Great Basin. Unlike the Intermountain Seismic Belt to the east, though, the zone of seismicity along the western border of the Great Basin does not coincide with large areas of dense population. However, the northern portion of this zone does pass near enough to Reno, Lake Tahoe, and surrounding communities to pose at least some concern. Moderate earthquakes are commonly felt in these areas, and there is the potential for major damage from a strong earthquake in the future.

There are several characteristics of the earthquakes in the Walker Lane–ECSZ area that distinguish them from the temblors generated along the Intermountain Seismic Belt. The majority of earthquake-producing faults in the western Great Basin are dextral-slip faults, rather than extensional faults with mostly vertical displacement. In addition to the Owens Valley Fault, other dextral-slip earthquake faults in the western Great Basin include the Lavic Lake and Bullion Faults (both involved in the 1999

Hector Mine earthquake) and the Johnson Valley Fault that slipped more than 15 feet during the 1992 Landers event. There are literally dozens of other dextral-slip faults in the Walker Lane area that have produced strong historic earthquakes, and may do so again in the future. In addition to the dextral-slip faults, there are many active normal faults along the base of the Sierra Nevada and in the Walker Lane belt immediately east. Some of these normal faults in the western Great Basin are slipping at rates estimated to be as high as several millimeters per year, suggesting that they may be even more active than the faults in the Intermountain Seismic Belt. Within the western Great Basin zone of seismicity, most of the earthquakes occur at relatively shallow depths, especially those generated by the dextral-slip faults. The main rupture during 1992 Lander earthquake, for example, occurred only slightly more than a kilometer (less than a mile) underground. No wonder that the surface vibrations from this earthquake were so powerful! In view of the extremely intense geological activity in the western Great Basin, verified by both the recent seismic history of the area and satellite measurements of ground motion, we can certainly expect to see many more powerful earthquakes in the California-Nevada region in the future.

Central Nevada Seismic Belt

The third major area of modern earthquake activity in the Great Basin is the Central Nevada Seismic Belt, a north-south-trending zone that extends for at least 200 miles, from Hawthorne to Winnemucca. Some geologists extend the Central Nevada Seismic Belt as far south as the Owens Valley fault zone, and as far north as the Nevada-Oregon border. The limits of this belt are, in fact, difficult to establish precisely because earthquake activity gradually diminishes northward from Winnemucca, while the southern end of the zone merges with the seismic activity related to the Walker Lane belt. However, there is little doubt about where the main portion of the Central Nevada Seismic Belt is located: it definitely coincides with the very strong earthquakes that occurred from

Figure 10.8 The prominent scarp of the 1954 Fairview Peak earthquake in west-central Nevada.

1915 to 1954 (see Table 10.1) in the Stillwater Range–Tobin Range–Dixie Valley area of Nevada. These earthquakes, the strongest historic temblors to strike Nevada, produced conspicuous ground ruptures that are the most distinctive topographic features of the Central Nevada Seismic Belt. Other regions of the Great Basin have experienced powerful earthquakes, but there are no better examples of modern surface ruptures than those produced during the historic earthquakes in central Nevada. The fault scarps are so well developed in this area that they have become marked tourist attractions along Highway 50 in western Nevada. The fault scarps resulting from the 1954 Fairview Peak earthquake about thirty miles southeast of Fallon, Nevada, are particularly prominent (Figure 10.8). In this area, an escarpment 10–20 feet high marks the actual ground break of the magnitude 7.2 earthquake. Careful measurements on the rock and soil displaced along this scarp indicate that the earthquake simultaneously produced about 7 feet of vertical offset (the mountain block moved up, the valley down) and about 11 feet of dextral-slip displacement. Thus, in 1954 the Fairview Peak Fault behaved much like the Owens Valley Fault did in 1872: blocks of rock along both faults jerked up and slipped to the right during their respective earthquake events. Several other faults in the Central Nevada Seismic Belt exhibit this same kind of composite motion—partly vertical, partly lateral—that geologists describe as **oblique slip**. Because the faults in this zone so commonly display oblique slip, some geologists consider the Central Nevada Seismic Belt to be

transitional between the eastern Great Basin, where seismicity is related to extensional faulting, and the Walker Lane belt, where the strongest earthquakes result primarily from sudden twitches along dextral-slip faults.

Despite the powerful historic earthquakes in the Central Nevada Seismic Belt, this interior zone of seismicity is less extensive than either the Walker Lane–ECSZ or the Intermountain Seismic Belt, situated along the margins of the Great Basin. No one knows exactly why there are so many earthquakes in the Central Nevada Belt, but it may have something to do with a rift zone that opened during the Miocene, when strong extensional forces were just beginning to sweep through the region. This ancient crack in the crust may constitute a north-trending zone of weakness that tends to localize the modern earthquakes in central Nevada. Nonetheless, on a regional scale, the modern seismicity of the Great Basin is clearly concentrated along the eastern and western boundaries of the province. This peripheral distribution of earthquakes, along with the modern measurements of extensional movement, is a sure sign that the Great Basin is still extending. The strongest seismicity occurs at the edges of the region because it is in those areas that the greatest variance in extensional velocities exists. Along the Intermountain Seismic Belt, the crust is being pulled to the west at rates that are low for the Great Basin but still higher than the nonextending crust to the east. On the western edge of the Great Basin, the combination of westward extension and northwest dextral-slip results in the highest velocities anywhere in the Great Basin and the most intense earthquake activity. It is ironic that in such a sparsely populated region, the two largest concentration of humans, the Wasatch Front and the western Nevada urban centers of Las Vegas and Reno, lie directly atop the most unstable ground.

MODERN VOLCANIC ACTIVITY IN THE GREAT BASIN

No one knows exactly when the last volcanic eruptions occurred in the Great Basin. We have already learned that it might have been the steam blast at Ubehebe Crater in northern Death Valley, the lava flow at Pisgah Crater in the Mojave region, the cinder cones near Lunar Crater in central Nevada, or Paoha Island in the Mono Craters chain. All of these volcanic events appear to have occurred less than 1,000 years ago, and many other eruptions probably took place elsewhere in the Great Basin over the past few millennia. Such recent volcanic activity is a strong indication that the volcanic fires are not yet extinguished beneath the Great Basin. If the region is still being stretched, as the seismic and geodetic information suggests, then eventually we should expect underground magma to reach the surface by squeezing through the widening faults and fractures. It is certain that future volcanic eruptions will occur somewhere in the Great Basin Province. But where?

Any of the areas of young volcanism in the Great Basin are good candidates for volcanic reignition in the future, but the Long Valley caldera region of eastern California stands above all others in posing the most serious threat of renewed activity. This area, the site of unimaginable volcanic violence 730,000 years ago, appears poised for eruptions that could conceivably begin at any moment. Recent geologic events in the Long Valley area have raised some warning flags of an impending eruption. In May of 1980, a swarm of small earthquakes began near Mammoth Mountain, a dormant volcano last active about 50,000 years ago. The vibrations continued for days in the southwest part of the Long Valley caldera, eventually producing several shocks in the magnitude 5–6 range. Shortly after the earthquake swarm, geologists from the U.S. Geological Survey reported that the resurgent dome within the Long Valley caldera had suddenly risen by approximately two feet. Another earthquake swarm occurred in 1989 and involved more than 1,500 small tremors. In 1990, carbon dioxide gas began escaping from several places around the flanks of Mammoth Mountain. On March 11, 1990, a forest ranger escaped a spring blizzard by taking refuge in a snowbound cabin near Horseshoe Lake at the foot of Mammoth Mountain. The ranger narrowly avoided suffocation from carbon dioxide that had accumulated in the cabin

from the soil beneath it. Eventually, this gas became so concentrated in the soils around the ancient volcano that it so damaged the roots of trees that kill zones began to develop in the forests. By 1994, the several patches of dead trees collectively amounted to more than seventy-five acres, and geological studies confirmed that the soils in the dead zones were nearly saturated with carbon dioxide. Today, between 50 and 150 tons of carbon dioxide leak from the soil around Mammoth Mountain each day, according to estimates by the U.S. Geological Survey. The only plausible source of the CO_2 emitted from the Mammoth area are the gases released directly from magma near the surface, and perhaps from older carbonate-rich metamorphic rocks that are in contact with the magma. Seismic profiling techniques have suggested that the magma pool under the western Long Valley caldera is now situated less than about 10 km (6 miles) below ground. Meanwhile, the current rate of uplift of the resurgent dome in the caldera is about 1 inch per year: the dome is swelling!

The Long Valley caldera is one of the most intensely monitored volcanic regions in the world. Geologists continuously gather information pertaining to the seismicity, inflation, geothermal activity, and gas emissions in the region. Most scientists agree that future volcanic eruptions *will* occur in the area, but no one can predict exactly *when* they will begin. However, some forecasts can be made for the types of eruptions we might expect, and where in the Long Valley area they are most likely to occur. The most recent eruptions in the Mono Lake–Long Valley area have occurred along a north-south zone aligned with the Mono Craters and Inyo Craters chains. This linear belt of volcanic activity probably reflects the opening of fissures under the influence of west-directed extensional stresses. Evidently, magma has moved up along this fissure zone and erupted from several places along the crack during late Holocene time. Each dome, crater, or volcano in the Inyo and Mono chains represents a point where magma rising as a sheet, or "wall," along the fissure reached the surface. The ascent of magma along such a vertical fissure is known as a **dike instrusion**. Because dike intrusions beneath the Mono and

Figure 10.9 Snow-dusted rhyolite flow (left) in Panum Crater was erupted about 600 years ago. Small eruptions of this type are likely in the Mono-Inyo region in the future.

Inyo craters have dominated the most recent phases of volcanic activity in the Long Valley area, and because we know that extensional forces are probably still pulling the fissure apart, future eruptions are most likely to occur along a north-south zone stretching from the western end of the Long Valley caldera to Mono Lake.

What type of eruption might occur? It is unlikely that a colossal volcanic cataclysm, such as the Pleistocene explosion that created the Long Valley caldera, will occur again. Such powerful volcanic events are very rare in the extensional plate tectonic setting that prevails in the modern Great Basin. Instead, we are probably much more likely to witness small blobs of thick rhyolitic magma ooze out from one or more sites along the Mono Craters–Inyo Craters trend. These eruptions could involve some groundwater explosions and perhaps even generate some localized ash flows. Small rhyolite domes, similar to those that dot the existing Long Valley terrain (Figure 10.9), are likely to be the final product of future eruptions in the area. This is not to suggest, of course, that the renewed volcanic activity will be harmless to humans. Depending on exactly where and when the eruptions occur, and how much magma they vent, thousands of lives could be lost in the ski resorts, campgrounds, and other recreational facilities clustered around Mammoth Mountain.

Beyond the Long Valley area, future volcanic activity in the Great Basin is either less likely or there is insufficient information to make a valid forecast. Two regions that appear to be primed for future eruptions are the Black Rock Desert of western Utah and the Amboy-Pisgah volcanic field of the Mojave region. However, the presumed likelihood of renewed eruptions in those areas is based primarily on the recency of Holocene eruptions, rather than on earthquake swarms, ground swelling, or gas emissions. Nonetheless, geologists would not be surprised to see magma belching from any of the volcanoes in the Mojave Desert or western Utah at any time, though there are no signs that such activity is imminent. Magma may also resurface from many other areas of Holocene volcanism in the Great Basin, including the Lunar Crater volcanic field, the many volcanic cones of southeastern Oregon, and the isolated fault-controlled volcanic centers that are widely scattered across the region. All of the geological evidence suggests that magma definitely exists at shallow levels under the Great Basin, though it is difficult to know exactly how much and where it is concentrated. Someday, perhaps soon, we may see the slumbering volcanoes awaken in the deserts of western Utah, or Nevada, or eastern California. In the meantime, even while it remains underground, the magma is affecting the crust above it and giving rise to another fascinating natural phenomenon in the Great Basin. Moreover, if our goal is a comprehensive geological model for the modern crust, one that embraces all of its geophysical characteristics, we must take into account an irrefutable consequence of the recent volcanic history: the Great Basin is still hot!

WHERE HOT WATERS FLOW: GEOTHERMAL ACTIVITY AND HEAT FLOW IN THE GREAT BASIN

Every miner, even the ones who prospected the Great Basin in the nineteenth century, soon becomes familiar with the warmth of the subsurface earth. The average rate of downward temperature increase for the entire planet is about 25°C per kilometer of depth, or roughly 70°F for each mile of descent. Known technically as the **geothermal gradient**, this rate of temperature increase is influenced by a number of local geologic factors, and varies considerably from place to place. The geothermal gradient exists because the center of the earth is very hot (about 6,000°C, almost as hot as the surface of the sun!), while the surface is cold. Heat energy always flows from a hot source to a cold "sink" until the temperature differences are equalized. When the temperature differences between the source and the sink vanish, so that neither is "hot" or "cold," equilibrium conditions exist and the flow of heat energy stops. Within the earth, heat flows from the center to the surface, where it is radiated through the atmosphere and eventually into the cold abyss of space. The earth, as a whole, is cooling down from this natural flow of heat, but only very slowly. This is because the heat of the inner earth is continuously (but not completely) replenished by the decay of radioactive elements, by heat generated from the crushing forces of gravity, and from the residual heat left over from the formation of the planet billions of years ago. The amount of heat generated in the interior of the earth does not match the amount radiated into space or absorbed elsewhere, but the difference is very small. Eventually, billions of years from now, the earth will be cold to the core, but until then, heat will continue to flow outward toward the crust. It is this inner heat, of course, that ultimately drives the entire global tectonic system.

The amount of heat flowing from subsurface sources through the crust is very small, but it is measurable. At the surface of the earth, geothermal heat is usually a negligible influence on the temperature of rocks, especially compared to the heat from the sun. A chunk of black basalt from the Lunar Crater Volcanic Field may be too hot to hold comfortably on a July day, but most of this heat was absorbed from the sun, not from the deep interior of the planet. Underground, however, the reverse is true: geothermal heat is much more important than solar heat. Even though it takes very sensitive instruments to measure heat moving through rock near the surface, underground geothermal heat has a great influence on the rigidity and density of rocks, the patterns of subsurface circulation, the rates of tectonic movements, and even the elevation of the surface. Nonetheless, the global aver-

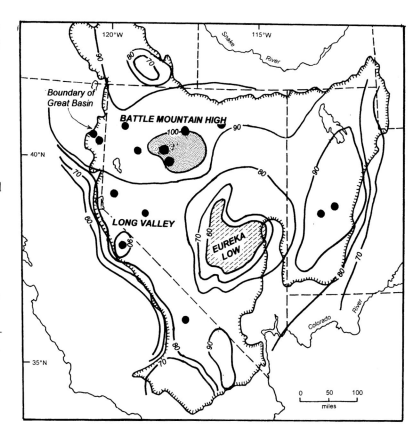

age heat flow at the surface is only about 60 mW/m², the standard units (milliwatts per square meter) reflecting the minuscule drift of heat energy through rocks of the outer earth. The amount of heat flowing through the crust at any location depends on such factors as the thickness of the crust, the presence and amount of magma, the age of young igneous bodies, the degree and style of deformation (fracturing), the concentration of radioactive elements, and the motion of groundwater and other subsurface fluids. It seems that nothing is ever *average* in the Great Basin, and heat flow is no exception. The heat flow of the Great Basin is variable, as it is everywhere, but generally hovers in the range of 90–95 mW/m², significantly higher than the global average. In places such as the "Battle Mountain High" of northwest Nevada, the heat flow exceeds 100 mW/m² (Figure 10.10). Only in a relatively small area near Eureka, Nevada, appropriately called the "Eureka Low," does the heat flow in the Great Basin dip slightly below the global average. The Eureka Low notwithstanding, the Great Basin is clearly a geologically hot region. The high heat flow

Figure 10.10 Contour map of heat flow in the Great Basin. Across the Great Basin, heat flow is significantly higher than the global average of about 60 mw/m2. The "hottest" area is the Battle Mountain High in northwest Nevada, while the "coolest" is the Eureka Low. Black dots represent major geothermal power-generating facilities.

Figure 10.11 The Night-- ingale geothermal area in northwest Nevada.

results in some interesting geological phenomena, but it also tells us something very important about the nature of the crust.

Hot springs are one of the most obvious expressions of the geothermal heat in the Great Basin. There are so many hot springs in the region that no one has ever made a comprehensive count of them all. Nevada has more officially designated thermal areas, 312, than any other state. Utah has an additional 116 official hot springs, most of which are located in the western part of the state. There are probably thousands of smaller, isolated, and unnamed hot springs scattered around the Great Basin. In recent years, the exploration of hot springs has become one of the more popular outdoor activities in the region, and it is now rare to encounter back-country hot springs without finding some sign of recreational use. Some of the geothermal areas in the Great Basin have temperatures high enough to vaporize groundwater, giving rise to roaring vents that discharge steam at temperatures exceeding 300°F. If there is enough steam at sufficiently high pressures, it can be circulated through turbines to generate electrical power. There are twenty-one such geothermal power plants situated in the Great Basin, with more planned in the future to ease the energy supply problems that all Americans will face in the coming years. Collectively, the geothermal plants in the Great Basin can now produce more than 200 megawatts of power, and projected expansions will greatly increase

this yield in the future. In addition to the geothermal power plants, hot water from other Great Basin thermal sites is used for space heating, aquaculture, and food dehydration in many places throughout the region. Each of the many thermal sites in the Great Basin has unique characteristics that reflect the local geological and hydrological framework. However, the overall abundance and diversity of geothermal resources in the area is a direct reflection of the heat energy flowing through the crust.

Why is the foundation of the Great Basin so hot? The high heat flow in the region arises from the combination of several different factors that are directly related to the relatively recent geologic history. Magma near the surface is obviously an important source of heat in areas of recent (or future?) volcanism. At the Casa Diablo power facility in the Long Valley area, for example, heat from magma at shallow depths drives a geothermal system that can produce a total power yield of more than 40 megawatts. This magmatic heat also results in many thermal springs, particularly the well-known pools along the appropriately named Hot Creek (Plate 17). In other areas, young bodies of igneous rock, usually less than a million years old, sometimes retain enough residual heat to generate significant geothermal phenomena and resources. In the Roosevelt Hot Springs and Cove Fort geothermal areas of western Utah, heat from young granitic intrusions can produce more than 35 megawatts of electrical power. Still other hot springs in the Great Basin are situated where no shallow magma or young igneous masses exist. In such cases, it is most likely the thin and fractured nature of the crust that results in the geothermal activity. For obvious reasons, more heat is transferred from the hot mantle through the thin crust of the Great Basin than through the thicker crust of adjacent regions. In a similar way, humans lose more body heat through a thin sheet than through a thick down comforter on a cold winter night. Recall that the crust of the Great Basin is not only thin but also has been highly fractured by extensional deformation over the past 30 million years. The faults in the stretched and broken crust allow water to trickle down to deep levels, where it is heated and driven back to the surface. The importance of fractures in localizing geothermal phenomena is underscored by the fact that 90 percent of the geothermal systems in the Great Basin are associated with late Pleistocene–Holocene faults. The continuous convective motion of deep hot water toward the surface greatly enhances the flow of heat through the crust, augmenting the high regional heat flow in the Great Basin. We have already learned that such hydrothermal systems, linked to Mesozoic and Cenozoic igneous activity, were important in the formation of rich ore bodies million of years ago. Many similar systems, driven by the earth's internal heat, exist across the Great Basin today. Thus, the high heat flow in the Great Basin is the consequence of three general factors: shallow magma or young igneous rock bodies, a relatively thin and broken crust, and the vigorous circulation of hydrothermal fluids through highly fractured subterranean rocks.

CONTEMPORARY STRUCTURE OF THE CRUST: THE CLOSING CIRCLE

The modern geologic fabric of the Great Basin is the cumulative result of billions of years of natural history. We have traced much of this history in the preceding chapters, and have a general understanding of how the delightfully heterogeneous and complex substrate of the region was assembled, piece by piece, one step at a time. Though the crust of the Great Basin today is heated, stretched thin by extension, and broken into myriad pieces, it has only existed in that state for less than 1 percent of its long history. In earlier chapters of this book, we have explored how utterly different were the various prehistoric landscapes of the Great Basin from the alternating mountains and basins that we know today. During the past 3 billion years microcontinents have collided and coalesced, megacontinents have split apart, volcanic mountain systems have come and gone, tropical seas have invaded and receded, and glaciers and gigantic lakes have appeared and vanished. All of these ancient events have left evidence in the geological foundation of the Great Basin, a

rocky chronicle so complex that it has taken scientists well over a century to decipher the clues and untangle the commingled stories. But, what about the future? What long-term changes can we anticipate from the current geologic activity with the Great Basin? Geologists, by nature, are generally much more comfortable reconstructing the past than predicting the future, because most ancient events leave some tangible evidence: rocks! Future events, of course, cannot now provide evidence upon which a testable model for what might happen next can be formulated. However, our understanding of the deep history of the Great Basin can be combined with the observed patterns of modern geologic activity to allow us to make some reasonable speculations about where we are headed in the millennia to come. Such forward projections of current phenomena are always somewhat conjectural, but if they are based on accurate reconstructions of the geological past, and guided by a sound understanding of contemporary tectonics, then the forecasts will be at least plausible, if not certain.

In addition to the wealth of direct information about the rocks comprising the crust in the Great Basin, studies of geophysical properties such as the propagation of seismic waves, variations in the intensity of gravitational forces, and deviations in the magnetic field can be used to develop models of the deep unseen portion of the North American Plate. Though this information on the deeper-level structure of the lithosphere is impossible to verify directly and with absolute certainty, the geophysical models provide valuable information about the composition and structure of the lower parts of the plate, the way it behaves, and the forces that are affecting it. For example, seismic profiling techniques suggest that the crust in the Great Basin is 20–30 km (12–19 miles) thick, about half the thickness of normal continental crust. Though scientists have not drilled through the crust to directly verify the geophysical determination of its thickness, the concept of relatively thin crust in the Great Basin is compatible with the geologic structures, rock types, heat flow, and other properties that can be measured directly on the surface. For these reasons, we accept the indirect

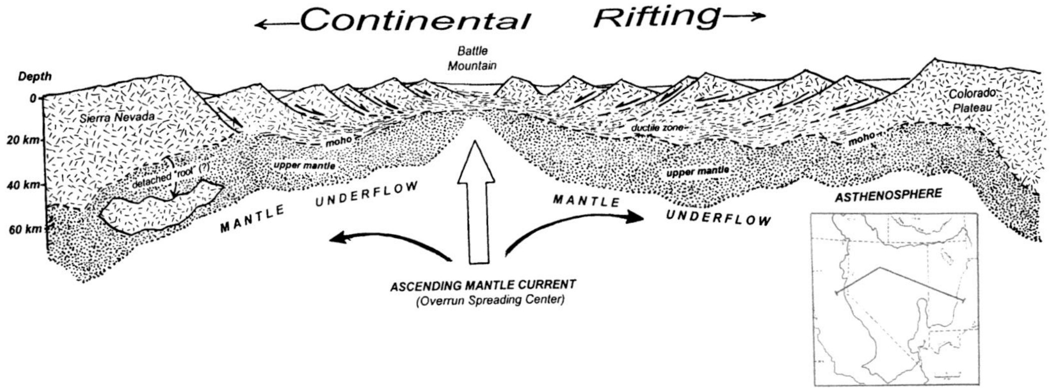

←Continental   Rifting→

Figure 10.12 Continental rifting in the Great Basin. Along the line of transect (inset map), the rifting appears to be centered near Battle Mountain, Nevada, where the crust is thinnest and heat flow is highest.

measurement of crustal thickness as generally accurate, even if it is not perfectly precise. The assumption of a thin crust in the Great Basin, in turn, suggests certain things about what events are most likely to occur in the future. Over the past two decades, several indirect means (using geomagnetic, electrical, and gravitational data) of probing the deeper levels of the Great Basin crust have resulted in a fairly consistent view of its overall characteristics. Consequently, geologists are in general agreement about the broad structure of the Great Basin lithosphere, although many uncertainties still exist and numerous details have yet to be resolved. We may not know everything about the deeper structure of the Great Basin, but we know a lot.

Across the widest part of the Great Basin, between the northern Sierra Nevada and the Wasatch Mountains, the crust of the Great Basin is somewhat variable in thickness, but it is relatively thin everywhere (Figure 10.12). The thinnest crust is less than 20 km (about 13 miles) thick and appears to be situated in northwest Nevada, where this "thin spot" may be related to the high heat flow of the Battle Mountain High. The crust in western Utah, near the boundary between the Great Basin and the Colorado Plateau, is 28–30 km (17–18 miles) thick, though in places it appears to be slightly less. Elsewhere in the region, the crust appears to be remarkably uniform in thickness, varying from 20 to 30 km almost everywhere. Crustal rocks abruptly thicken beyond the western and eastern margins of the Great Basin. In the Sierra Nevada, the crust is as much as 55 km (34 miles) thick, but this value varies along the northwest-southeast trend of the mountain range. The crust in the

southern Sierra Nevada appears to be relatively thin, only 30–40 km (19–25 miles) thick, perhaps by virtue of having lost a portion of the deep "root" of the Sierra Nevada batholith through a process known as "de-lamination": the breaking away, and sinking, of a slab of rock from the lower crust. East of the Great Basin, the undeformed crust beneath the Colorado Plateau is 40–50 km (25–31 miles) thick. The highly deformed crust in the Rocky Mountain region is, for the most part, comparable in thickness to that of the Colorado Plateau. However, the crust under the southern Wasatch Mountains, along the northeastern boundary of the Great Basin, may be thinned slightly by a localized "mantle upwarp" in that region.

As we have already discovered, the thin crust of the Great Basin is amazingly heterogeneous and, of course, it is also highly deformed. Thick sequences of varied Precambrian, Paleozoic, and Mesozoic sedimentary rock have been bent, crumpled, and shattered during the numerous waves of mountain building that have swept through the region. The contorted rock sequences were later intruded by magma that cooled to form the many Mesozoic plutons. Then the whole mangled mess was intruded by magma again in mid-Cenozoic time, and ripped apart by extensional forces in the late Cenozoic Era. Even now, the seismic evidence suggests the presence of "soft" zones in the crust, probably the result of pooled residual magma or partial melting induced by localized high temperatures. In at least some places, the high regional heat flow and tensional stresses in the Great Basin cause the crust to become "soft" at a depth of about 15 km (9 miles). At this depth, the rigid and brittle rocks of the crust are transformed into more ductile material. Ductile deformation ("bending and flowing") in the soft zone in the lower crust explains why earthquakes are generally so shallow in the Great Basin: rocks *break* only when they are rigid and brittle. Below about 15 km, the crust of the Great Basin accommodates stress primarily by flowing rather than breaking, and earthquake-spawning ruptures are rare. On either side of the Great Basin, the thicker crust is composed of more uniform, more rigid, and generally less deformed assemblages of rock. To the west, the Sierra Nevada crust is dominated by a large composite batholith associated with inclusions of Paleozoic and Mesozoic metamorphic rock. East of the Great Basin, the crust is characterized by either the relatively thick and undeformed sedimentary sequences of the Colorado Plateau or the folded and faulted Precambrian-Paleozoic rocks of the Middle Rocky Mountains. The striking variability of the Great Basin crust, compared to the relative homogeneity of the crust in adjacent areas, reflects its unique geological heritage. No other part of western North America has undergone a similar succession of geologic events spanning such as enormous chasm of time.

Immediately beneath the crust in the Great Basin is a variable thickness of solid mantle rock. Small pieces of this dense rock appear in some of the younger Quaternary basalt flows of the Great Basin, indicating that the magma in those volcanic areas originated from upper mantle or lower crust source rocks. The comparatively high density of mantle rocks accelerates seismic waves across a surface known as the **moho** (named in honor of its Croatian discoverer, Andrija Mohorovičić), the boundary between the crust and the mantle (Figure 10.12). The depth to the moho is the same as the depth to the top of the mantle, and, therefore, is generally equal to the thickness of the crust. Thus, in the Great Basin, the moho is usually situated 20–30 km (12–19 miles) below the surface. As we learned in an earlier chapter, far below the moho mantle rocks soften from heat and pressure to form the weak asthenosphere, which marks the base of the lithosphere. It is very difficult to establish the precise thickness of the lithosphere in the Great Basin, but it probably ranges from 50 to 70 km (30–43 miles) in thickness, compared the Colorado Plateau lithosphere, where it is estimated to be about 100 km (60 miles) thick.

So, what does all this mean? From the foregoing discussion it is clear that the crust of the Great Basin today is thin, broken, hot, and subject to powerful stretching forces. These considerations all point toward the same intriguing conclusion about the contemporary geologic setting of western North America: the Great Basin appears to be a zone of continental rifting! The modern interactions between lithospheric plates in western North America are evidently tearing the edge of the continent apart, and the Great Basin is a broad rift zone separating the westernmost pieces from the more stable tract to the east. This rifting is driven by a combination of several different tectonic forces related to the interactions between the North American and Pacific Plates. Recall that a segment of the East Pacific Rise, an oceanic spreading ridge, has been overrun by North America and is now probably situated somewhere beneath the Great Basin. Along the subterranean axis of the buried oceanic ridge, hot and partially molten rock still rises from the asthenosphere, welling up under the central Great Basin. The ascending mantle material under the Great Basin partially accounts for the high heat flow of the region, and may also be responsible for melting the rock of the uppermost mantle to produce the very young basaltic magmas that were erupted during Holocene time. The softening of crustal rocks to produce the shallow ductile zone 15 km (9 miles) deep may also be related to the heat carried upward by the rising mantle currents. The ascending mantle plume is probably deflected laterally where it meets the base of the crust, and slowly oozes outward beneath the ductile zone in the overlying slab. This lateral underflow results in some of the extensional stress that is tearing the Great Basin apart. In addition, the drag between the northwest-moving Pacific Plate and the North American Plate contributes to the rifting in the Great Basin by enhancing the stretching forces in the

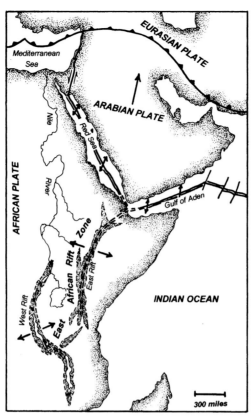

Figure 10.13 The Arabian Plate and the East African rift zone both originated through the rifting of fragments of continental crust from Africa. The rifting in northeast Africa began about the same time that extension was initiated in the Great Basin.

region and introducing the dextral slip that characterizes the Walker Lane belt. The modern rifting of the Great Basin may thus be linked to at least two principal mechanisms: the lateral mantle underflow in the central region, and the shearing related to the San Andreas system along the western edge of the province. In addition, it recently has been proposed that gravitational forces also may play an important role in the rifting by triggering the collapse and spreading of the crust when the subcrustal material that supports it either is removed by lateral flow, weakened by heating, or thinned by stretching (or all three!). Though geologists continue to debate the details of the mechanisms of rifting, the reality of the process in the Great Basin is undeniable. The Broken Land heralds the breaking of the continent.

The best place in the world to study the process of modern continental rifting is Africa, a continent that has been self-destructing for more than 20 million years. In the northeast corner of Africa (Figure 10.13), the Arabian Plate was separated from the rest of the continent about 5 million years ago, when the modern Red Sea

and Gulf of Aden opened as narrow ocean basins. As sea-floor spreading continues in these narrow oceanic troughs, the severed Arabian Plate is forced farther to the northeast, where it continually collides with Asia. This rifting event was evidently driven by a plume of hot mantle material rising upward beneath a volcanically active area known as the Afar Triangle in the Ethiopian highlands. More pertinent to the tectonic setting of the Great Basin is the great East African rift zone, which extends south from the Afar Triangle for more than 2,000 miles. In this linear zone of continental rifting, several slivers of East Africa have been broken from the continental interior along prominent north-trending extensional faults. Two great linear valleys, the Eastern Rift and the Western Rift, have developed where the African crust has collapsed as the rock to either side has been stretched and broken. Rivers, including a portion of the upper Nile and its tributaries, tend to flow into and through the rifts, depositing thick sequences of gravelly sediment on the basin floors. Several large lakes have developed in the lowest parts of the rift valleys, where enclosed basins have developed. Under the influence of the warm and dry East African climate, these lakes lose great amounts of water through evaporation. This causes the water of the larger rift-valley lakes to be strongly mineralized and alkaline. Some of the smaller lakes evaporate entirely during the dry season, as do many of the rivers and streams. Shallow earthquakes occur continuously in the East African rift zone as the extensional faults yield to the persistent stretching forces. Recent volcanic activity in East Africa is related to mantle-derived magma, and is very similar to the bimodal eruptions that have occurred in the Great Basin during the past several million years. Eventually, the extensional tectonics in East Africa will result in the complete separation of elongated slivers of the continental crust. These splinters of Africa will drift away from the mainland while narrow ocean basins, similar to the modern Red Sea and Gulf of Aden, open in the places now occupied by the rift valleys. Thus, the modern geology of the East African zone is dominated by extensional faulting, linear rift systems, collapsing valleys, ephemeral lakes and rivers, dry playas, salt lakes, shallow earthquakes, and bimodal volcanism. Sound familiar? Though they are not geologically identical, the contemporary geology of the East African rift zone and the Great Basin share many common themes. There is a very good reason for the geological similarities: both regions are situated in areas of active continental rifting.

Pondering the Great Basin as a nascent continental rift zone fosters some interesting questions. Unlike the narrow rift valleys of East Africa, the Great Basin is very broad. Why is this? As we have seen, the extension and collapse of the Great Basin began about the same time that the extension of the East African rift zone commenced. So, the unusual width of the Great Basin in comparison to the East African rift cannot be attributed to any significant differences in the timing of the rifting. It seems that the rate of extension in the Great Basin has been much higher than the rate in Africa, though the magnitude of subsidence and collapse has been about the same in the two areas. This may reflect the overrun oceanic spreading ridge that may now simmer underneath the Great Basin. No such feature exists beneath the East African rift, and the lateral mantle underflow associated with it in western North America may be accelerating the extension of the Great Basin. Thus, what would have been a narrow rift valley in the absence of the overrun spreading ridge is now more than 500 miles wide. It is also important to understand that the rifting in Africa is occurring within a single plate, while in the Great Basin it proceeds under the influence of the sliding motion between the North American and Pacific Plates. The northwest movement of the Pacific Plate along the western edge of North America may further enhance the extension of the Great Basin. In contrast, nothing comparable to the San Andreas system and the Walker Lane belt exists in the vicinity of the East African rift.

Assuming that continental rifting is in progress in the Great Basin, we might ask, where in this broad rift zone is the center located? In East Africa, the narrow rift valleys clearly define the principal zones of continental fragmentation,

the places that will someday be transformed into narrow ocean basins. Because there is really no principal rift valley anywhere in the Great Basin, it is much less certain where the axis of extension is located, or even if there is one. However, the crust in the vicinity of Battle Mountain in northwest Nevada appears to be the thinnest in the region, based on heat-flow measurements and seismic studies. In addition, the Central Nevada Seismic Belt bisects this area of thin crust, indicating that the faults in northwestern Nevada rupture more frequently than do faults in other areas in the central Great Basin. Finally, northwest-aligned igneous dikes of middle Tertiary age in this same general area suggest that extensional forces may have opened fissures earlier (in late Oligocene time) in the northwest Great Basin than elsewhere in the region. However, if the Battle Mountain area was the axis of extension in the Tertiary Period, it seems that the most intense geological activity has since shifted to the eastern and western periphery of the Great Basin. Some geologists have suggested that the Death Valley region, in the seismically active Walker Lane belt, and with a basin floor nearly 300 feet below sea level and still dropping, may be the most actively extending part of the Great Basin. Older geological trends, such as the rifted edge of the Precambrian basement (the 0.706 Line discussed in Chapter 2), may help localize the stretching forces in the modern Great Basin, creating several discreet zones of rapid extension. Such zones may separate the Great Basin into several "microplates," each of which responds differently to the omnipresent extensional stresses. Compared to the relatively simple East Africa rift zone, extension in the Great Basin appears to be more widely dispersed, locally more intense, and superimposed on a more variable geological foundation. Despite the complexities and uncertainties of modern Great Basin geology, it is clear that the crust in the region has been so thinned, weakened, and stretched that it appears ready to pull apart completely in the geological future. No matter how ageless and enduring the Great Basin terrain seems, we are undoubtedly on the threshold of a new landscape.

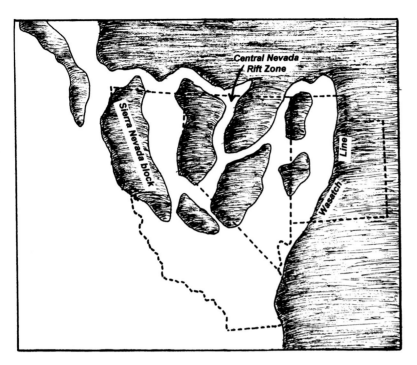

Figure 10.14 A sketch map of the future Great Basin. Continental rifting in the Great Basin will ultimately lead to the fragmentation of western North America over the next 30 million years. Though it is impossible to predict the precise details of the future geography, it seems clear that the modern landscape of the Great Basin is slowly, but surely, being transformed into a new rifted continental margin.

THE FUTURE LANDSCAPE: FORWARD INTO THE PAST

It is almost certain that extensional forces currently at work in the Great Basin will ultimately result in the rifting of several fragments from the western edge of North America. Between the severed blocks of crust and the stable interior of the continent, new oceanic lithosphere will be generated along embryonic spreading centers similar to those on the floor of the Red Sea and Gulf of Aden. Taking into consideration the regional geological patterns in the Great Basin and adjacent areas, and incorporating the current estimates of the rate of extension, we can make some reasonable speculations about how the future landscape will evolve through the next 30 million years or so. Modern geodetic measurements have shown that the Sierra Nevada and the Great Valley of California are moving together, as a coherent block, to the northwest about 12 millimeters (one-half inch) each year. This motion is facilitated by both the extension of the Great Basin and the dextral shearing in the Walker Lane belt. West of the Sierra Nevada–Great Valley block, a narrow sliver of coastal California is also moving to the northwest with the Pacific Plate, but in a slightly different direction and three times faster than the Sierra Nevada block. In the future, the Sierra

Nevada–Great Valley block will probably remain unified as a large island severed from North America. Thirty million years from now, this island will be 220 miles offshore from the broken edge of the continent, with the coastal California slice well to the north (Figure 10.14). If the two cities still exist in the year 30,000,000 C.E., Los Angeles will be north of San Francisco and situated on a different island mass.

On the east side of the Great Basin, the "Wasatch line" today marks the transition from fractured, thin crust in the Great Basin to the thick and relatively undeformed crust of the Colorado Plateau. This zone, essentially the eastern boundary of the modern Great Basin, will probably constitute the eastern limit of continental rifting. The new coastal margin for North America will probably form along the western edge of the Colorado Plateau as the unstable tracts in the Great Basin separate from the continent. Persistent faulting and powerful earthquakes in the Central Nevada Seismic Belt will probably lead to another major zone of rifting in northwestern Nevada. This central rift may likely propagate south to merge with, or become superimposed on, the edge of Precambrian continental crust (the 0.706 Line) in central Nevada. The future rifting may bend sharply to follow the 0.706 Line into what was the Walker Lane region. Meanwhile, the ocean will creep north from the modern Gulf of California, submerging the basins between the rifted masses that will have dropped below sea level. Several other island masses, all relatively small, may arise in the area between the Central Nevada Rift (the current seismic belt) and the Wasatch line. These islands might evolve from areas in the eastern Great Basin that have relatively thick Precambrian continental crust. Candidates for such rifted island masses would include the Sheeprock Mountains of western Utah, the Deep Creek and Snake Ranges astride the Utah-Nevada border, and the Ruby Mountains in eastern Nevada. Though it is impossible to predict the impending landscape of the Great Basin with much detail, it is difficult to argue against all the hard evidence that points to continental rifting now and in the future.

It is ironic that both the earliest recognized

events and the most recent phases in the inconceivably long natural history of the Great Basin involve continental rifting. As we learned in Chapter 2, the foundation for billions of years of geologic history was established when the core of primeval North America was rifted in the Great Basin region during Proterozoic time. Upon that foundation, an amazingly varied rock succession was built in a series of upheavals, each resulting in profound changes in the prehistoric landscapes and environments of the region. If the postulated continental rifting in the modern Great Basin is real, then we have come full-circle to a new episode of fragmentation. It appears that we are now witnessing a slow rerun of the late Precambrian events in the region, as primordial history has returned from the tectonic grave to become contemporary crustal dynamics. Perhaps the most significant distinction between the ancient and modern rifting events is that humans, now and in the future, will be affected by the breaking of North America. We will not see the entire process, of course, but even tiny increments of geological change can have dramatic impacts on creatures as puny as ourselves. In Lone Pine, California, no one that survived the event ever forgot how a small piece of crust was suddenly jerked north along the edge of North America on March 26, 1872. And so it will continue into the unimaginable future, far beyond the age of humans, until a new landscape emerges where the sagebrush now grows. The majestic Great Basin is slowly, and inexorably, vanishing.

In the preceding pages, we have explored the history of a special place and glimpsed, though a geological lens, the changes that have transformed the region over billions of years. The Great Basin we know today—the dry, fallen land of sage-covered mountains and valleys—may be young, but it is already becoming something else. As reverent as we might feel in the face of the immense history of the Great Basin, we should resist thinking of the modern landscape as "finished." Until the inner earth cools completely, billions of years from now, nothing will attain a truly final state on this dynamic and evolving planet. The geological legacy of the Great Basin is one of continuous change driven by the shifting arrangements of lithospheric plates and the variable interactions between them. Such interactions created the Great Basin, and they are also destroying it at this very moment. Reflecting on the billions of years summarized in the preceding chapters, it becomes apparent that change has been the only constant. Nothing is permanent in the Great Basin except impermanence, and the region will continue to change in response to the rhythmic undulations from the molten-iron heart of the planet.

If the story of the Great Basin resonates within us, it may be because there is a deep, almost spiritual, nexus between us and the barren sagebrush expanse. Like the desert landscape, each of us is also a dynamic entity, continually responding to external influences, and never remaining the same for very long. Each of us responds to the turbulent convulsions of life in our own way, eventually blazing our own unique path of personal transformation and growth. We become what we are in increments, as each unforeseen trauma, challenge, or triumph results in new personal dimensions and new thresholds of spiritual transformation. For as long as life endures, this process never stops; we steadily evolve through the various stages of life, constantly redefining ourselves as people, and reshaping our essential humanity as the world swirls around us. And yet, as natural as is this process of personal growth and change, it seems that there is an almost universal tendency among humans to seek permanence in *something*, even if we know that such constancy in nature is as illusory as a watery mirage shimmering across a dusty Great Basin playa. Perhaps it is the uncontrollable chaos of the natural world that provokes this innate human quest for stability. No matter how well we may comprehend the capriciousness of nature, we still struggle to embrace it in our own lives. Change, particularly when we cannot direct, control, or command it, scares us.

As the pace of global change accelerates, the impulse to find stability seems to be intensifying into a fundamental heartfelt longing in many people. The human population of the earth is now rapidly approaching the level at which it will force—indeed, it *is* already forcing—rapid changes in the living conditions for people on a planetary scale. It has become increasingly clear that traditional habits and customs, particularly the wasteful and resource-intensive practices of "developed" societies such as ours, cannot be sustained very much longer anywhere on the earth. Humans are now poised on the threshold of great self-inflicted changes in the way each of us interacts with the natural cycles that sustain the entire global ecosystem. Unless we are prepared to accept a future of rapid decline in the quality of life preceding our ultimate extinction, some revolutionary changes in the way we all live will be necessary. Though we may be able to understand these changes intellectually, our deep yearning for permanence can make the prospect of such fundamental revolutions in life disconcerting. In this regard, the story of endlessly changing landscapes in the Great Basin can help us to understand that such changes should not be feared or resisted. They are merely the consequences of being a part of the irrepressible biosphere that has evolved on a dynamic planet. If there is any message for humanity in the geological story of the Great

Basin, it is that we inhabit a living, sometimes convulsive, planet on which landscapes shift relentlessly under the influence of continuous environmental change. For billion of years, life in the Great Basin, and everywhere else on the earth, has adapted to new ecological challenges. The mountains and valleys of the Great Basin deserts stand as historical monuments to such natural rhythms, and the history of this stark land can inspire us to accommodate and embrace the changes that the future will certainly bring. For, as disconcerting as it may be, we *must* change to survive. We must find ways to apply our innate creativity, coupled with our unique brainpower, to transcend the coming environmental crises. Times of transition are never easy, but we must learn to look ahead to the next stage in the immense history of land and life on the earth, rising to meet challenges of survival that loom on the near horizon. For more than three billion years, life in the Great Basin has always responded that way: adapting to circumstances, shifting with the cadence of nature, and turning ecological threats into evolutionary opportunities.

With any luck, we will too.

Glossary

Acanthodians an extinct group (Subclass *Acanthodii*) of primitive fish characterized by paired fins supported by bony spines along one edge.

Accreted terranes also known as "exotic" or "suspect" terranes, these represent large blocks of crust, or "micro-plates," that are added to the edge of a larger continental mass. Most of North America west of the Rocky Mountains consists of such accreted terranes, principally sutured to the continent during the past 300 million years.

Accommodation zone a more or less linear zone that separates the regional tilt domains in the Great Basin.

Age a formal unit of geologic time smaller than an epoch; the time equivalent of a *stage*.

Allochthon a mass of generally deformed rock that has been transported from its original position by some tectonic (mountain-building) process.

Alluvium any sediment—sand, gravel, silt, etc.—deposited by running water in the relatively recent geologic past.

Alpine glaciers elongated "rivers" of ice that develop in mountainous regions and flow downhill through preexisting canyons. Also known as valley glaciers, these glaciers are generally much smaller than the expansive ice sheets that cover flat regions during Ice Age conditions.

Antler Orogeny a compressive mountain-building event that affected the central Great Basin, principally during the late Devonian and early Mississippian Periods.

Archean the formal eon of geological time, between the Hadean and Proterozoic, that began about 4.0 billion years ago and ended 2.5 billion years ago.

Asthenosphere a soft weak zone within the earth's mantle, immediately below the lithosphere, extending from about 100 km to 350 km or more beneath the surface.

Batholith a large pluton that is exposed on the surface for more than 40 square miles (100 square kilometers).

Breccia a granular sedimentary rock composed primarily of angular rock fragments larger than 2 mm (about 1/10 inch). The term **megabreccia** is sometimes used when the rock fragments are extremely large.

Bryozoan an aquatic invertebrate organism belonging to the Phylum Bryozoa, characterized by the construction of delicate coral-like colonies and commonly referred to as "moss animals."

Caldera a large depression, generally circular or elliptical in shape, that develops as a consequence of volcanic eruptions.

Cinder cone a small, conical volcano composed primarily of loosely consolidated cinders and other volcanic rubble.

Coquina a variety of limestone consisting chiefly of the shells and shell fragments of invertebrate organisms such as clams, brachiopods, or crinoids.

Cordilleran derived from the Spanish term for "mountain chain" or "range of mountains," this term is most commonly used to describe the mountainous part of western North America, generally from the Rocky Mountains to the West Coast. Some geologists restrict the use of the term to the region *between* the Rockies and the coast. In either case, the term includes all of the Great Basin.

Crinoids a group of echinoderms belonging to the Class Crinoidea, distantly related to starfish and sea urchins, and known informally as "sea lilies."

Cross-bedding a style of rock layering in which small sets of strata are arranged at an angle to a larger overall stratification.

Décollement a detachment surface, commonly a thrust fault, along which one mass of rock moves independently over another.

Deflation the lifting and removal of small rock particles by wind.

Desert pavement a durable mantle of closely packed pebbles and stones that results from the selective removal of small rock grains from the surface by wind and water.

Detachment faults in geological usage, this term generally describes a gently tilted or curved plane of separation between an upper mass of mobile rock that moves over a stationary underlying foundation.

Dextral slip a term describing the lateral motion of blocks along a fault in which each rock mass moves to the right, relative to the other.

Dike a tabular mass of igneous rock that cuts across the layering in surrounding rock. Dikes generally represent magma that moves along some plane of weakness, such as a fault, to solidify in a sheetlike form.

Diorite a crystalline plutonic igneous rock, commonly of medium gray color.

Dipnoans a group (Order Dipnoi) of lobe-finned fish that includes the modern lungfish and their prehistoric relatives.

Dolomite a carbonate mineral of the composition $CaMg(CO_3)_2$.

Dolostone a nongranular sedimentary rock consisting primarily of dolomite.

Ductile a term describing the nonbrittle flow of rock under high temperature and pressure.

Epoch an increment of geologic time smaller than a period and larger than an age; the time equivalent of a *series*.

Erg an extensive tract of sand dunes; a "sand sea," such as the modern Sahara Desert, where the term originated.

Evaporites a term designating chemical sediments that result from the evaporation of large bodies of water. Evaporite deposits are dominated by such minerals as halite (sodium chloride, or salt), gypsum (calcium sulfate), and lime (calcium carbonate).

Extrusive a genetic term that describes the eruptive origin of volcanic rocks; a synonym for "volcanic."

Fault a fracture in the earth's crust along which the blocks of rock on either side have been displaced relative to each other by tectonic forces.

Fenestral a textural term that is applied to rocks that have openings or pockets that may or may not be filled with secondary minerals.

Flood basalt dark-colored volcanic rocks that result from the eruption of fluid, low-silica magma. The low viscosity of such magma allows it to flow over vast areas, as a flood, before it cools and hardens.

Flysch a German term for thinly bedded shale and impure sandstone, comingled with conglomerate and coarse sandstone, that accumulates in a marine basin adjacent to an area experiencing uplift.

Foredeep a deep, troughlike depression on the sea floor immediately bordering an orogenic belt or other land mass.

Gabbro a crystalline igneous rock similar to granite in origin but dominated by dark-colored minerals such as pyroxene, calcium feldspar, and olivine.

Gniess a crystalline metamorphic rock, typically with alternating bands of dark- and light-colored minerals, that results from alteration under extreme temperature and pressure conditions.

Gondwana a late Paleozoic supercontinent of the Southern Hemisphere, consisting of the central regions of modern Antarctica, Africa, Australia, South America, and India.

Graded bedding a style of layering in sediment and sedimentary rocks in which each individual layer is defined by a progressive decrease in particle size from bottom to top.

Granite a light-colored crystalline igneous rock dominated by the minerals quartz, potassium-rich feldspar, and muscovite mica.

Graptolite a fossil of a primitive colonial marine organism thought to be related to the corals, echinoderms, or chordates.

Greenstone a dense metamorphic rock composed predominately of greenish-colored minerals (such as epidote, chlorite, or actinolite) that forms through the alteration of igneous rocks such as basalt and gabbro.

Hadean the first formal eon of the geological time scale encompassing the period from the origin of the earth about 4.6 billion years ago to the beginning of the Archean Eon, around 4.0 billion years ago.

Hirnantian the last age of the Ordovician Period, spanning the interval from approximately 440–438 million years ago.

Horst-and-Graben a term describing a landscape characterized by alternating valleys and mountains that result from widespread normal faulting.

Hydrothermal of, or pertaining to, hot water, either underground or as thermal springs on the suface.

Hypersaline excessively saline or salty; possessing a salinity substantially greater than normal; seawater.

Ichthyosaur extinct marine reptiles of the Mesozoic Era that bore a superficial resemblance to modern dolphins and porpoises.

Ignimbrite a sequence of volcanic rocks that results from the cooling of hot ash explosively discharged from an erupting volcano as a *nuee ardente*. Ignimbrite sequences include a variety of specified types of volcanic rocks, such as crystal tuff, lithic tuff, and vitrophyres.

Intraformational conglomerate a coarse sedimentary rock consisting of limestone fragments and pebbles cemented together by nongranular limestone.

Intrusive a term describing the underground origin of granitic igneous rocks; a synonym for *plutonic*.

Jasperoid a dense, brittle, and commonly fractured rock in which silica has replaced carbonate minerals in rocks such as limestone and dolomite. Jasperoid in the Great Basin is most common along fault planes that cut through Paleozoic strata.

Laurasia an ancient supercontinent of the Northern Hemisphere that included North America, Greenland, and Eurasia. Laurasia resulted from the initial fragmentation of Pangaea about 250 million years ago, and was the counterpart of *Gondwana* in the Southern Hemisphere. By the end of the Mesozoic Era, Laurasia had broken apart into the predecessors of the modern northern continents.

Listric fault a normal fault characterized by a curved, concave-up plane that changes from steeply inclined at the surface to nearly horizontal in the subsurface.

Lithic tuff a pyroclastic volcanic rock consisting primarily of consolidated volcanic ash particles and larger rock fragments.

Lithosphere the outer rocky shell of the earth, including all of the crust and a portion of the rigid upper mantle, that is supported by the soft underlying asthenosphere.

Lycopods a general term for primitive treelike plants belonging to the Class Lycopsida, characterized by the spiral arrangement of scalelike leaves and rootlets.

Maar a broad pit or crater resulting from the explosive discharge of steam generated when rising magma comes into contact with groundwater.

Mafic a term describing the composition of igneous rocks dominated by dark-colored, iron- and magnesium-rich silicate minerals such as biotite, hornblende, or olivine.

Magma molten rock generated under high pressure and temperature conditions under the earth's surface.

Magmatic, as in *magmatic arc* or *magmatic history*, refers to any process or phenomena involving magma.

Metamorphism the process of chemical and physical modification of rocks by the combined effects of heat, pressure, and reactive fluids or vapors in the subsurface earth.

Metavolcanic a term applied to volcanic rocks that become metamorphosed through exposure to heat, pressure, or chemically active fluids and vapors. Metavolcanic rocks in the Great Basin are common hard, brittle, compact, and dark-colored.

Monzonite a light-colored plutonic igneous rock similar to granite but with less quartz and a different dominant feldspar mineral.

Mylonite a platy metamorphic rock, usually streaked or banded, that forms where moving rock masses grind or shear against each other.

Nena an ancient (Archean) continent that comprised the core of North America and consisted of older basement blocks assembled through accretionary tectonic events.

Oblique slip any movement along a fault that includes both vertical and lateral displacement of adjacent blocks.

Ostracoderms a varied group of extinct jawless fishlike animals related to the modern lampreys and hagfish.

Oreodont a group of extinct ungulates that bore crude similarities to modern pigs, sheep, and goats.

Orogenic of, or pertaining to, an orogeny.

Orogeny an episode of mountain building; a geological disturbance that results in the development of a mountain system.

Pangaea a supercontinent that existed about 250 million years ago and encompassed nearly all of the continental crust and land surface on the earth. The term was first applied in the early 1900s and is derived from Latin roots meaning "all earth."

Pegmatite an igneous rock consisting of exceptionally large crystals of silicate minerals. Roughly similar to granite in composition, pegmatite commonly contains unusual minerals such as lepidolite, tourmaline, and spodumene.

Phaneritic a textural term that is applied to igneous rocks consisting of large (easily visible to the unaided eye) and completely intergrown mineral crystals. Phaneritic rocks are *plutonic* or *intrusive* in origin.

Placoderms a group of primitive and heavily armored jawed fish (Class Placodermi) that were particularly abundant in Devonian time.

Playa a flat surface marking the lowest portion of a desert basin. Derived from a Spanish term for "beach" or "shore," playas commonly have salt-encrusted surfaces covering fine-grained sediments deposited during infrequent floods.

Playa lake a temporary desert lake, usually relatively shallow, that covers the surface of a playa following heavy rainfall.

Pluton any mass of magma that cools underground to form a solid body of intrusive igneous rock embedded in the surrounding host rock.

Plutonic a genetic term that describes the underground origin of rocks, such as granite, that form from the slow, prolonged cooling of magma.

Porphyry a general term for an igneous rock that contains distinctly large crystals embedded in a mass of smaller crystals.

Porphyry copper deposit a mass of intrusive igneous rock (a porphyry) that contains enough copper-bearing minerals (usually sulfides) to allow profitable extraction of the metal.

Precambrian an informal term designating all time preceding the Phanerozoic Eon, and encompassing the Hadean, Archean, and Proterozoic Eons.

Proterozoic the formal eon of geological time that spans the interval from about 2.5 to 0.54 billion years ago; the eon of "primitive life."

Pyroclastic a textural term that describes any volcanic rock, such as tuff or volcanic breccia, that is composed dominantly of rock particles or fragments.

Quartzite a hard, compact rock consisting of sand-sized quartz grains firmly cemented and compacted into a durable mass. Quartzites may result from metamorphism of sandstone (**meta-quartzite**) or from the thorough cementation of sand grains in the sedimentary process (**ortho-quartzite**).

Resurgent dome a low dome of volcanic rock, built through the effusion of viscous silicic magma, that develops in a caldera following an explosive eruption.

Rhipidistians a group of lobe-finned fish thought to be ancestral to the first terrestrial vertebrates (amphibians).

Rhyolite a quartz-rich, generally light-colored volcanic rock; the extrusive equivalent of granite.

Rodinia an ancient supercontinent that, during Proterozoic time, contained virtually all the world's land masses.

Roof pendants an inclusion of older rock, generally metamorphosed, into younger plutonic igneous rock. Roof pendants are most abundant near the margins of a batholith, where fragments of the enclosing rock were engulfed by the magma while it was still fluid.

Sacroptyergians the "lobe-finned" fish (of the Subclass Sarcopterygii, Class Osteichthyes), characterized by an fleshy lobe that connects the fins with the body wall.

Scarp a low cliff or embankment resulting from either faulting or erosion; an abbreviated form of the term *escarpment,* that generally applies to higher and more prominent cliffs.

Series the rocks formed during a specific *epoch* of geologic time; a subdivision of a *system.*

Silica a general term for any form of silicon dioxide, SiO_2, including crystalline quartz, microcrystalline chalcedony, noncrystalline forms such as opal, and other minerals.

Sinistral indicates any movement, orientation, or position to the left.

Skarn a complex assemblage of rocks derived from the metamorphism of limestone and/or dolomite. Skarns commonly contain calcium-rich silicate minerals such as garnet, epidote, and wollastonite, along with a variety of valuable ore minerals.

Sonomia a name for an ancient microcontinent that collided with the western margin of North America during the Permian and Triassic Periods.

Stage a sequence of rock layers that formed during a particular *age* of geological time; a subdivision of a *series.*

Still stand as applied to pluvial lakes, a term designating a time of relative stability in lake level; sometimes simply referred to as a "stand."

Stock a pluton with a surface exposure less than 40 square miles (100 square kilometers); similar to a *batholith,* but smaller.

Stratigraphy the study of layered sequences of rock.

Stratotype a sequence of rock layers designated as the standard of reference for a stratigraphic unit such as a system, series, or stage.

Subduction the process by which a lithospheric plate is driven, pulled, or sinks into the deeper portions of the mantle; the sinking of one slab of the lithosphere beneath another.

Supergene enrichment the process of enriching an ore deposit through the downward leaching of minerals near the surface to a zone of reprecipitation in the subsurface; supergene enrichment leads to the development of a rich underground ore body within, or from, an original mineral deposit of lower grade.

Synorogenic any process or phenomena that is contemporaneous with *orogeny,* or mountain building.

System in stratigraphic usage, a sequence of rock layers that formed during a specific period of geologic time.

Tectonism any movement of the crust that results from the interaction of lithospheric plates, including uplift, subsidence, and lateral motion.

Terrace a narrow, relatively flat, surface bounded by steeper descending and ascending slopes. A terrace is narrower, and more elongated, than a plain; is commonly situated as a "notch" or "bench" on hillsides; and may result from several different mechanisms of slope erosion.

Thrust fault, or **"Thrust"** a low-angle fracture along which one block of rock is forced over an underlying mass during compressive deformation.

Titanotheres an extinct group (Family Brontotheriidae) of large odd-toed ungulates with rhinoceros-like bodies and multiple blunt horns. Titanotheres are also known as *brontotheres,* and are distant relatives of the modern rhinos, tapirs, and horses.

Transpression a type of stress that represents a combination of *compression* (or "squeezing") and lateral, or *transcurrent,* motion.

Triple Junction the common meeting point of three different lithospheric plates.

Tsunami a large sea wave produced by any sudden displacement of mass on the sea floor. Commonly (and erroneously) known as "tidal waves," tsunamis may be generated by volcanic eruptions, earthquakes, asteroid impacts, or any other event that abruptly displaces water.

Tufa a term applied to massive, bulbous, or spongy calcium carbonate deposited from lake water, around springs and seeps, and along the banks of mineralized streams; not to be confused with tuff (see below).

Tuff a general term for volcanic rocks consisting primarily of consolidated volcanic ash.

Turbidite a term applied to sedimentary rocks representing sediment transported during, and deposited from, a dense flowing mixture of sediment and water known as a *turbidity current* or submarine mudflow.

Unconformity a gap, or break, in the rock record signifying an increment of time during which either no rocks formed or the rocks were subsequently removed by erosion.

Ventifact a general term for any rock abraded, polished, and faceted by wind-blown rock particles. Ventifacts most commonly have several smooth, flat surfaces that meet to form roughly triangular facets.

Vitrophyre a special type of pyroclastic volcanic rock in which the constituent particles are thoroughly fused into a glasslike material. Vitrophyres are also known as **vitric tuffs**.

Volcanic breccia a pyroclastic igneous rock consisting of particles and blocks larger than 64 mm (2.5 inches); essentially, a consolidated mass of volcanic rubble.

Wallrock screens relatively thin partitions, generally of metamorphic rock, that separate adjacent plutons in a composite batholith. Wallrock screens represent drapes of rock that were not engulfed by ascending bodies of magma.

Xerophytic a term that is applied to plants adapted to dry soils and arid climates with infrequent and/or irregular rainfall.

References

Chapter One: The Fallen Land

Christiansen, R.L., and Yeats, R.S. 1992. Post-Laramide geology of the U.S. Cordilleran region. In *The Cordilleran Orogen: Contiminous U.S.,* Burchfiel, B.C., Lipman, P.W., and Zoback, M.L., eds., 261–406. *The Geology of North America,* v. G-3. Boulder, CO: Geological Society of America.

Cline, G.G. 1988. *Exploring the Great Basin.* Reno: University of Nevada Press.

Eaton, G.P. 1982. The Basin and Range province: Origin and tectonic significance. *Annual Review of Earth and Planetary Sciences* 8:409–40.

Fenneman, N.M. 1928. Physiographic divisions of the United States. *Annals of the Association of American Geographers* 18, no. 4:263–353.

Frémont, J.C. 1845. *Report of the exploring expedition to the Rocky Mountains in the year 1842, and to Oregon and California in the years 1843–1844.* Washington, D.C.: Goles and Seaton.

Grayson, D.K. 1993. *The desert's past: A natural prehistory of the Great Basin.* Washington and London: Smithsonian Institution Press.

Houghton, S.G. 1976. *A trace of desert waters.* Reno: University of Nevada Press.

Hunt, C.B. 1974. *Natural regions of the United States and Canada.* San Francisco: W.H. Freeman and Company.

McPhee, John. 1980. *Basin and Range.* New York: Farrar, Straus and Giroux.

Nolan, T.B. 1943. The Basin and Range province in Utah, Nevada, and California. *U.S. Geological Survey Professional Paper* 197-D:141–96.

Raymond, C. Elizabeth. 1989. Sense of place in the Great Basin. In *East of Eden, west of Zion: Essays on Nevada,* Shepperson, W.S., ed., 17–29. Reno: University of Nevada Press.

Smith, R.B., Nagy, W.C., Julander, K.A., Viveiros, J.J., Barker, C.A., and Gants, D.G. 1989. Geophysical and tectonic framework of the eastern Basin and Range–Colorado Plateau–Rocky Mountain transition. In *Geophysical framework of the continental United States,* Pakiser, L.C., and Mooney, W.D., eds., 205–33. Memoir 172. Boulder, CO: Geological Society of America.

Stokes, W.L. 1988. *Geology of Utah.* Utah Museum of Natural History Occasional Paper 6. Salt Lake City: Utah Museum of Natural History/Utah Geological and Mineral Survey.

Wolfe, Jack A., Schorn, Howard E., Forest, Chris E., and Molnar, P. 1997. Paleobotanical evidence for high altitudes in Nevada during the Miocene. *Science* 276:1672–75.

Chapter Two: The Breaking of a Continent

Bennett, V.C., and DePaolo, D.J. 1987. Proterozoic crustal history of the western United States as determined by neodymium isotopic mapping. *Geological Society of America Bulletin* 99:674–85.

Bond, G.D., and Kominz, M.A. 1984. Construction of tectonic subsidence curves for the early Paleozoic miogeocline, southern Canadian Rocky Mountains: Implications for subsidence mechanisms, age of breakup, and crustal thinning. *Geological Society of America Bulletin* 95:155–73.

Bond, G.D., Nickerson, P.A., Kominz, M.A. 1984. Breakup of a supercontinent between 625 Ma and 555 Ma: New evidence and implications for continental histories. *Earth and Planetary Science Letters* 70:325–45.

Christie-Blick, N. 1997. Neoproterozoic sedimentation and tectonics in west-central Utah. *Brigham Young University Geology Studies* 42, part 1:1–30.

Condie, K.C. 1982. Plate-tectonics model for Proterozoic continental accretion in the southwestern United States. *Geology* 10:37–42.

———. 1992. Proterozoic terranes and continental accretion in southwestern North America. In *Proterozoic Crustal Evolution*, Condie, K.C., ed., 447–80. Amsterdam: Elsevier Publishing Company.

Condie, K.C., and Chomiak, B. 1996. Continental accretion: Contrasting Mesozoic and early Proterozoic tectonic regimes in North America. *Tectonophysics* 265:101–26.

Crittenden, M.D., Jr., Schaeffer, F.E., Trimble, D.E., and Woodward, L.E. 1971. Nomenclature and correlations of some upper Precambrian and basal Cambrian sequences in western Utah and southeast Idaho. *Geological Society of America Bulletin* 82:581–602.

Dalziel, I.W.D. 1991. Pacific margins of Laurentia and East Antarctica–Australia as a conjugate rift pair: Evidence and implications for an Eocambrian supercontinent. *Geology* 19:598–601.

Ehlers, T.A., Chan, M.A., and Link, P.K. 1997. Proterozoic tidal, glacial, and fluvial sedimentation in Big Cottonwood Canyon, Utah. *Brigham Young University Geology Studies* 42, part 1:31–58.

Farmer, G.L., and Ball, T.T. 1997. Sources of middle Proterozoic to early Cambrian siliciclastic sedimentary rocks in the Great Basin: A Nd isotope study. *Geological Society of America Bulletin* 109 (9): 1193–1205.

Frost, C.D., Frost, B.R., Chamberlain, K.R., and Hulsebosch, T.P. 1998. The late Archean history of the Wyoming province as recorded by granitic magmatism in the Wind River Range, Wyoming. *Precambrian Research* 89:145–73.

Hedge, C.E., Houston, R.S., Tweto, O.L., Peterman, Z.E., Harrison, J.E., and Reid, R.R. 1986. The Precambrian of the Rocky Mountain region. *U.S. Geological Survey Professional Paper* 2141-D.

Karlstrom, K.E., Ahäll, K.-I., Harlan, S.S., Williams, M.L., McLelland, J., and Geissman, J.W. 2001. Long-lived (1.8–1.0 Ga) convergent orogen in southern Laurentia, its extensions to Australia and Baltica, and implication for refining Rodinia. *Precambrian Research* 111:5–30.

Karlstom, K.E., Williams, M.L., McClelland, J., Geissman, J.W., Ahäll, K.-I. 1999. Refining Rodinia: Geologic evidence for the Australian–western U.S. connection in the Proterozoic. *GSA Today* 9 (10): 1–7.

Karlstrom, K.E., et al. 2000. Chuar group of the Grand Canyon: Record of breakup of Rodinia, associated change in the global carbon cycle, and ecosystem expansion by 740 Ma. *Geology* 28 (7): 619–22.

Levy, M., and Christie-Blick, N. 1991. Late Proterozoic paleogeography of the eastern Great Basin. In *Paleozoic paleogeography of the western United States–II*, Cooper, J.D., and Stevens, C.H., eds., 371–86. Pacific Section, Society of Economic Paleontologists and Mineralogists (SEPM), v. 67.

Lush, A.P., and Snoke, A.W. 1987. Allochthonous Archean basement in the northern East Humboldt Range, Nevada. *Geological Society of America Abstracts with Programs* 19:752.

Miller, J.M.G. 1985. Glacial and syntectonic sedimentation: The upper Proterozoic Kingston Peak Formation, southern Panamint

Range, eastern California. *Geological Society of America Bulletin* 96:1537–53.

Moores, E. M. 1991. Southwest U.S.–Antarctica (SWEAT) connection: A hypothesis. *Geology* 19:425–28.

Ramo, O.T., and Calzia, J.P. 1998. Nd isotopic composition of cratonic rocks in the southern Death Valley region: Evidence for a substantial Archean source component in Mojavia. *Geology* 26 (10): 891–94.

Rogers, J.J. 1996. A history of continents in the past three billion years. *Journal of Geology* 104:91–107.

Sears, J.W., and Price, R.A. 1978. The Siberian connection: A case for Precambrian separation of the North American and Siberian cratons. *Geology* 6:267–70.

———. 2000. New look at the Siberian connection: No SWEAT. *Geology* 28 (5): 423–26.

Sears, J.W., Graff, P.J., and Holden, G.S. 1982. Tectonic evolution of lower Proterozoic rocks, Uinta Mountains, Utah and Colorado. *Geological Society of America Bulletin* 93:990–97.

Snoke, A.W., Howard, K.A., McGrew, A.J., Burton, B.R., Barnes, C.G., Peters, M.T., and Wright, J.E. 1997. The grand tour of the Ruby-East Humboldt metamorphic core complex, northeastern Nevada: Part 1—introduction and road log. *Brigham Young University Geology Studies* 42, part 1: 225–69.

Stewart, J.H. 1972. Initial deposits in the Cordilleran Geosyncline: Evidence of a late Precambrian (<850 m.y.) continental separation. *Geological Society of America Bulletin* 83:1345–60.

Stewart, J.H., and Gehrels, G.E., Barth A.P., Link, P.K., Christie-Blick, N., and Wrucke, C.T. 2001. Detrital zircon provenance of Mesoproterozoic to Cambrian arenites in the western United States and northwestern Mexico. *Geological Society of America Bulletin* 113 (10): 1343–56.

Unrug, R. 1997. Rodinia to Gondwana: The geodynamic map of Gondwana supercontinent assembly. *GSA Today* 7 (1): 1–6.

Wooden, J.L., and Miller, D.M. 1990. Chronologic and isotopic framework for early Proterozoic crustal evolution in the eastern Mojave Desert region, SE California. *Journal of Geophysical Research* 95:20133–46.

Wright, L.A., Troxel, B.W., Williams, E.G., Roberts, M.T., and Diehl, P.E. 1976. Precambrian sedimentary environments on the Death Valley region, eastern California. In *Geologic features of Death Valley*, Troxel, B.W., and Wright, L.A., eds., 7–15. *California Division of Mines and Geology Special Report* 106.

Chapter Three: Water Bugs in Stone Houses

Armstrong, R.L. 1968. The Cordilleran miogeosyncline in Nevada and Utah. *Utah Geological and Mineralogical Survey Bulletin* 78:1–58.

Brady, M.J., and Koepnick, R.B. 1979. A middle Cambrian platform-to-basin transition, House Range, west-central Utah. *Brigham Young University Geology Studies* 26:1–7.

Brady, M.J., and Rowell, A.J. 1976. An upper Cambrian subtidal blanket carbonate, eastern Great Basin. *Brigham Young University Geology Studies* 23:153–64.

Brett, C.E., Baird, G.C., and Speyer, S.E. 1997. *Fossil Lagerstätten*: Stratigraphic record of paleontological and taphonomic events. In *Paleontological events: Stratigraphic, ecological, and evolutionary implications*, Brett, C.E., and Baird, G.C., eds., 1–40. New York: Columbia University Press.

Butterfield, N.J. 1995. Secular distribution of Burgess-Shale-type preservation. *Lethaia* 28:1–13.

Church, S.B., Rigby, J.K., Gunther, L.F., and Gunther, V.G. 1999. A large *Protospongia hicksi* Hinde, 1887, from the middle Cambrian Spence Shale of southeastern Idaho. *Brigham Young University Geology Studies* 44:17–25.

Eddy, J.D., and McCollum, L.B. 1998. Early middle Cambrian *Albertella* biozone trilobites of the Pioche shale, southeastern Nevada. *Journal of Paleontology* 72 (5): 864–89.

Foster, J.R. 1994. A note on depositional environments of the lower-middle Cambrian Cadiz formation, Marble Mountains, California. *Mountain Geologist* (Rocky Mountain Association of Geologists) 31 (1): 29–36.

Gunthur, L.F., and Gunthur, V.G. 1981. Some middle Cambrian fossils of Utah. *Brigham Young University Geology Studies* 28:1–79.

Gevitsman, D.A., and Mount, J.F. 1986. Paleoenvironments of an earliest Cambrian (Tommotian) shelly fauna in the southwestern Great Basin. *Journal of Sedimentary Petrology* 56:412–21.

Hintze, L.F. 1988. *Geologic history of Utah.* Provo, UT: Brigham Young University Geology Studies, Special Publication 7.

Hintze, L.F., and Palmer, A.R. 1976. Upper Cambrian Orr formation: Its subdivisions and correlatives in western Utah. *U.S. Geological Survey Bulletin* 1405-G:G1–G25.

Hintze, L.F., and Robison, R.A. 1975. Middle Cambrian stratigraphy of the House, Wah Wah, and adjacent ranges in western Utah. *Geological Society of America Bulletin* 86:881–91.

Li, Xing, and Droser, M.L. 1997. Nature and distribution of Cambrian shell concentrations: Evidence from the Basin and Range province of the western United States (California, Nevada, and Utah). *Palaios* 12:111–26.

McCollum, L.B., and Miller, D.M. 1991. Cambrian stratigraphy of the Wendover area, Utah and Nevada. *U.S. Geological Survey Bulletin* 1948.

Mount, J.F., and Bergk, K.J. 1998. Depositional sequence stratigraphy of lower Cambrian grand cycles, southern Great Basin, U.S.A. In *Integrated earth and environmental evolution of the southwestern United States*, Ernst, W.G., and Nelson, C.A., eds., 180–202. Columbia, MD: Bellwether Publishing, Ltd. (for the Geological Society of America).

Nelson, C.A. 1978. Late Precambrian–early Cambrian stratigraphic and faunal succession of eastern California and the Precambrian boundary. *Geological Magazine* 115 (2): 121–26.

———. 1980. *Guidebook to the eastern Sierra Nevada, Owens Valley, White-Inyo Range.* Los Angeles: Department of Earth and Space Sciences, UCLA.

Palmer, A.R. 1971. The Cambrian of the Great Basin and adjacent areas, western United States. In *Cambrian of the new world*, Holland, C.H., ed., 1–78. New York: Wiley-Interscience.

Palmer, A.R., and Halley, R.B. 1979. Physical stratigraphy and trilobite biostratigraphy of the Carrara formation (lower and middle Cambrian) in the southern Great Basin. *U.S. Geological Survey Professional Paper* 1047.

Rees, M.N. 1986. A fault-controlled trough through a carbonate platform: The middle Cambrian House Range embayment. *Geological Society of America Bulletin* 97:1054–69.

Rigby, J.K. 1980. The new middle Cambrian sponge *Vauxia magna* from the Spence Shale of northern Utah and taxonomic position of the Vauxiidae. *Journal of Paleontology* 54 (1): 234–40.

———. 1983. Sponges of the middle Cambrian Marjum Limestone from the House Range and Drum Mountains of western Millard County, Utah. *Journal of Paleontology* 57 (2): 240–70.

Robison, R.A. 1971. Additional middle Cambrian trilobites from the Wheeler Shale of Utah. *Journal of Paleontology* 45:796–804.

———. 1985. Affinities of *Aysheaia* (Onychophora), with description of a new Cambrian species. *Journal of Paleontology* 59 (1): 226–35.

———. 1976. Middle Cambrian trilobite biostratigraphy of the Great Basin. *Brigham Young Geology Studies* 23, part 2:93–109.

Robison, R.A., and Richards, B.C. 1981. Larger bivalve arthropods from the middle Cambrian of Utah. *University of Kansas Paleontological Contributions*, Paper 106.

Robison, R.A., and Rowell, A.J., eds. 1976. *Paleontology and depositional environments: Cambrian of western North America. Brigham Young University Geology Studies* 23, part 2.

Salak, M., and Lescinsky, H.L. 1999. *Spygoria zappania* new genus and species, a *Cloudina*-like biohermal metazoan from the lower Cambrian of central Nevada. *Journal of Paleontology* 73 (4): 571–76.

Saltzman, M.R., Runnegar, B., and Lohmann, K.C. 1998. Carbon isotope stratigraphy of upper Cambrian (Steptoean stage) sequences of the eastern Great Basin: Record of a global oceanographic event. *Geological Society of America Bulletin* 110 (3): 285–97.

Schweickert, R.A., and Lahren, M.M. 1991. Age and tectonic significance of metamorphic rocks along the axis of the Sierra Nevada batholith: A critical reappraisal. In *Paleozoic paleogeography of the western United States–II*, Cooper, J.D., and Stevens, C.H., eds., 653–76. Pacific Section, Society of Economic Paleontologists and Mineralogists (SEPM), v. 67.

Seilacher, A., Reif, W.-E., and Westphal, F. 1985. Sedimentological, ecological, and temporal patterns of fossil Lagerstätten. *Philosophical Transactions of the Royal Society of London* B311:5–23.

Stewart, J.H., and Poole, F.G. 1974. Lower Paleozoic and uppermost Precambrian Cordilleran miogeosyncline, Great Basin, western United States. In *Tectonics and sedimentation*, Dickinson, W.R., ed., 28–58. Society of Economic Paleontologists and Mineralogists Special Publication 22.

Sumrall, C.D., and Sprinkle, J. 1999. *Ponticulocarpus*, a new cornute-grade stylophoran from the middle Cambrian Spence Shale of Utah. *Journal of Paleontology* 73 (5): 886–91.

Taylor, M.E. 1976. Indigenous and redeposited trilobites from late Cambrian basinal environments of central Nevada. *Journal of Paleontology* 50:668–70.

Taylor, M.E., and Robison, R.A. 1976. Trilobites in Utah folklore. In *Paleontology and depositional environments: Cambrian of western North America*, Robison, R.A., and Rowell, A.J., eds., 1–5. *Brigham Young University Geology Studies* 23, part 2.

Chapter Four: The Sun Always Shines on the Ordovician

Berndan, J.M. 1988. Middle Ordovician (Whiterockian) palaeocopid and podpcopid ostracods from the Ibex area, Millard County, western Utah. In *Contributions to Paleozoic paleontology and stratigraphy in honor of Rousseau H. Flower*, Wolberg, D.L., comp., 273–301. New Mexico Bureau of Mines and Mineral Resources Memoir 44.

Braithwaite, L.F. 1976. *Graptolites from the lower Ordovician Pogonip Group of western Utah.* Geological Society of America Special Paper 166.

Budge, D.R., and Sheehan, P.M. 1980. *The upper Ordovician through middle Silurian of the eastern Great Basin: Part 1—Introduction: Historical perspective and stratigraphic synthesis.* Milwaukee Public Museum Contributions in Biology and Geology 28.

Church, S.B. 1991. A new lower Ordovician species of *Calathium*, and skeletal structure of western Utah calathids. *Journal of Paleontology* 65 (4): 602–10.

———. 1974. Lower Ordovician patch reefs in western Utah. *Brigham Young University Geology Studies* 21, part 3:41–62.

Cooper, J.D., and Edwards, J.C. 1991. Cambro-Ordovician craton-margin carbonate section, southern Great Basin: A sequence-stratigraphic perspective. In *Paleozoic paleogeography of the western United States–II*, Cooper, J.D., and Stevens, C.H., eds., 237–52. Pacific Section, Society of Economic Paleontologists and Mineralogists (SEPM), v. 67.

Datillo, B.F. 1993. The lower Ordovician Fillmore Formation of western Utah: Storm-dominated sedimentation on a passive margin. *Brigham Young University Geology Studies* 39:71–99.

Denison, R.H. 1952. Early Devonian fishes from Utah, part I, Osteostaci, and part II, Heterostraci. *Fieldiana Geology* 11 (6): 264–55.

———. 1958. Early Devonian fishes from Utah, part III, Arthrodira. *Fieldiana Geology* 11 (9): 459–551.

Dunne, G.C., and Suczek, D.A. 1991. Early Paleozoic eugeoclinal strata in the Kern Plateau pendants, southern Sierra Nevada, California. In *Paleozoic paleogeography of the western United States–II*, Cooper, J.D., and Stevens, C.H., eds., 677–92. Pacific Section, Society of Economic Paleontologists and Mineralogists (SEPM), v. 67.

Finney, S.C., and Perry, B.D. 1991. Depositional setting and paleogeography of Ordovician Vinini Formation, central Nevada. In *Paleozoic paleogeography of the western United States–II*, Cooper, J.D., and Stevens, C.H., eds., 747–66. Pacific Section, Society of Economic Paleontologists and Mineralogists (SEPM), v. 67.

Finney, S.C., Berry, W.B.N., Cooper, J.D., Ripperdan, R.L., Sweet, W.C., Jacobsen, S.R., Soufiane, A., Achab, A., and Noble, P.J. 1999. Late Ordovician mass extinction: A new perspective from stratigraphic sections in central Nevada. *Geology* 27 (3): 215–18.

Fortney, R.A., and Droser, M.L. 1999. Trilobites from the base of the type Whiterockian (middle Ordovician) in Nevada. *Journal of Paleontology* 73 (2): 182–201.

Gil, A.V. 1988. Whiterock (lower middle Ordovician) cephalopod fauna from the Ibex area, Millard County, western Utah. In *Contributions to Paleozoic paleontology and stratigraphy in honor of Rousseau H. Flower*, Wolberg, D.L., comp., 27–59. New Mexico Bureau of Mines and Mineral Resources Memoir 44.

Gregory, J.T., Morgan, T.G., and Reed, J.W. 1977. Devonian fishes in central Nevada. In *Western North America Devonian*, Murphy, M.A., Berry, W.B.N., and Sandberg, C.A., eds., 112–19. Riverside: University of California, Riverside (Campus Museum Contribution 4).

Harris, M.T., and Sheehan, P.M. 1997. Carbonate sequences and fossil communities from the upper Ordovician–lower Silurian of the eastern Great Basin. *Brigham Young University Geology Studies* 42, part 1:105–28.

Hintze, L.F. 1987. Exceptionally fossiliferous lower Ordovician strata in the Ibex area, western Millard County, Utah. In *Geological Society of America Centennial Field Guide*, v. 2:261–64.

———. 1951. *Lower Ordovician detailed stratigraphic sections for western Utah*. Utah Geological and Mineral Survey Bulletin 39.

———. 1960. Ordovician of the Utah-Nevada Great Basin. *Intermountain Association of Geologists 11th Annual Field Conference*, 59–62.

———. 1979. Preliminary zonation of lower Ordovician of western Utah by various taxa. *Brigham Young University Geology Studies* 35, part 2:13–19.

Jensen, R.G. 1967. Ordovician brachiopods from the Pogonip Group of Millard County, western Utah. *Brigham Young University Geology Studies* 14:67–100.

Johns, R.A. 1994. Ordovician lithistid sponges of the Great Basin. *Nevada Bureau of Mines and Geology Open-File Report* 94-1.

Johnson, J.G., and Klapper, G. 1990. Lower and middle Devonian brachiopod-dominated communities of Nevada, and their position in a biofacies-province-realm model. *Journal of Paleontology* 64 (6): 902–41.

Johnson, J.G., and Murphy, M.A. 1984. Time-rock model for Siluro-Devonian continental shelf, western United States. *Geological Society of America Bulletin* 95:1349–59.

Kim, J.C., and Lee, Y.I. 1995. Flat-pebble conglomerate: A characteristic lithology of upper Cambrian and lower Ordovician shallow-water carbonate sequences. In *Ordovician odyssey: Short papers for the seventh international symposium on the Ordovician system*, Cooper, J.D., Droser, M.L., and Finney, S.L., eds., 371–74. Pacific Section, Society for Sedimentary Geology.

Lane, N.G. 1970. Lower and middle Ordovician crinoids from west-central Utah. *Brigham Young University Geology Studies* 17:3–17.

Li, X., and Droser, M.L. 1999. Lower and middle Ordovician shell beds from the basin and range province of the western United States (California, Nevada, and Utah). *Palaios* 14:215–33.

Matti, J.C., and McKee, E.H. 1977. Silurian and lower Devonian paleogeography of the outer continental shelf of the Cordilleran miogeosyncline, east-central Nevada. In *Paleozoic paleogeography of the western United States*, Stewart, J.H., Stevens, C.H., and Fritsche, A.E., eds., 181–215. Pacific Coast Paleogeography Symposium 1, Pacific Basin Section, Society of Economic Paleontologists and Mineralogists.

Matti, J.C., Murphy, M.A., and Finney, S.C. 1975. Silurian and lower Devonian basin and basin-slope limestones, Copenhagen Canyon, Nevada. *Geological Society of America Special Paper* 159.

McDowell, R.R. 1995. Depositional history of a middle Ordovician, mixed carbonate-clastic unit—the Kanosh Formation, eastern Great Basin, U.S.A. In *Ordovician odyssey: Short papers for the seventh international symposium on the Ordovician system*, Cooper, J.D., Droser, M.L., and Finney, S.C., eds., 367–70. Pacific Section, Society for Sedimentary Geology.

Miller, J.F., and Taylor, M.E. 1995. Biostratigraphic significance of *Iapetognathus* (Conodonta) and *Jujuyaspis* (Trilobita) in the House Limestone, Ibex area, Utah. In *Ordovician odyssey: Short papers for the seventh international symposium on the Ordovician system*, Cooper, J.D., Droser, M.L., and Finney, S.C., eds., 109–12. Pacific Section, Society for Sedimentary Geology.

Miller, R.H., and Zilinsky, G.H. 1981. Lower Ordovician through lower Devonian cratonic margin rocks of the southern Great Basin. *Geological Society of America Bulletin* 92:255–61.

Murphy, M.A., and Gronberg, E.C. 1970. Stratigraphy of the lower Nevada group (Devonian) north and west of Eureka, Nevada. *Geological Society of America Bulletin* 81:127–36.

Murphy, M.A., Morgan, T.G., and Dineley, D.L. 1976. *Astrolepis* sp. from the upper Devonian of central Nevada. *Journal of Paleontology* 50 (3): 467–71.

Poole, F.G., Stewart, J.H., Palmer, A.R., Sandberg, C.A., Madrid, R.J., Ross, R.J., Jr., Hintze, L.F., Miller, M.M., and Wrecke, C.T. 1992. Latest Precambrian to latest Devonian time: Development of a continental margin. In *The Cordilleran Orogen: Conterminous U.S.*, Burchfiel, B.C., Lipman, P.W., and Zoback, M.L., eds., 9–56. *The Geology of North America*, v. G-3. Boulder, CO: Geological Society of America.

Reed, J.W. 1985. Devonian dipnoans from Red Hill, Nevada. *Journal of Paleontology* 59 (5): 1181–93.

Ross, R.J., Jr. 1972. Fossils from the Ordovician bioherm at Meikeljohn Peak, Nevada. *U.S. Geological Survey Professional Paper* 685.

———. 1964. Middle and lower Ordovician formations in southernmost Nevada and adjacent California. *U.S. Geological Survey Bulletin* 1180-C.

———. 1970. Ordovician brachiopods, trilobites, and stratigraphy in eastern and central Nevada. *U.S. Geological Survey Professional Paper* 639.

———. 1967. Some middle Ordovician brachiopods and trilobites from the basin ranges, western United States. *U.S. Geological Survey Professional Paper* 523D.

Ross, R.J., Jr., Ethington, R.L., and Mitchell, C.E. (appen.). 1991. Stratotype of the Ordovician Whiterock Series. *Palaios* 6:156–73.

Ross, R.J., Jr., Hintze, L.F., Ethington, R.L., Miller, J.F., Taylor, M.E., and Repetski, J.E. 1997. The Ibexian series in the North American Ordovician. *U.S. Geological Survey Professional Paper* 1579-A:1–50.

Ross, R.J., Jr., James, N.P., Hintze, L.F., and Ketner, K.B. 1991. Early middle Ordovician (Whiterock) paleogeography of basin ranges. In *Paleozoic paleogeography of the western United States–II*,

Cooper, J.D., and Stevens, C.H., eds., 39–50. Pacific Section, Society of Economic Paleontologists and Mineralogists, v. 67.

Sheehan, P.M. 1979. Silurian continental margin in northern Nevada and northwestern Utah. *University of Wyoming Contributions to Geology* 17 (1): 25–35.

———. 1990. Late Ordovician and Silurian paleogeography of the Great Basin. *University of Wyoming Contributions to Geology* 27:41–54.

Sheehan, P.M., and Boucot, A.J. 1991. Silurian paleogeography of the western United States. In *Paleozoic paleogeography of the western United States–II*, Cooper, J.D., and Stevens, C.H., eds., 51–82. Pacific Section, Society of Economic Paleontologists and Mineralogists, v. 67.

Sheehan, P.M., and Harris, M.T. 1997. Upper Ordovician-Silurian macrofossil biostratigraphy of the eastern Great Basin, Utah and Nevada. *U.S. Geological Survey Professional Paper* 1579-C:89–115.

Sweet, W.C., and Tolbert, Celeste M. 1997. An Ibexian (lower Ordovician) reference section in the southern Egan Range, Nevada, for a conodont-based chronostratigraphy. *U.S. Geological Survey Professional Paper* 1579-B:53–84.

Turner, Susan and Murphy, M.A. 1988. Early Devonian vertebrate microfossils from the Simpson Park Range, Eureka County, Nevada. *Journal of Paleontology* 62 (6): 959–64.

Webb, G.W. 1958. Middle Ordovician stratigraphy in eastern Nevada and western Utah. *American Association of Petroleum Geologists Bulletin* 42 (10): 2335–77.

Chapter Five: Mississippian Meyhem

Brew, D.A. 1971. Mississippian stratigraphy of the Diamond Peak area, Eureka County, Nevada. *U.S. Geological Survey Professional Paper* 661.

Bues, S.S., and Lane, N. Gary. 1969. Middle Pennsylvanian fossils from Indian Springs, Nevada. *Journal of Paleontology* 43 (4): 986–1000.

Burchfiel, B.C., and Davis, G.A. 1975. Nature and controls of Cordilleran orogenesis, western United States: Extensions of an earlier hypothesis. *American Journal of Science* 275-A:363–96.

———. 1972. Structural framework and evolution of the southern part of the Cordilleran orogen, western United States. *American Journal of Science* 272:97–118.

Carroll, R.L., Bybee, P., and Tidwell, W.D. 1991. The oldest microsaur (Amphibia). *Journal of Paleontology* 65 (2): 314–22.

Chamberlain, A.K. 1990. *Stigmaria*: Indicator of erosional surface of low sea level stands in the Mississippian Antler basin, Utah and Nevada. *American Association of Petroleum Geologists Bulletin* 74 (8): 1319.

Churkin, M., Jr. 1974. Paleozoic marginal ocean basin volcanic arc systems in the Cordilleran foldbelt. In *Modern and ancient geosynclinal sedimentation*, Dott, R.H., Jr., and R.H. Shaver, eds., 174–92. Society of Economic Paleontologists and Mineralogists Special Publication 19.

Dickinson, W.R. 1977. Paleozoic plate tectonics and the evolution of the Cordilleran continental margin. In *Paleozoic paleogeography of the western United States*, Stewart, J.H., Stevens, C.H., and Fritsche, A.E., eds., 137–55. Pacific Coast Paleogeography Symposium 1, Pacific Section, Society of Economic Paleontologists and Mineralogists (SEPM).

Goebel, K.A. 1991. Paleogeographic setting of late Devonian to early Mississippian transition from passive to collisional margin, Antler foreland, eastern Nevada and western Utah. In *Paleozoic paleogeography of the Western United States–II*, Cooper, J.D., and Stevens, C.H., eds., 401–18. Pacific Section, Society of Economic Paleontologists and Mineralogists (SEPM), v. 67.

Gutschick, R.C., and Rodriguez J. 1979. Biostratigraphy of the Pilot Shale (Devonian- Mississippian) and contemporaneous strata in Utah, Nevada, and Montana. *Brigham Young University Geology Studies* 26, part 1:37–63.

Gutschick, R.C., Sandberg, C.A., and Sando, W.J. 1980. Mississippian shelf margin and carbonate platform from Montana to Nevada. In *Paleozoic paleogeography of the west-central United States*, Fouche, T.D., and Magathan, E.R., eds., 111–28. Rocky Mountain Paleogeography Symposium 1, Rocky Mountain Section, Society of Economic Paleontologists and Mineralogists (SEPM).

Harbaugh, D.W., and Dickinson, W.R. 1981. Depositional facies of Mississippian clastics, Antler foreland basin, central Diamond Range, Nevada. *Journal of Sedimentary Petrology* 51:1223–34.

Hose, R.K. 1966. Devonian stratigraphy of the Confusion Range, west-central Utah. *U.S. Geological Survey Professional Paper* 550-B:B36–B41.

Hose, R.K., Wrucke, C.T., and Armstrong, A.K. 1979. Mixed Devonian and Mississippian conodont and foraminiferal faunas and their bearing on the Roberts Mountains Thrust, Nevada. *Geological Society of America Abstracts with Programs* 11:446.

Ingersoll, R.V. 1998. Phanerozoic tectonic evolution of central California and environs. In *Integrated earth and environmental evolution of the southwestern United States*, Ernst, W.G., and Nelson, C.A., eds., 349–64. Columbia, MD: Bellwether Publishing, Ltd. (for the Geological Society of America).

Johnson, J.G. 1972. Antler effect equals Haug effect. *Geological Society of America Bulletin* 83 (8): 2497–98.

Johnson, J.G., and Pendergast, A. 1981. Timing and mode of emplacement of the Roberts Mountains allochthon, Antler Orogeny. *Geological Society of America Bulletin* 92:648–58.

Johnson, J.G., Sandberg, C.A., and Poole, F.G. 1991. Devonian lithofacies of western United States. In *Paleozoic paleogeography of the western United States–II*, Cooper, J.D., and Stevens, C.H., eds., 83–105. Pacific Section, Society of Economic Paleontologists and Mineralogists, v. 67.

Kay, M., and Crawford, J.P. 1964. Paleozoic facies from the miogeosynclinal to the eugeosynclinal belt in thrust slices, central Nevada. *Geological Society of America Bulletin* 86:425–54.

Ketner, K.B. 1970. Limestone turbidite of Kinderhook age and its tectonic significance, Elko County, Nevada. *U.S. Geological Survey Professional Paper* 700-D:D18–D22.

———. 1998. The nature and timing of tectonism in the Western Facies Terrane of Nevada and California—An outline of evidence and interpretations derived from geologic maps of key areas. *U.S. Geological Survey Professional Paper* 1592.

Marcantel, J. 1975. Late Pennsylvanian and early Permian sedimentation in northeast Nevada. *American Association of Petroleum Geologists Bulletin* 59 (11): 2079–98.

Merriam, C.W. 1963. Paleozoic rocks of Antelope Valley, Eureka and Nye Counties, Nevada. *U.S. Geological Survey Professional Paper* 423.

Miller, W.E. 1981. Cladont shark teeth from Utah. *Journal of Paleontology* 55 (4): 894–95.

Murchey, B.L. 1990. Age and depositional setting of siliceous sediments in the upper Paleozoic Havallah sequence near Battle Mountain, Nevada: Implications for the paleogeography and structural evolution of the western margin of North America. In *Paleozoic and early Mesozoic paleogeographic relations: Sierra Nevada, Klamath Mountains, and related terranes*, Harwood, D.S., and Miller, M.M., eds., 137–55. *Geological Society of America Special Paper* 255.

Nelson, C.R., and Tidwell, W.D. 1987. *Brodioptera stricklandi* n. sp. (Megasecoptera: Brodioptera), a new fossil insect from the Upper Manning Canyon Shale Formation, Utah (lowermost Namurian B). *Psyche* 94:309–16.

Nichols, K.M., Silberling, N.J., Cashman, P.H., and Trexler, J.H., Jr. 1992. Extraordinary synorogenic and anoxic deposits amidst sequence cycles of the late Devonian–early Mississippian carbonate shelf, Lakeside and Stansbury Mountains, Utah. In *Field guide to geologic excursions in Utah and adjacent areas of Nevada, Idaho, and Wyoming*, Wilson, J.R., ed., 123–45. *Utah Geological Survey Miscellaneous Publication* 92-3.

Nilsen, T.H., and Stewart, J.H. 1980. The Antler Orogeny—Mid-Paleozoic tectonism in western North America. *Geology* 8:298–302.

Poole, F.G. 1974. Flysch deposits of the Antler foreland basin, western United States. In *Tectonics and sedimentation*, Dickinson, W.R., ed., 58–82. *Society of Economic Paleontologists and Mineralogists Special Publication* 22.

Poole, F.G., and Sandberg, C.A. 1977. Mississippian paleogeography and tectonics of the western United States. In *Paleozoic paleogeography of the western United States*, Stewart, J.H., Stevens, C.H., and Fritsche, A.E., eds., 181–215. Pacific Coast Paleogeography Symposium 1, Pacific Section, Society of Economic Paleontologists and Mineralogists.

Poole, F.G., and Sandberg, C.A. 1991. Mississippian paleogeography and conodont biostratigraphy of the western United States. In *Paleozoic paleogeography of the western United States–II*, Cooper, J.D., and Stevens, C.H., 107–36. Pacific Section, Society of Economic Paleontologists and Mineralogists, v. 67.

Rich, M. 1962. Mississippian stigmarian plant fossils from southern Nevada. *Journal of Paleontology* 36 (2): 347–49.

Rigby, J.K., and Washburn, A.T. 1972. A new hexactinellid sponge from the Mississippian-Pennsylvanian Diamond Peak Formation in eastern Nevada. *Journal of Paleontology* 46 (2): 266–70.

Roberts, R.J. 1972. Evolution of the Cordilleran foldbelt. *Geological Society of America Bulletin* 83:1989–2003.

Rodriguez, J., and Gutschick, R.C. 1978. A new shallow water *Schizophoria* from the Leatham Formation (late Famennian), northeastern Utah. *Journal of Paleontology* 52 (6): 1346–55.

Rose, P.R. 1976. Mississippian carbonate shelf margins, western United States. *U.S. Geological Survey Journal of Research* 4:449–66.

Saller, A.H., and Dickinson, W.R. 1982. Alluvial to marine facies transitions in the Antler overlap sequence, Pennsylvanian and Permian of north-central Nevada. *Journal of Sedimentary Petrology* 52 (3): 925–40.

Saltzman, M.R., Gonzalez, L.A., and Lohmann, K.C. 2000. Earliest carboniferous cooling step triggered by the Antler Orogeny? *Geology* 28 (4): 347–50.

Sandberg, C.A., and Gutschick, R.A. 1980. Sedimentation and biostratigraphy of Osagean and Merimecian starved basin and foreslope, western United States. In *Paleozoic paleogeography of the west-central United States*, Fouche, T.D., and Magathan, E.R., eds., 129–47. Rocky Mountain Paleogeography Symposium 1, Rocky Mountain Section, Society of Economic Paleontologists and Mineralogists.

Sandberg, C.A., Morrow, J.R., and Warme, J.E. 1997. Late Devonian Alamo impact event, global Kellwasser events, and major eustatic events, eastern Great Basin, Nevada and Utah. *Brigham Young University Geology Studies* 42, part 1:129–60.

Sandberg, C.A., Poole, F.G., and Gutschick, R.C. 1980. Devonian and Mississippian stratigraphy and conodont zonation of the Pilot and Chainman Shales, Confusion Ranges, Utah. In *Paleozoic paleogeography of the west-central United States*, Fouche, T.D., and Magathan, E.R., eds., 71–79. Paleozoic Paleogeography Symposium 1, Rocky Mountain Section, Society of Economic Paleontologists and Mineralogists.

Schultze, H.-P. 1990. A new acanthodian from the Pennsylvanian of Utah, U.S.A., and the distribution of otoliths in gnathostomes. *Journal of Vertebrate Paleontology* 10 (1): 49–57.

Silberling, N.J., and Nichols. K.M. 1992. Petrology and regional significance of the Mississippian Delle phosphatic member, Lakeside Mountains, northwestern Utah. In *Field guide to geologic excursions in Utah and adjacent areas of Nevada, Idaho, and Wyoming,*

Utah, Wilson, J.R., ed., 147–69. *U.S. Geological Survey Miscellaneous Publication* 92-3.

Silberling, N.J., Nichols, K.M., Macke, D.L., and Trappe, J. 1997. Upper Devonian–Mississippian stratigraphic sequences in the distal Antler foreland of western Utah and adjoining Nevada. *U.S. Geological Survey Bulletin* 1988-H.

Silberling, N.J., Nichols, K.M., Trexler, J.H., Jr., Jewell, P.W., and Crosbie, R.A. 1997. Overview of Mississippian depositional and paleotectonic history of the Antler foreland, eastern Nevada and western Utah. *Brigham Young University Geology Studies* 42, part 1:161–96.

Smith, J.F., Jr., and Ketner, K.B. 1968. Devonian and Mississippian rocks and the date of the Roberts Mountains thrust in the Carlin-Pinyon Range area. *U.S. Geological Survey Bulletin* 1251-I: I1–I18.

———. 1975. Stratigraphy of Paleozoic rocks in the Carlin-Pinyon Range area, Nevada. *U.S. Geological Survey Professional Paper* 867-A.

Speed, R.C., and Sleep, N.H. 1980. Antler Orogeny and foreland basin: A model. *Geological Society of America Bulletin* 93:815–28.

Stevens, C.H., Stone, P., and Belasky, P. 1991. Paleogeography and structural significance of an upper Mississippian facies boundary in southern Nevada and east-central California. *Geological Society of America Bulletin* 103:876–85.

Stewart, J. H. 1980. *Geology of Nevada*. Nevada Bureau of Mines and Geology Special Publication 4.

Wallin, E.T. 1990. Provenance of selected lower Paleozoic siliciclastic rocks in the Roberts Mountains allochthon, Nevada. In *Paleozoic and early Mesozoic paleogeographic relations: Sierra Nevada, Klamath Mountains, and related terranes*, Harwood, D.S., and Miller, M.M., eds., 17–32. *Geological Society of America Special Paper* 255.

Warme, J.E., and Kuehner, H-C. 1998. Anatomy of an anomaly: The Devonian catastrophic Alamo impact breccia of southern Nevada. In *Integrated earth and environmental evolution of the southwestern United States*, Ernst, W.G., and Nelson, C.A., eds., 80–107. Columbia, MD: Bellwether Publishing, Ltd. (for the Geological Society of America).

Welsh, J.E. 1979. Paleogeography and tectonic implications of the Mississippian and Pennsylvanian in Utah. In *Basin and Range symposium*, Newman, G.W., and Goode, H.D., eds., 93–106. Rocky Mountain Association of Geologists–Utah Geological Association Field Conference and Symposium 1979.

Whiteford, W.B. 1990. Paleographic setting of the Schoonover sequence, Nevada, and implications for the late Paleozoic margin of western North America. In *Paleozoic and early Mesozoic paleogeographic relations: Sierra Nevada, Klamath Mountains, and related terranes*, Harwood, D.S., and Miller, M.M., eds., 115–36. *Geological Society of America Special Paper* 255.

Wilson, B.R., and Laule, S.W. 1979. Tectonics and sedimentation along the Antler orogenic belt of central Nevada. In *Basin and Range symposium*, Newman, G.W., and Goode, H.D., eds., 81–92. Rocky Mountain Association of Geologists–Utah Geological Association Field Conference and Symposium 1979.

Chapter Six: The Sonoma Event and Other Mesozoic Convulsions

Armstrong, R.L. 1968. Sevier orogenic belt in Nevada and Utah. *Geological Society of America Bulletin* 79:429–58.

Axen, G.J. 1984. Thrusts in the eastern Spring Mountains, Nevada: Geometry and mechanical implications. *Geological Society of America Bulletin* 95 (10): 1202–7.

Burchfiel, B.C., Cowan, D.S., and Davis, G.A. 1992. Tectonic overview of the Cordilleran Orogen in the western United States. In *The Cordilleran Orogen: Conterminous U.S.*, Burchfiel, B.C., Lipman, P.W., and Zoback, M.L., eds., 407–79. *The Geology of North America*, v. G-3. Boulder, CO: Geological Society of America.

Camilleri, P.A., and Chamberlain, K.R. 1997. Mesozoic tectonics and metamorphism in the Pequop Mountains and Wood Hills region, northeast Nevada: Implications for the architecture and evolution of the Sevier Orogen. *Geological Society of America Bulletin* 109 (1): 74–94.

Camilleri, P., Yonkee, A., Coogan, J., DeCelles, P., McGrew, A., and Wells, M. 1997. Hinterland to foreland transect through the Sevier Orogen, northeast Nevada to north-central Utah: Structural style, metamorphism, and kinematic history of a large contractional wedge. *Brigham Young University Geology Studies* 42, part 1:297–324.

Camp, C.L. 1980. Large ichthyosaurs from the upper Triassic of Nevada. *Palaeontographica* 170:139–200.

Cowen, D.S., and Bruhn, R.L. 1992. Late Jurassic to early late Cretaceous geology of the U.S. Cordillera. In *The Cordilleran Orogen: Conterminous U.S.*, Burchfiel, B.C., Lipman, P.W., and Zoback, M.L. eds., 169–203. *The Geology of North America*, v. G-3. Boulder, CO: Geological Society of America.

DeCelles, P.G, and Mitra, G. 1995. History of the Sevier orogenic wedge in terms of critical taper models, northeast Utah and southwest Wyoming. *Geological Society of America Bulletin* 107 (4): 454–62.

Fleck, R.J. 1970. Tectonic style, magnitude, and age of deformation in the Sevier orogenic belt in southern Nevada and eastern California. *Geological Society of America Bulletin* 81:1705–20.

Hose, R.K. 1977. Structural geology of the Confusion Range, west-central Utah. *U.S. Geological Survey Professional Paper* 971.

Jones, A.E. 1991. Sedimentary rocks of the Golconda terrane: Provenance and paleogeographic implications. In *Paleozoic paleogeography of the western United States*, Cooper, J.D., and Stevens, C.H., eds., 783–800. Pacific Section, Society of Economic Paleontologists and Mineralogists, v. 67.

Kelsey, Michael R. 1997. *Hiking, climbing, and exploring western Utah's Jack Watson Ibex Country*. Provo, Utah: Kelsey Publications.

Ketner, K.B., Day, W.C., Elrick, M., Vaag, M.K., Zimmerman, R.A., Snee, L.W., Saltus, R.W., Repetski, J.E., Wardlaw, B.R., Taylor, M.E., and Harris, A.G. 1998. An outline of tectonic, igneous, and metamorphic events in the Goshute-Toano Range between Silver Zone Pass and White Horse Pass, Elko County, Nevada: A history of superposed contractional and extensional deformation. *U.S. Geological Survey Professional Paper* 1593.

Longwell, C.R. 1949. Structure of the northern Muddy Mountain area, Nevada. *Geological Society of America Bulletin* 90:923–68.

Lore, D. 1941. *Leptolepis nevadensis*, a new Cretaceous fish. *Journal of Paleontology* 15 (3): 318–21.

Lucas, S.G., and Marzolf, J.E. 1993. Stratigraphy and sequence stratigraphic interpretation of upper Triassic strata in Nevada. In *Mesozoic paleogeography of the western United States–II*, Dunn, G., and McDougall, K., eds., 375–88. Pacific Section, Society of Economic Paleontologists and Mineralogists, v. 71.

MacNeil, F.S. 1939. Fresh-water invertebrates and land plants of Cretaceous age from Eureka, Nevada. *Journal of Paleontology* 13 (3): 355–60.

Marzolf, J.E. 1993. Plainspastic reconstruction of early Mesozoic sedimentary basin near the latitude of Las Vegas: Implications for the early Mesozoic Cordilleran cratonal margin. In *Mesozoic paleogeography of the western United States–II*, Dunn, G., and McDougall, K., eds., 433–62. Pacific Section, Society of Economic Paleontologists and Mineralogists, v. 71.

Miller, D.M., and Hoisch, T.D. 1995. Jurassic tectonics of northeastern Nevada and northwestern Utah from the perspective of barometric studies. In *Jurassic magmatism and tectonics of the North American cordillera*, Miller, D.M., and Busby, C., eds., 267–94. *Geological Society of America Special Paper* 299.

Nolan, T.B., Merriam, C.W., and Williams, J.S. 1956. The stratigraphic section in the vicinity of Eureka, Nevada. *U.S. Geological Survey Professional Paper* 276.

Oldow, J.S. 1984. Evolution of a late Mesozoic back-arc fold and thrust belt, northwestern Great Basin, USA. *Tectonophysics* 102:245–74.

———. 1981. Structure and stratigraphy of the Luning allochthon and the kinematics of allochthon emplacement, Pilot Mountains, west-central Nevada. *Geological Society of America Bulletin* 92, part 1:888–911.

Oldow, J.S., and Bartel, R.L. 1987. Early to middle (?) Jurassic extensional tectonism in the western Great Basin: Growth faulting and synorogenic deposition of the Dunlap Formation. *Geology* 15:740–43.

Paulsen, T., and Marshak, S. 1998. Charleston transverse zone, Wasatch Mountains, Utah: Structure of the Provo salient's northern margin, Sevier fold-thrust belt. *Geological Society of America Bulletin* 110 (4): 512–22.

Presnell, R., and Parry, W.T. 1995. Evidence of Jurassic tectonism from the Barneys Canyon gold deposit, Oquirrh Mountains, Utah. In *Jurassic magmatism and tectonics of the North American Cordillera*, Miller, D.M., and Busby, C., eds., 313–26. *Geological Society of America Special Paper* 299.

Roberts, R.J., and Crittenden, M.D. 1973. Orogenic mechanisms, Sevier orogenic belt, Nevada and Utah. In *Gravity and tectonics*, DeJong, K.A., and Scholten, S., eds., 409–28. New York: John Wiley and Sons.

Saleeby, J.B., and Busby-Serpa, C. 1992. Early Mesozoic teectonic evolution of the western U.S. cordillera. In *The Cordilleran Orogen: Conterminous U.S.*, Burchfiel, B.C., Lipman, P.W., and Zoback, M.L., eds., 107–68. *The Geology of North America*, v. G-3. Boulder, CO: Geological Society of America.

Silberling, N.J, and Roberts, R.J. 1962. Pretertiary stratigraphy and structure of northwestern Nevada. *Geological Society of America Special Paper* 72.

Smith, D.L., Miller, E.L., Wild, S.J., and Wright, J.E. 1993. Progression and timing of Mesozoic crustal shortening in the northern Great Basin, western U.S.A. In *Mesozoic paleogeography of the western United States–II*, Dunn, G., and McDougall, K., eds., 389–406. Pacific Section, Society of Economic Paleontologists and Mineralogists, v. 71.

Snoke, A.W., Howard, K.A., McGrew, A.J., Burton, B.R., Barnes, C.G., Peters, M.T., and Wright, J.E. 1997. The grand tour of the Ruby-East Humboldt metamorphic core complex, northeastern Nevada: Part 1—Introduction and road log. *Brigham Young University Geology Studies* 42, part 1:225–69.

Speed, R.C. 1979. Collided Paleozoic microplate in the western United States. *Journal of Geology* 87:279–92.

Taylor, W.J., Bartley, J.M., Fryxll, J.E., Schmitt, J.G., and Vandervoort, D.S. 1993. Tectonic style and regional relations of the central Nevada thrust belt. In *Crustal evolution of the Great Basin and Sierra Nevada*, Lahren, M.M., Trexler, J.H., Jr., and Soinosa, C., eds., 57–96. Geological Society of America Guidebook. Reno: Department of Geological Sciences, University of Nevada.

Taylor, W.J., Bartley, J.M., Martin Mark W., Geissman, J.W., Walker, J.D., Armstrong, P.A., and Fryxell, J.E. 2000. Relations between hinterland and foreland shortening: Sevier Orogeny, central North American Cordillera. *Tectonics* 19 (6): 1123–43.

Ward, P.L. 1995. Subduction cycles under western North America during the Mesozoic and Cenozoic eras. In *Jurassic magmatism and tectonics of the North American Cordillera*, Miller, D.M., and Busby, C., eds., 1–39. *Geological Society of America Special Paper* 299.

Wiltscchko, D.V., and Door, J.A., Jr. 1983. Timing of deformation in overthrust belt and foreland of Idaho, Wyoming, and Utah. *American Association of Petroleum Geologists Bulletin* 67 (8): 1304–22.

Wyld, S.J., and Wright, J.E. 1997. Triassic-Jurassic tectonism and magmatism in the Mesozoic continental arc of Nevada: Classic relations and new developments. *Brigham Young University Geology Studies* 42, part 1:197–224.

Yonkee. W.A., Evans, J.P., and DeCelles, P.G. 1992. Mesozoic tectonics of the northern Wasatch Range, Utah. In *Field guide to geologic excursions in Utah and adjacent areas of Nevada, Idaho, and Wyoming*, Wilson, J.R., ed., 429–59. *Utah Geological Survey Miscellaneous Publication* 92-3.

Zamudio, J.A., and Atkinson, W.W., Jr. 1995. Mesozoic structures of the Dolly Varden Mountains and Currie Hills, Elko County, Nevada. In *Jurassic magmatism and tectonics of the North America Cordillera*, Miller, D.M., and Busby, C., eds., 292–312. *Geological Society of America Special Paper* 299.

Chapter Seven: The Fires Below

Armstrong, R.L., and Suppe, J. 1973. Potassium-argon geochronometry of Mesozoic igneous rocks in Nevada, Utah, and southern California. *Geological Society of America Bulletin* 84:1375–92.

Bateman, P.C. 1992. Plutonism in the central part of the Sierra Nevada batholith, California. *U.S. Geological Survey Professional Paper* 1483.

Burchfiel, B.C., Cowan, D.S., and Davis, G.A. 1992. Tectonic overview of the Cordilleran Orogen in the western United States. In *The Cordilleran Orogen: Conterminous U.S.*, Burchfiel, B.C., Lipman, P.W., and Zoback, M.L., eds., 407–79. *Geology of North America*, v. G-3. Boulder, CO: Geological Society of America.

Busby-Spera, C.J. 1988. Speculative tectonic model for the early Mesozoic arc of the southwest Cordilleran United States. *Geology* 16:1121–25.

Busby-Spera, C.J., Mattinson, J.M., Riggs, N.R., and Schermer, E.R. 1990. The Triassic-Jurassic magmatic arc in the Mojave-Sonoran deserts and the Sierra-Klamath region: Similarities and differences in paleogeographic evolution. In *Late Paleozoic and Mesozoic paleogeographic relations, Klamath-Sierra and adjacent regions*, Harwood, D., and Miller, M., eds., 325–37. *Geological Society of America Special Paper* 225.

Coleman, D.S., and Glazer, A.F. 1998. The Sierra crest magmatic event: Rapid formation of juvenile crust during the late Cretaceous in California. In *Integrated earth and environmental evolution of the southwestern United States*, Ernst, W.G., and Nelson, C.A., eds., 253–72. Columbia, MD: Bellwether Publishing, Ltd. (for the Geological Society of America).

Dunne, G.C., and Walker, J.D. 1993. Age of Jurassic volcanism and tectonism, southern Owens Valley region, east-central California. *Geological Society of America Bulletin* 105:1223–30.

Garside, L.J. 1998. Mesozoic metavolcanic and metasedimentary rocks of the Reno–Carson City area, Nevada and adjacent California. *Nevada Bureau of Mines and Geology Report* 49.

Marolf, J.E., and Cole, R.D. 1987. Relationship of the Jurassic volcanic arc to backarc stratigraphy, Cowhole Mountains, San Bernardino County, California. In *Centennial field guide, Volume I, Cordilleran section*, Hill, M.L., ed., 115–20. Boulder, CO: Geological Society of America.

McGrew, A.J., Peters, M.T., and Wright, J.E. 2000. Thermobarometric constraints on the tectonothermal evolution of the East Humboldt Range metamorphic core complex, Nevada. *Geological Society of America Bulletin* 112 (1): 45–60.

Miller, D.M., Nakata, J.K., and Glick, L.L. 1990. K-Ar ages of Jurassic to Tertiary plutonic and metamorphic rocks, northwestern Utah and northeastern Nevada. *U.S. Geological Survey Bulletin* 1906.

Quinn, M.J., Wright, J.E., and Wyld, S.J. 1997. Happy Creek igneous complex and tectonic evolution of the early Mesozoic arc in the Jackson Mountains, northwest Nevada. *Geological Society of America Bulletin* 107 (4): 461–82.

Saleeby, J.B. 1981. Ocean floor accretion and volcano-plutonic arc evolution of the Mesozoic Sierra Nevada. In *The geotectonic development of California*, Ernst, W.G., ed., 132–81. Englewood Cliffs, NJ: Prentice-Hall.

Schweickert, R.A. 1978. Triassic and Jurassic paleogeography of the Sierra Nevada and adjacent region, California and western Nevada. In *Mesozoic paleogeography of the western U.S.*, Howell, D.G., and McDougall, K.A., eds., 361–84. Pacific Coast Paleogeography Symposium 2, Pacific Section, Society of Economic Paleontologists and Mineralogists.

Schweickert, R.A., and Lahren, M.M. 1999. Triassic caldera at Tioga Pass, Yosemite National Park, California: Structural relationships and significance. *Geological Society of America Bulletin* 111 (11): 1714–22.

———. 1993. Triassic-Jurassic magmatic arc in eastern California and western Nevada: Arc evolution, cryptic tectonic breaks, and significance of the Mojave–Snow Lake Fault. In *Mesozoic paleogeography of the western U.S–II*, Dunne, G., and McDougall, K., eds., 227–46. Pacific Section, Society of Economic Paleontologists and Mineralogists, v. 71.

Sorensen, S.S., Dunne, G.C., Hanson, R.B., Barton, M.D., Becker, J., Tobish, O.T., and Fiske, R.S. 1998. From Jurassic shores to Cretaceous plutons: Geochemical evidence for paleoalteration environments of metavolcanic rocks, eastern California. *Geological Society of America Bulletin* 110 (3): 326–43.

Wyld, S.J., and Wright, J.E. 1997. Triassic-Jurassic tectonism and magmatism in the Mesozoic continental arc of Nevada: Classic relations and new developments. In *Proterozoic to recent stratigraphy, tectonics, and volcanology, Utah, Nevada, southern Idaho, and central Mexico*, Link, P.K., and Kowallis, B.J., eds., 197–224. *Brigham Young University Geology Studies* 42, part 1.

Chapter Eight: The Conflagration and the Army of Caterpillars

Armstrong, R.L. 1970. Geochronology of Tertiary igneous rocks, eastern Basin and Range Province, western Utah, eastern Nevada, and vicinity, U.S.A. *Geochimica et Cosmochimica Acta* 34:203–32.

Axen, G.J., Taylor, W.J., and Bartley, J.M. 1993. Space-time patterns and tectonic controls of Tertiary extension and magmatism in the Great Basin of the western United States. *Geological Society of America Bulletin* 105:56–76.

Best, M.G., and Grant, S.K. 1987. Stratigraphy of the volcanic Oligocene Needles Range group in southwestern Utah and eastern Nevada. *U.S. Geological Survey Professional Paper* 1433-A.

Best, M.G., Christiansen, E.H., and Blank, H.R., Jr. 1989 Oligocene caldera complex and clac-alkaline tuffs and lavas of the Indian Peak volcanic field, Nevada and Utah. *Geological Society of America Bulletin* 101:1076–90.

Best, M.G., Scott, R.B., Rowley, P.D., Swadley, W.C., Anderson, R.E., Gromme, C.S., Harding, A.E., Deino, A.L., Christiansen, E.H., Tingey, D.G., and Sullivan, K.R. 1993. Oligocene-Miocene caldera complexes, ash-flow sheets, and tectonism in the central and southeastern Great Basin. In *Crustal evolution of the Great Basin and Sierra Nevada, Cordilleran–Rocky Mountain section*, Lahren, M.M., Trexler, J.H., Jr., and Spinosa, C., eds., 285–311. Geological Society of America Guidebook. Reno: Department of Geological Sciences, University of Nevada.

Burke, K.J., and Axen, G.J. 1997. Structural geometry resulting from episodic extension in the northern Chief Range area, eastern Nevada. In *Geologic studies in the Basin and Range–Colorado Plateau transition in southeastern Nevada, southwestern Utah, and northwestern Arizona*, Maldonado, F., and Nealy, L.D., eds., 267–88. *U.S. Geological Survey Bulletin* 2153.

Chadwick, R.A. 1985. Overview of Cenozoic volcanism in the west-central United States. In *Cenozoic paleogeography of west-central United States*, Florcs, R.W., and Kaplan, S.S., eds., 359–76. Rocky

Mountain Section, Society of Economic Paleontologists and Mineralogists.

Christiansen, E.H., Sheridan, M.F., and Burt, D.M. 1986. The geology and geochemistry of Cenozoic topaz rhyolite from the western United States. *Geological Society of America Special Paper* 205.

Coney, P.J. 1980. Cordilleran metamorphic core complexes: An overview. In *Cordilleran metamorphic core complexes*, Crittenden, M.D., Jr., Coney, P.J., and Davis, G.H., eds., 7–31. *Geological Society of America Memoir* 153.

Coogan, J.C., and DeCelles, P.G. 1996. Extensional collapse along the Sevier Desert reflection, northern Sevier Desert basin, western United States. *Geology* 24 (10): 933–36.

duBray, E.A. 1995. Geochemistry and petrology of Oligocene and Miocene ash-flow tuffs of the southeastern Great Basin, Nevada. *U.S. Geological Survey Professional Paper* 1559.

Gans, P.B., Mahood, G.A., and Schermer, E. 1989. Synextensional magmatism in the Basin and Range Province: A case study from the eastern Great Basin. *Geological Society of America Special Paper* 233.

Gans, P.B., Seedorff, E., Fahey, P.L., Hasler, R.W., Maher, D.J., Jeanne, R.A., and Shaver, S.A. 2001. Rapid Eocene extension in the Robinson district, White Pine County, Nevada: Constraints from $^{40}Ar/^{39}Ar$ dating. *Geology* 29 (6): 475–78.

Gregory-Wodzicki, K.M. 1997. The late Eocene House Range flora, Sevier Desert, Utah: Paleoclimate and paleoelevation. *Palaios* 12:552–67.

Henry, C.D., and Boden, D.R. 1998. Eocene magmatism: The heat source for Carlin-type gold deposits of northern Nevada. *Geology* 26 (12): 1067–70.

Henry, C.D., and Ressel, M.W. 2000. Interrelation of Eocene magmatism, extension, and Carlin-type gold deposits in northeastern Nevada. In *Great Basin and Sierra Nevada: Geological Society of America Field Guide 2*, Lageson, D.R., Peters, S.G., and Lahren, M.M., eds., 165–87. Boulder, Colo.: Geological Society of America.

Heylum, E.B. 1965. Reconnaissance of the Tertiary sedimentary rocks in western Utah. *Utah Geological and Mineralogical Survey Bulletin* 75.

Humphreys, E.D. 1995. Post-Laramide removal of the Farallon slab, western United States. *Geology* 23:987–90.

McGrew, A.J., Peters, M.T., and Wright, J.E. 2000. Thermobarometric constraints on the tectonothermal evolution of the East Humboldt Range metamorphic core complex, Nevada. *Geological Society of America Bulletin* 112 (1): 45–60.

McKee, E.H. 1996. Cenozoic magmatism and mineralization in Nevada. In *Geology and ore deposits of the North American Cordillera*, Coyner, A.R., and Fahey, P.L., eds., 581–88. Geological Society of Nevada Symposium Proceedings, Reno/Sparks, Nevada, April 1995.

McKee, E.H., and Silberman, M.L. 1970. Geochronology of Tertiary igneous rocks in central Nevada. *Geological Society of America Bulletin* 81:2317–28.

Merriam, J.C. 1919. Tertiary mammalian faunas of the Mohave Desert. *University of California, Department of Geology Bulletin* 11:437–585.

Murphy, J.B., Opplinger, G.L., Brimhall, G.H., Jr., and Hynes, A. 1998. Plume-modified orogeny: An example from the western United States. *Geology* 26 (8): 731–34.

Nelson, S.T., and Tingey, D.G. 1997. Time-transgressive and extension-related basaltic volcanism in southwest Utah and vicinity. *Geological Society of America Bulletin* 109 (10): 1249–65.

Rowley, P.D., and Siders, M.A. 1988. Miocene calderas of the Caliente caldera complex, Nevada-Utah. *U.S. Geological Survey Professional Paper* 1149.

Smith, R.B., Nagy, W.C., Julander, K.A., Viveiros, J.J., Barker, C.A., and Gants, D.G. 1989. Geophysical and tectonic framework of the eastern Basin and Range–Colorado Plateau–Rocky Mountain transition. In *Geophysical framework of the continental United States*, Pakiser, L.C., and Mooney, W.D., eds., 205–33. *Geological Society of America Memoir* 172.

Stewart, J.H., and Diamond, D.S. 1990. Changing patterns of extensional tectonics: Overprinting of the basin of the middle and upper Miocene Esmeralda Formation in western Nevada by younger structural basin. In *Basin and Range extensional tectonics near the latitude of Las Vegas, Nevada*, Wernicke, B.P., ed., 447–67. *Geological Society of America Memoir* 176.

Stock, C. 1949. Mammalian fauna from the Titus Canyon Formation, California. *Carnegie Institution of Washington Publication* 584.

———. 1936. Titanotheres from the Titus Canyon Formation, California. *Proceedings of the National Academy of Science* 229 (11): 656–61.

Taylor, W.J., and Switzer, D.D. 2001. Temporal changes in fault strike (to 90°) and extension directions during multiple episodes of extension: An example from eastern Nevada. *Geological Society of America Bulletin* 113 (6): 743–59.

Wernicke, B. 1992. Cenozoic extensional tectonics of the U.S. Cordillera. In *The Cordilleran Orogen: Conterminous U.S.*, Burchfiel, B.C., Lipman, P.W., and Zoback, M.L., eds., 553–81. *The Geology of North America*, v. G-3. Boulder, CO: Geological Society of America.

Wernicke, B., and Snow, J.K. 1998. Cenozoic tectonism in the central Basin and Range: Motion of the Sierran–Great Valley block. In *Integrated earth and environmental evolution of the southwestern United States*, Ernst, W.G., and Nelson, C.A., eds., 111–18. Columbia, MD: Bellwether Publishing, Ltd. (for the Geological Society of America).

Whistler, D.P., and Burbank, D.W. 1992. Miocene biostratigraphy and biochronology of the Dove Spring Formation, Mojave Desert, California, and characterization of the Clarendonian mammal age (late Miocene) in California. *Geological Society of America Bulletin* 104:644–58.

Wolfe, J.A., Forest, C.E., and Molnar, P. 1998. Paleobotanical evidence of Eocene and Oligocene paleoaltitudes in midlatitude western North America. *Geological Society of America Bulletin* 110 (5): 664–78.

Wolfe, J.A., Schorn, H.E., Forest, C.E., and Molnar, P. 1997. Paleobotanical evidence for high altitudes in Nevada during the Miocene. *Science* 276:1672–75.

Woodburne, M.O., Tedford, R.H., Stevens, M.S., and Taylor, B.E. 1974. Early Miocene mammalian faunas, Mojave Desert, California. *Journal of Paleontology* 48:6–26.

Woodburne, M.O., Tedford, R.H., and Swisher, C.C., III. 1990. Lithostratigraphy, biostratigraphy, and geochronology of the Barstow Formation, Mojave Desert, southern California. *Geological Society of America Bulletin* 102 (4): 459–77.

Chapter Nine: Fire and Ice: The Modern Great Basin Emerges

Adams, K.D., and Wesnousky, S.G. 1998. Shoreline processes and the age of the Lake Lahontan highstand in the Jessup embayment, Nevada. *Geological Society of America Bulletin* 110 (10): 1318–32.

Adams, K.D., Wesnousky, S.G., and Bills, B.G. 1999. Isostatic rebound, active faulting, and potential geomorphic effects in the Lake Lahanton basin, Nevada and California. *Geological Society of America Bulletin* 111 (12): 1739–56.

Anderson, D. 1999. Latest Quaternary (≥30 ka) lake high-stand fluctuations and evolving paleohydrology of Death Valley. *U.S. Geological Survey Open-File Report* 99-153:124–31.

Blair, T.C. 2001. Outburst flood sedimentation on the proglacial Tuttle Canyon alluvial fan, Owens Valley, California, U.S.A. *Journal of Sedimentary Research* 71 (5): 657–79.

Benson, L.V., and Thompson, R.S. 1987. Lake-level variation in the Lahontan basin for the past 50,000 years. *Quaternary Research* 28:69–85.

Benson, L.V., Currey, D.R., Dorn, R.I., Lajoie, K.R., Oviatt, C.G., Robinson, S.W., Smith, G.I., and Stine, S. 1990. Chronology of expansion and contraction of four Great Basin lake systems during the past 35,000 years. *Paleolakes and Paleo-oceans, Palaeogeography, Palaeoclimatology, Palaeoecology* (Meyers, P.A., and Benson, L.V., eds.) 78:241–86.

Blackwelder, E. 1954. Pleistocene lakes and drainage in the Mojave region, southern California. *California Division of Mines Bulletin* 170, part 5:35–40.

Brattstrom, B.H. 1976. A Pleistocene herpetofauna from Smith Creek Cave, Nevada. *Bulletin of the Southern California Academy of Science* 57:5–13.

Currey, D.R. 1990. Quaternary paleolakes in the evolution of semidesert basins, with special emphasis on Lake Bonneville and the Great Basin, U.S.A. *Palaeogeography, Palaeoclimatology, Palaeoecology* 76:189–214.

Currey, D.R., Atwood, G., and Mabey, D. 1983. Major levels of Great Salt Lake and Lake Bonneville. Utah Geological and Mineral Survey Map 73.

Currey, D.R., Oviatt, C.G., and Czarnomski, J.E. 1984. Late Quaternary geology of Lake Bonneville and Lake Waring. *Utah Geological Association Publication* 13:227–38.

Gilbert, G.K. 1890. Lake Bonneville. Washington, D.C.: U.S. Geological Survey.

Grayson, D.K. 1987. The biogeographic history of small mammals in the Great Basin: Observation on the last 20,000 years. *Journal of Mammalogy* 68 (2): 359–75.

———. 1993. *The desert's past.* Washington and London: Smithsonian Institution Press.

———. 1982. Toward a history of Great Basin mammals during the past 15,000 years. In *Man and Environment in the Great Basin*, Madsen, D.B., and O'Connell, J.F., eds., 82–101. *Society for American Archaeology Paper* 2.

Heaton, T.H. 1990. Quaternary mammals of the Great Basin: Extinct giants, Pleistocene relics, and recent immigrants. In *Causes of evolution: A paleontological perspective*, Ross, R.M., and Allmon, W.D., eds., 422–65. Chicago: University of Chicago Press.

———. 1985. Quaternary paleontology and paleoecology of Crystal Ball Cave, Millard County, Utah: With emphasis on mammals and description of a new species of fossil skunk. *Great Basin Naturalist* 45 (3): 337–90.

Kelly, T.S. 1997. Additional late Cenozoic (latest Hemphillian to earliest Irvingtonian) mammals from Douglas County, Nevada. *PaleBios* 18 (1): 1–31.

Lowenstein, T.K., Li, J., Brown, C., Roberts, S.M., Ku T.-L., Luo, S., and Yang, W. 1999. 200 k.y. plaeoclimate record from Death Valley salt core. *Geology* 27 (1): 3–6.

Machette, Michael N., Johnson, Margo L., and Slate, J.L. 2001. Quaternary and late Pliocene geology of the Death Valley region: Recent observations on tectonics, stratigraphy, and lake cycles (guidebook for the 2001 Pacific Cell-Friends of the Pleistocene field trip). *U.S. Geological Survey Open-File Report* 01-51:A5–Q246.

Morrison, R.B. 1965. Quaternary geology of the Great Basin. In *The Quaternary of the United States*, Wright, H.E., Jr., and Frey, D.G., eds., 265–85. Princeton, NJ: Princeton University Press.

———. 1991. Quaternary stratigraphic, hydrologic, and climatic history of the Great Basin, with emphasis on Lakes Lahontan, Bonneville, and Tecopa. In *Quaternary nonglacial geology: Conterminous U.S.*, Morrison, R.B., ed., 283–317. *The Geology of North America*, v. K-2. Boulder, CO: Geological Society of America.

Morrison, R.B., and Frye, J.C. 1965. Correlation of the middle and late Quaternary successions of the Lake Lahontan, Lake Bonneville, Rocky Mountain (Wasatch Range), southern Great

Plains, and eastern Midwest areas. *Nevada Bureau of Mines Report* 9 (Mackay School of Mines, University of Nevada).

Oviatt, C.G. 1997. Lake Bonneville fluctuations and global climate change. *Geology* 25 (2): 155–58.

Oviatt, C.G., and McCoy, W.D. 1992. Early Wisconsin lakes and glaciers in the Great Basin, U.S.A. In *The last interglacial-glacial transition in North America*, Clark, P.U., and Lea, P.D., eds., 279–87. *Geological Society of America Special Paper* 270.

Reheis, M., and Morrison, R. 1997. High, old pluvial lakes of western Nevada. *Brigham Young University Geology Studies* 42, part 1:459–92.

Russell, I.C. 1885. Geological history of Lake Lahontan. Washington, D.C.: U.S. Geological Survey.

———. 1889. Quaternary history of the Mono Valley, California. *Annual Report of the U.S. Geological Survey* 8:267–394.

Smith, G.I. 1984. Paleohydrologic regimes in the southwest Great Basin, 0–3.2 m.y. ago, compared with other long records of "global" climate. *Quaternary Research* 22:1–27.

Smith, G.I., Benson, L., and Currey, D.R. 1989. Quaternary geology of the Great Basin. *28th International Geological Congress, Field Trip Guidebook* T117. Washington, D.C.: American Geophysical Union.

Snyder, C.T., Hardman, G., and Zdenek, F.F. 1964. Pleistocene lakes in the Great Basin. U.S. Geological Survey Map I-416.

Spaulding, W.G. 1999. Paleoclimate and active tectonics: Middle to late Quaternary environmental changes in Death Valley and vicinity. In *Proceedings of conference on status of geological research and mapping, Death Valley National Park*, Slate, J.L., ed., 121–23. *U.S. Geological Survey Open-File Report* 99-153.

Thompson R.S., and Mead, J.I. 1982. Late Quaternary environments and biogeography in the Great Basin. *Quaternary Research* 17:39–55.

Chapter Ten: Closing Circle, Unfinished Story

Beck, C., and Jones, G.T. 1997. The terminal Pleistocene/early Holocene archaeology of the Great Basin. *Journal of World Prehistory* 11:161–236.

Bennett, R.A., Davis, B.P., and Wernicke, B.P. 1999. Present-day pattern of cordilleran deformation in the western United States. *Geology* 27:371–74.

Caskey, S.J., Bell, J.W., Slemmons, D.B., and Ramelli, A.R. 2000. Historical surface faulting and paleoseismology of the central Nevada seismic belt. In *Great Basin and Sierra Nevada*, Lageson, D.R., Peters, S.G., and Lahren, M.M., eds., 23–44. *Geological Society of America Field Guide* 2.

Grayson, D.K. 1993. *The desert's past: A natural prehistory of the Great Basin.* Washington, D.C.: Smithsonian Institution Press.

Lee, J., Spencer, J., and Owen, L. 2001. Holocene slip rates along the Owens Valley Fault, California: Implications for the recent evolution of the Eastern California Shear Zone. *Geology* 29 (9): 819–22.

Miller, C.D. 1985. Holocene eruptions at the Inyo volcanic chain: Implications for possible eruptions in Long Valley caldera. *Geology* 13:14–17.

Oldow, J.S., Kohler, G., and Donelick, R.A. 1994. Late Cenozoic extensional transfer in the Walker Lane strike-slip belt, Nevada. *Geology* 22:637–40.

Rhode, D., Adams, K.D., and Elston, R.G. 2000. Geoarchaeology and Holocene landscape history of the Carson Desert, western Nevada. In *Great Basin and Sierra Nevada*, Lageson, D.R., Peters, S.G., and Lahren, M.M., eds., 45–74. *Geological Society of America Field Guide* 2.

Smith, R.B., and Eaton, G.P. 1978. Cenozoic tectonics and regional geophysics of the western cordillera. *Geological Society of America Memoir* 152.

Stewart, J.H. 1998. Regional characteristics, tilt domains, and extensional history of the late Cenozoic Basin and Range Province, western North America. In *Accommodation zones and transfer zones: The regional segmentation of the Basin and Range Province,* Faulds, J.E., and Stewart, J.H., eds., 47–74. *Geological Society of America Special Paper* 323.

———. 1988. Tectonics of the Walker Lane belt, western Great Basin: Mesozoic and Cenozoic deformation in a zone of shear. In *Metamorphism and crustal evolution of the western United States,* Ernst, W.G., ed., 683–713. Englewood Cliffs, NJ: Prentice-Hall.

Thatcher, W., Foulger, G.R., Julian, B.R., Svarc, S.L., Quilty, E., and Bawden, G.W. 1999. Present-day deformation across the Basin and Range Province, Nevada. *Geological Society of America Bulletin* 84:627–32.

Thenhaus, P.C., and Barnhard, T.P. 1998. Insights from Quaternary relations for segmentation of the Great Basin by regional, transverse accommodation zones. In *Accommodation zones and transfer zones: The regional segmentation of the Basin and Range Province,* Faulds, J.E., and Stewart, J.H., eds., 229–38. *Geological Society of America Special Paper* 323.

Wallace, R.E. 1984. Patterns and timing of late Quaternary faulting in the Great Basin Province and relation to some regional tectonic features. *Journal of Geophysical Research* 89, no. B7: 5763–69.

Wernicke, B., Friedrich, A.M., Niemi,N.A., Bennett, R.A., and Davis, J.L. 2000. Dynamics of plate boundary fault systems from Basin and Range Geodetic Network (BARGEN) and geologic data. *GSA Today* 10 (11): 1–3.

Zandt, G., Myers, S., and Wallace, T. 1995. Crust and mantle structure across the Basin and Range-Colorado Plateau boundary at 37° north latitude and implications for Cenozoic extensional mechanism. *Journal of Geophysical Research* 100 (10):10,529–48

brachiopods: and Cambrian reefs, 63; and extinction
event at end of Permian, 132; and Mississippian strata,
117; and Ordovician fossil assemblages, 89, 91; and Silurian dolomite, 95
breccia, 111
bristlecone pine (*Pinus longaeva*), 10
brittle fracturing, 185–86
bryozoans, 91
Buenaventura, Rio, 4
Bullion Fault, 233
Burgess Shale, 66–67, 70
Burro Canyon Formation (Colorado), 148
Butterfield, Nicholas, 70

Caesar Canyon Formation, 87–88
Calamites, 116, 117
calcium carbonate, and limestone, 57
caldera complexes, 176–77
calderas, 158, 176–77, *178*
Calico Mining District, 179
Caliente caldera complex, 177
California: and boundaries of Great Basin, 8; and
collapsed borderland model, 113; and Cretaceous
dinosaur fossils, 148, 149. *See also* Death Valley; San
Andreas Fault
Cambrian Eon: and fossils, 45, 60; and Ordovician tran-
sition, 90–92; and seas of Great Basin, 52–71
Cambrian explosion, in evolution of marine organisms,
45, 60
camels, 222
Camp, Charles, 133
Canadian Shield, and Archeon Eon, 26
Caninia, 117
carbonate belt, 68, 69
carbonate rocks, 57
carbon dioxide, and volcanic activity, 235
Cardipeltis, 96
Carrara Formation, *58*, 62
Carson Desert volcanic field, 219
Carson River, 210
Casa Diablo power facility, 238
Cascade Range, and climate of Great Basin, 15–16
Cedar Mountain Formation, 148
Cedarville (town), 11
Cenozoic Period: middle of and rocks of Great Basin,
171–80; and transition from Mesozoic in Great Basin,
166–68
Central Nevada–Eureka Thrust Belt, 139–41, 145
Central Nevada Seismic Belt, 233–34, 243, 244
Cephalaspis utahensis, 96
cephalopods, 89, *117*
Chainman Shale, 108, 114
chert, 59, 102
China Lake, 214
Chinle Formation, 131
chitinozoans, 93
Christmas Lake, 216
Cima volcanic field, 218
cinder cones, 180
cities, and earthquakes in Great Basin, 231. *See also*
Reno; Salt Lake City
clams, 132, 147, 221
Clark Mountain Thrust, 141
clastic rocks, 58
climate: and Cambrian cycles, 60; and drainage pattern
of Great Basin, 8, 14–17; and Cretaceous Period, 147,

149; and lakes in East Africa, 242; and Ordovician
Period, 91–94; Quaternary Period and Ice Age,
198–201, 221, 222
Climatius, 96
Cloverly Formation (Wyoming), 148
coastal margin, of North America, 244
Cocos Plate, 189
collapsed borderland model, 113
collared lizard (*Crotaphytus bicinctores*), 6
Colorado Plateau, 240, 244
Colorado River, 14, 208
Comet Hale-Bopp, 170. *See also* asteroids
Composita, 117
composite pluton, 159
Comstock Lode, 179
Condor Canyon Formation, 174
Confusion Mountains: and field geology, 74; naming of,
126; and Ordovician strata, 79, 80, *83*; and unconformity, 127, 145–46
conifers, 147, 221
conodonts, 89, 93
continental rifting: and contemporary structure of crust
in Great Basin, 240, 241–42, 243; and late Proterozoic,
39–43; in modern Africa, 30, 42, 242–43. *See also* continents; Farallon Plate; North American Plate; Pacific
Plate; plate tectonics
continental slope, 56–57, *98*
continents: and Archean Eon, 26, 27; and Ordovician
Period, 91; and Pennsylvanian Period, 121–22; and
Proterozoic Eon, 27–30, 34. *See also* continental rifting; microcontinents; plate tectonics; supercontinents
copper, 164–65
coquinas, 89
corals, 89, 91, 95, 117. *See also* reefs
Cordilleran Geosyncline, 55–71
Cordilleran magmatic arc, 157–60, 161
Coso Range (California), 213
Cottonwood Wash Tuff, 174
Cove Fort geothermal area, 238
Crater Lake (Oregon), 158, 176
Cravenoceras, 117
Crawford Thrust, 141
Cretaceous Period: and plate tectonics, 160; and rock
sequences, 136; and Sevier Orogeny, 141
crinoids: and Cambrian reefs, 65; and extinction event at
end of Permian, 132; and Mississippian strata, *117*; and
Ordovician fossil assemblages, 91
Cronise Lake, 216
cross-bedding, 84, 139
crust, of earth, 239–46
Cruziana, 64
Crystal Peak, 174
Crystal Spring Formation, 35
Crystal tuff, 174–75
cyanobacteria, 44, 45
cycads, 147
Cymbospondylus, 134–35

dacite, 172
Death Valley: and early Oligocene, 191; extension in
modern, 243; Ice Age innundation of, 215; and late
Proterozoic rocks, 35
Death Valley National Park, *44*
décollement, 141–42
deflation, 16
Delta (town), 11

hydrated minerals, 155
hydrology. *See* drainage; rivers and streams
hydrothermal systems, 164
hyolithids, 70
hypersaline, 87

Ibapah Peak, 11
Ibex Hills, 125, 126
Ibexian Series, 77
ice ages: in Great Basin, 201–16; and megafauna, *150G*;
    Pleistocene Epoch and concept of, 197–98; Quaternary
    Period and climate change, 198–201. *See also* glaciers
    and glaciation
ice cores, analysis of, 200
ichthyosaurs, 133
Idaho, and boundaries of Great Basin, 8
igneous rocks: and Cordilleran magmatic arc, 157–60;
    and granite in Great Basin, 151–52; subduction,
    magma, and minerals, 152–57. *See also* basalt; granite;
    magma; metamorphism
ignimbrites, 175–76
Independence Range: and Antler Orogeny, 108; and
    Central Nevada–Eureka Thrust Belt, 140, *141*; and
    Cenozoic volcanic rocks, 172–73
Indian Peak caldera, 177
Indian Wells Valley, 214
insects: and Miocene Epoch, 193; and Mississippian Pe-
    riod, 117; and Quaternary environment, 221. *See also*
    fauna; fossils
Intermountain Seismic Belt (ISB), 231–32
intraformational conglomerate, 80–82
intrusive volcanic rock, 153
inverted topography, 177
Inyo Craters (California), 219–20
iridium, 112
Iron Springs Mining District, 179
isostacy, 56, 113–14
Ivanpah Orogeny, 33

Jackson Mountain, 159
Jarbridge Range, 140, 172–73
jasperoids, 102
Jefferson, Mount, 11
jellyfish, 70
Johnnie Formation, 35
Johnson Valley Fault, 233
Juan de Fuca Plate, 189
Jurassic Period, 135–43

Kane Spring caldera, 176
Kanosh Shale, *89*
Kawich caldera, 176
Kelsey, Michael, 125
Keystone thrust fault, 142
King, Clarence, 196
King Lear Formation, 146, 148
Kingston Peak Formation, 36–39
Kinnikinic Formation, 83
Klamath Falls earthquake (Oregon, 1993), 226
Klamath Mountains, 104, 106, 129
Koipato Group, 127, 128
Krakatoa (volcanic eruption), 177
Kula Plate, 144

Lagerstätten, and fossils, 66–71
LaHontan, Baron de, 196

Lahontan, Lake, 12, 196, 209–12
Lahren, Mary, 161
lakes: and climate in East Africa, 242; and domed shape
    of Great Basin, 12; and Miocene Epoch, 193; and Pleis-
    tocene glaciation, 201, 203–16. *See also* Bonneville,
    Lake; Great Salt Lake; Lahonton Lake; Mono Lake;
    Pyramid Lake
Laketown Dolomite, 95
Lamoille Canyon, 20–21, *150F*
Landers earthquake (1992), 226, 233
landscape: and future of Great Basin, 244–46; and mid-
    Tertiary, 189–93. *See also* environment; fauna; plants
Lapworth, Charles, 75, 76
Laramide Orogeny, 187
Latham Shale, 62
Laurasia, 121–22
Lava Mountains, 218
Lavic Lake Fault, 223
Leach Canyon Tuff, 174
Leidy, Joseph, 133
Leptolepis nevadensis, 147
limestone: and Antler foreland basin, 115; and Cambrian
    seas, 53, 54–55, 57–58; and dolomite, 86; and late Pro-
    terozoic, 35
lions, 222. *See also* mountain lion
listric faults, 181, 183
lithic tuff, 174, *175*
lithosphere, and oceans, 155
lithostatic pressure, 156
Little Cottonwood Canyon, *37, 203*
Little Willow Formation, 31, 32
lobe-fin fish, 97
Lone Mountain Dolomite, 95
Lone Pine earthquake (California, 1872), 225–26, 227
Lone Pine fault (California), 225, 233
longshore currents, 84
Long Valley caldera, *150H*, 220, 235–36
Long Valley–Mono Craters area, 219–20
Long Valley region, and ancestral Owens River, 213
Lunar Crater, 219, 235
lunar eclipse, 170
Lund Formation, 174
lungfish, 97
Luning Formation, 133
Luning-Fencemaker Belt, 136–39, 145
lycopods, 116, 117
Lyell, Charles, 197

maars, 219
magma, 137, 152–57, 162–66. *See also* volcanoes and
    volcanism
magmatic arc, 136–37
magnesium, in dolomite, 86–87
magnetic field, and reversals in polarity, 198
Malheur, Lake, 216
mammals: and Cenozoic Period, 170, 171; and late
    Mesozoic Era, 148; and middle Tertiary, *190*. *See also*
    animals; fauna; fossils
Mammoth Mountain (California), 220, 235, 236
mammoths, 222
Manix Lake, 216
Manly, Lake, 215
Manning Canyon Shale, 115, 116, 117, 118
mantle upwarp, 240
maps and mapping, and graphic images of Great Basin,
    11–13